COMMUNICATION
LIFE
MEGA-EVOLUTION

Decrypting Life's Nature

COMMUNICATION
LIFE
MEGA-EVOLUTION

Decrypting Life's Nature

Arnold De Loof

Leuven University Press
2002

Contact:
Arnold De Loof
Department of Biology
Zoological Institute
Naamsestraat 59, 3000 Leuven, Belgium
arnold.deloof@bio.kuleuven.ac.be

© 2002 Leuven University Press / Universitaire Pers Leuven / Presses Universitaires de Louvain
Europahuis, Blijde-Inkomststraat 5, B-3000 Leuven (Belgium)
Tel. +32-16-325345
Fax: +32-16-325352
E-mail: university.press@upers.kuleuven.ac.be
http://www.kuleuven.ac.be/upers

All rights reserved. Except in those cases expressly determined by law, no part of this publication may be multiplied, saved in an automated data file or made public in any way whatsoever without the express prior written consent of the publishers.

ISBN 90 5867 236 0

D / 2002 / 1869 / 43

NUR: 922

Cover : Julie Puttemans

To my good friend

Howard Bern

at the University of California, Berkeley,

as a tribute to his pioneering research in

*Comparative Endocrinology and
Integrative Biology*

PREFACE

Life is all around us, in an overwhelming abundance and in a seemingly endless diversity of forms. We ourselves are living creatures. Notwithstanding its omnipresence and the recent breakthroughs made in the biological sciences, the very nature of life continues to remain largely hidden for us. Is our brain still too underdeveloped to grasp life's complexity? Are we overlooking something? Is there a dimension that we cannot experience with our sense organs? Are we not asking the right questions? Or have researchers not yet found the biological equivalent of the Rosetta Stone that will enable us to decipher nature's system and put things in the right order?

The Rosetta Stone contains three versions of the same text, each written in a different language: hieroglyphics (used for official texts), demotic (used for everyday communication) and Greek. As J.F. Champollion, the final decipherer, showed in 1899, one does not need very long texts or sets of data to succeed in decryption. The trick was to search for repeated and matching elements. Once these were discovered, he managed to decipher the whole writing system by logical deduction. If this methodology is to be applied to biology, the relevant question is: "What is always repeated in all living systems and what is not?" By systematically eliminating all that is not essential, the very essence should become apparent. Furthermore, as the ancient Greek philosophers pointed out centuries ago, a good way to adequately describe a property of something is to compare it to its opposite. 'Cold' acquires meaning only when it is opposed to 'warm', 'light' to 'dark', etc. Taking this approach, I have asked the question, "What do we see changing when we say that something is 'still alive' at a given moment but 'no longer alive' thereafter?" In other words, we deduce the definition of 'life' in conjunction with the definition of 'death'.

In this book I argue that the ability to solve problems is *the* repeated element that makes the difference, again and again, between living and non-living matter. Problem solving is done by 'communicating compartments', the definition, properties and hierarchical organization of which are outlined in this book. This overall approach has paved the way to my definitions of 'life' and 'death'.

It seems to me that the time has come to incorporate into contemporary biology the insights into the basic mechanisms of communication that have been acquired in recent decades. This field of study offers a vocabulary that is also very useful in biology. It allows us to address basic questions that traditional biology has failed to answer, mainly because the right terminology did not yet exist. If 'life' is viewed from the point of view of communication and problem solving, and if the *morphological* unit, the 'cell', is replaced by the *functional* unit, the 'communicating compartment', then all the different pieces of the puzzle called "What is life?" begin falling nicely into their respective places. A very logical system becomes apparent, with some quite unexpected consequences, in particular regarding the theory of evolution and the biological foundations of certain ethical principles.

The original idea of the expanding universe, now commonly known as the big bang theory, came from the late Professor Georges Lemaître, whose laboratory was located within a few yards of the present site of my own developmental physiology laboratory.

In analogy with the 'big bang' (in fact a term introduced by Fred Hoyle to denigrate the expansion idea), I call the realization of the first act of communication, the first sign of life on Earth, the 'big hello'. In the history of the universe, this 'big hello' remains the single most important event subsequent to the big bang. It marked the onset of an era in which fossil stardust started to solve problems with ever-increasing degrees of complexity. This event transformed the passive universe into an active universe. Some 3.7 billion years after the big hello, specific aggregates of fossil stardust that we call human beings are using communication to master not only all the known forces of physics, but also to reflect on their own origin and destiny and the meaning of their existence.

'Life' has many aspects and it is difficult for one person to keep up-to-date on all of them. I have tried to keep up as much as possible, but this is obviously more difficult for those specialties that are not really closely related to my own field of research but that I felt needed some coverage to make the story as comprehensive as possible. My own experimental research activities have been concentrated on the identification of insect and crustacean hormones and the determination of their physiological effects in a comparative invertebrate-vertebrate context. This means that I see evolution through the eyes of a physiologist.

Both teachers and students know that a good picture or figure is worth a thousand words. Over the years I have asked both collaborators and students in some of my undergraduate classes to try to visually illustrate some of the abstract ideas featured in this book. This yielded the drawings and cartoons featured in the successive chapters. My sincere thanks to all who have contributed in one way or another to the development of both this book and its original version in Dutch ('Wat is Leven? - 1996) into their final form and content. I am particularly indebted to my collaborators Julie Puttemans (who designed the cover and who made nearly all the drawings), Maria Van der Eeken, Marijke Christiaens, Hendrik Van den bergh (who passed away in 1998) and Michael Breuer, as well as to Hilde Lens from Leuven University Press. Not being a native speaker of English, I am also very much indebted to Richard Sundahl for the final text correction, as well as for coauthoring the dynamic version of the Story of Creation at the end of this book.

I dedicate this book to my good friend Howard Bern, professor at the University of California at Berkeley. His pioneering work in the field of 'integrative biology', in particular its physiological and endocrinological aspects, has greatly influenced my way of thinking in integrative and comparative biology.

I hope that this book may serve as a tool to better familiarize biologists with the principles of communication and students in the humanities with the truly relevant principles of biology.

Leuven, June 2002

CONTENTS

Preface   7

Chapter 1:   17
The hardware of living matter: disguised fossil stardust

1.1	"Remember that you are dust and to dust you shall return"	19
1.2	The origin of the universe according to the standard cosmological model, also known as the "Big Bang" theory. Inflation	20
1.3	Sequence of events leading to the formation of atoms	21
1.4	The formation of our solar system and the planet Earth. The formation of planets is quite different from that of stars	25
1.5	The nascent atmosphere of the Earth. The origin of water and free oxygen gas	26
1.6	The formation of continents and oceans. Plate tectonics	27
1.7	The oldest fossils	28
1.8	Will our universe expand forever? The 'brane theory': Big Bang's new rival?	30
1.9	How old are you?	31
1.10	The static concept of the world as formulated in the opening account of creation in the book of Genesis	32
Essentials		34

Chapter 2:   37
Hardware biodiversity after 3.7 billion years of micro-, macro-, and mega-evolution. Genes, proteins, and the (first) central dogma

2.1	Introduction	39
2.2	The five-Kingdom classification system of Whittaker and Margulis (1978). More Kingdoms?	40
2.3	The origin of the eukaryotic cell type according to Lynn Margulis: unmatched natural 'genetic' engineering of cells	44
2.4	Elementary chemistry of living matter	45
2.5	Genes and heredity	48
2.6	A gene invariably codes for a chain of amino acids, and never for a (poly)saccharide or for a lipid	48
2.7	"One gene, one protein" is too simplistic. Introns and exons	53
2.8	How can genetic information change? Mutation versus modification	55
2.9	Micro-, macro- and mega-evolution. Evolution as a process of alienation resulting in a huge Babel-like confusion of tongues	56
2.10	Complexity. The human genome as compared to that of other organisms	59
2.11	Degree of kinship: man with other organisms	61
2.12	The descent of man: Paleontology versus the 'Eve – theory'	62
2.13	Philosophical consequences of the common descent view. *Homo* and his soul	63
Essentials		65

Chapter 3: 67
The nature and purpose of communication, the main activity of all biological systems. Software. Unconscious and conscious problem solving.

3.1	Introduction	69
3.2	A few 'simple' questions	69
3.3	Definitions of communication, learning, memory, hardware and software	70
3.4	Basic architecture of a communication system. Why do we have a name?	71
3.5	The boundary of a communicating compartment must have 'functional holes'	72
3.6	Sources and sinks. The importance of gradient formation	72
3.7	Remembering. The need for memory capacity	73
3.8	Prerequisites for a reliable memory system	74
3.9	Some reflections and speculations on the molecular nature of the carrier of long-lasting memory	74
3.10	Characteristics and evolution of software	76
3.11	The energy question	78
3.12	Communication can master any force of physics	79
3.13	When is something information? A few definitions of 'information'	79
3.14	The purpose of communication: Make others work for you in order to increase your own degree of contentment!	80
3.15	'Communication', a synonym for '(un)conscious problem solving'? Automation	81
3.16	The difference between communication and interaction. The (first) central dogma again	82
3.17	Communication and the anticipation of progressing time	82
3.18	Life as a double continuum. Two central dogmas?	83
3.19	Emotions, feelings, contentment, motivation	85
3.20	The basic architecture of a communication system in relation to the origin of the living state	86
3.21	Definitions of 'communicating compartment' and 'act of communication'	87
Essentials		89

Chapter 4: 91
Diversity in communicating compartments. Principles of mega-evolution. Revolutions

4.1	Introduction: Why should Mega-evolution be introduced to complement Micro- and Macro-evolution?	93
4.2	Compartment. Complexity	94
4.3	Definition of Mega-evolution. Revolutions. Continuous versus discontinuous evolution	94
4.4	Why is a novel classification system needed that is not based on kinship? Bringing order to the multitude of communicating compartments. Principles of the 16-level classification system	95
4.5	Compartments restricted to one and the same individual organism: mono-organismal compartments	97
4.6	Compartments which consist of more than one individual of the same species: polyorganismal-monospecies compartments	105

4.7	Compartments consisting of individuals belonging to different species: heterospecies compartments	108
4.8	The mechanisms instrumental to Mega-evolution. Some common denominators in the mechanisms for creating higher levels of compartmentalization	110
4.9	Newly formed compartments expand. Need for maintaining order. Installation of headquarters	112
4.10	Mechanisms for generating daughter compartments	113
4.11	General conclusions	114
Essentials		115

Chapter 5:
Traditional attempts to define 'life'. Literature survey
117

5.1	The necessity of defining 'life'	119
5.2	'Life': of extraterrestrial origin?	119
5.3	The 'circular definition' approach	119
5.4	The approach based on listing the 'properties of living matter'	120
5.5	The approach based on the 'classical laws of thermodynamics'	121
5.6	The approach based on the far-from-equilibrium thermodynamics of Ilya Prigogine: "Living systems are 'dissipative systems', as is shown by their heat production"	122
5.7	"Living beings are characterized by the fact that they are continually self producing. Autopoiesis and cell suicide (apotposis)"	123
5.8	"A system in which information carriers can duplicate themselves is alive"	124
5.9	"A self-correcting system is alive"	124
5.10	"Life is a machine"	125
5.11	Other approaches: "Feeling alive"	126
5.12	A list of traditional approaches to the definition of life	127
5.13	Still something missing? Where is the 'intellectual dimension' of 'life'?	128
Essentials		129

Chapter 6:
Alive, no longer alive. The deduction of a definition of 'death' as a key to that of 'life'
131

6.1	Accidental death (necrosis) and programmed cell death (apoptosis or cell suicide)	133
6.2	The deduction of a definition of 'death', the master key to a plausible definition of 'life'?	134
6.3	Ethical considerations	138
Essentials		140

Chapter 7: 141
Life defined in terms of communication and problem solving. L =ΣC

7.1	Requirements for an acceptable definition of 'Life'	143
7.2	Underlying logic	143
7.3	Definition of an 'act of communication'. Example: control of muscle contraction	143
7.4	The multitude of possible messengers and responses in biological communication	146
7.5	Only 'active' acts of communication are important	152
7.6	The total sum of acts of communication	152
7.7	A symbolic notation of 'Life' (as an activity)	152
7.8	Definitions of 'Life' (as an activity)	155
7.9	Properties of 'life'	156
Essentials		157

Chapter 8: 159
Making choices and the resulting variability and freedom in communication and behavior. Life's basic drive: "Solve problems to feel comfortable and contented!"

8.1	An example of the interrelationship between the environment on the one hand and human communication and behavior on the other	161
8.2	The origin of complexity and variability in communication: illustration with simple whistle music	161
8.3	Translation of the whistle music principles into the chemical language of a simple biological cell system	163
8.4	More holes: the complexity of recorder and accordion music	165
8.5	Bifurcation points, the often overlooked companions of mutations in generating communicational variability, offer possibilities but no certitudes. The (un)predictability of communication	171
8.6	Life's basic drive: "Solve problems if you want to feel comfortable and contented!"	173
8.7	Instinctive versus learned behavior. Not only humans have free will	174
8.8	Summary: freedom, responsibility and suffering	175
Essentials		177

Chapter 9: 179
No life without electricity. The principles of biological electricity as relevant to communication. The primordial gradient of the living state

9.1	Electric fishes are at the origin of the discovery of electricity	181
9.2	The nature of 'biological electricity'	182
9.3	Every living cell maintains a potential difference over its plasma membrane on the order of tens of thousands of volts per centimeter	183
9.4	All living cells actively generate and maintain differences in concentrations of specific ions over their plasma membrane. Ion pumps	184

9.5	There is also passive 'leaking' of ions through the plasma membrane. Ion channels	185
9.6	Cells manage to transduce a difference in concentration of some simple ions on the two sides of their plasma membrane into a potential gradient over this membrane. The principle of the concentration chamber	186
9.7	Calculation of the membrane potential. Depolarization and hyperpolarization	188
9.8	Ion pumps and channels act in concert	189
9.9	Transcellular ion fluxes and extracellular electric fields	191
9.10	Self-electrophoresis: a means for generating intracellular potential gradients and gradients of charged macromolecules under specific conditions. The concept of "the cell as a miniature electrophoresis chamber"	192
9.11	Actin filaments in the cytoskeleton: electricity-conducting wires in between the plasma membrane, the cytoplasm and the nucleus? DNA and electricity	193
9.12	Excitable membranes. Impulse conduction along axons. Electrical control of gene expression	194
9.13	Inorganic ions: one of the tools for generating order out of 'chaos'	195
9.14	General conclusion: to a large extent, life is an electrical phenomenon	196
Essentials		197

Chapter 10: 199
Viruses, prions, toolization, computers, man-made life

10.1	Introduction	201
10.2	Viruses are not alive	201
10.3	Prions	202
10.4	Man-made or artificial Life	203
10.5	Toolization. Self-communication	210
10.6	Refining the symbolic notation of 'life as an activity'	210
10.7	Electronic life and tool-aided (mechanical) reproduction	211
10.8	The future?	212
Essentials		213

Chapter 11: 215
No life without time. But time: what exactly is it?

11.1	Introduction	217
11.2	Time, space, space-time: an ultra-brief historical overview	218
11.3	The key questions	221
11.4	What do we mean when we say "duration"? Parameters influencing "duration"	221
11.5	A definition of time	223
11.6	Consequences of the definition	223
11.7	How to measure time? Time and motion	224
11.8	The speed of light	225
11.9	Bringing order into the multitude of different "times"	227
11.10	The arrow of time	227

11.11	Conclusions: time as special kind of 'elasticity coefficient'?	228
Essentials		229

Chapter 12: 231
The big hello: the very moment that life began. From passive to active universe. The Gradient Provoked - Triple S Principle

12.1	Life came into being the moment the first communication system started to function	233
12.2	What was first: the DNA-, the RNA-, or the peptide world?	233
12.3	The *pregradient* era: the problem of generating "order" in the primordial conditions	234
12.4	The *cytoskeleton*: as important as nucleic acids for the emergence of the living state?	235
12.5	The importance of the achievement of the *transmembrane ionic/voltage gradient* and of transcellular ion fluxes	236
12.6	Requirements for the perpetuation of the first communication system	240
12.7	The problem of osmosis. The Gradient-Provoked Swelling and Self-Selection Principle (the GP-Triple S Principle) and how it might apply to social and economic compartments	240
12.8	Conclusions	244
Essentials		246

Chapter 13: 247
Reproduction. Gender and Sex. How 'it' may have given birth to 'him' and 'her'

13.1	Introduction	249
13.2	The Progenote: why did it engage in cell division? Immortality as a reward	249
13.3	Mitotic cell division in the Urkaryote: a problem of coordination	251
13.4	Asexual reproduction in multicellular organisms: by regeneration and from stem cells	253
13.5	The sexual mode of reproduction: by means of gametes resulting from meiotic cell division	253
13.6	Formation of the germ cell line. Solitarization of the presumptive germ cells in insects as a model	254
13.7	What made the primordial germ cells come into existence? An ancient bacterial infection?	257
13.8	How to increase the number of gametes? Rounds of mitosis precede meiosis in higher animals	258
13.9	How do bipotential germ cells get committed to either the oocyte or the sperm cell scenario? Sex steroids	260
13.10	Gamete formation in the gonads: a benign form of cancer? The 'fight' between somatoplasm and germ plasm	260
13.11	The 'idea' behind yolk accumulation in eggs: based upon an ancient innate immunological mechanism?	261
13.12	Oviposition and sperm ejaculation: again 'hostile' acts by the somatoplasm	263

13.13	Gender: why only two isoforms, males and females? Gender-determining genes	263
13.14	Cholesterol, sex hormones and gender. Again the bacterial infection route. Parthenogenesis and sex-reversal	265
13.15	Males: from superfluous to indispensable?	267
13.16	DNA methylation and imprinting	267
13.17	Death of the somatoplasm: the price to be paid for meiosis. Sexual pleasure in return. The progeny: goal or free surprise bonus?	268
13.18	Sperm competition. The choosing female	269
13.19	Conclusions: the origin of reproduction is not what you thought it was	270
Essentials		272

Chapter 14: 273
Embryonic development. The acquisition of higher levels of compartmentalization and corresponding communication systems in the course of the lifetime

14.1	The key questions in developmental biology	275
14.2	Model organisms for the study of development	275
14.3	The importance of asymmetry	276
14.4	Successive morphological stages in animal development: *Branchiostoma lanceolatum* as a model	279
14.5	The key morphological principle underlying embryonic development in animals is centered on epithelium formation	279
14.6	Mechanisms instrumental to the acquisition of ever higher levels of compartmentalization as relevant to mega-development	279
14.7	The molecular biological approach in (micro)-development: *Drosophila* as a model	280
14.8	*Drosophila* complemented with Dolly: the scientific importance of cloning	282
14.9	The role of the environment: the example of the identical twins again	284
14.10	The macromolecular environment around the genes: fine tuning of gene expression. Transcription factors	284
14.11	The ionic environment around the genes. The complexity of the 'internal ionic environment'. Six levels of control	285
14.12	Processes that can be controlled by inorganic ions	286
14.13	The universal principle upon which differentiation in animals is based	287
14.14	Self-selection during development	288
14.15	The 'vital force' of Hans Driesch	289
14.16	Changes in the level of compartmentalization during development *In vitro* fertilization, cloning, abortion: ethical aspects	289
Essentials		292

Chapter 15: 293
Evolution. The *'hardware-software'* or *'double continuum'* theory

15.1	A sound theory of evolution should explain not only the formation of new species but also of 'life' at all its levels of compartmental organization	295
15.2	Micro-evolution. A brief overview of Darwinism	296

15.3	Elements in neo-Darwinism subject to improvement. Life is a double continuum	297
15.4	Thinking about evolution with insights from the computer era: not only the hardware evolves, but the software as well	299
15.5	Is Evolution based merely upon software possible? The example of the identical twins for the third time	299
15.6	'Cultural evolution' is software-driven, being based on teaching-learning processes. Abstract thinking. Consciousness. The role of tools	301
15.7	Definitions of 'natural selection'	303
15.8	Natural selection: cause or result? The driving force(s) of evolution Mutations and changes in the cognitive memory system	304
15.9	Summary of the essentials of the double continuum - or the hardware-software - theory of evolution	309
15.10	Input, through the software-aspect, of the organism's own input or free will in evolution? Creationism	315
15.11	Signal transduction pathways and the ongoing dispute between 'Phyletic gradualism' and 'Punctuated equilibrium'	316
15.12	An upgraded adage	316
Essentials		317

Chapter 16: 319
Properties of 'life'. 'Integrative' versus 'classical' biology. *Communico, ergo vivo*

16.1	Summary of the general properties of 'Life'	321
16.2	The understanding of 'Life' requires holistic thinking. 'Integrative biology' versus 'classical biology'	322
16.3	'Good' and 'evil'	323
16.4	The place of the *Homo sapiens* in the biosphere as a whole	323
16.5	Biology, ethics and philosophy. *Communico, ergo vivo*	326
16.6	If Darwin had had a computer	326
16.7	$L = \Sigma C$: Towards a biological paradigm?	327
Essentials		329

The Story of Creation: a contemporary dynamic version   331

References   335

Index   343

CHAPTER I

THE HARDWARE OF LIVING MATTER: DISGUISED FOSSIL STARDUST

God said: "Let there be a point of asymmetry so a Universe can come into being and life can arise in the fullness of time."

Contents

1.1 "Remember that you are dust and to dust you shall return"
1.2 The origin of the universe according to the standard cosmological model, also known as the "Big Bang" theory. Inflation
1.3 Sequence of events leading to the formation of atoms
1.4 The formation of our solar system and the planet Earth. The formation of planets is quite different from that of stars
1.5 The nascent atmosphere of the Earth. The origin of water and free oxygen gas
1.6 The formation of continents and oceans. Plate tectonics
1.7 The oldest fossils
1.8 Will the universe expand forever? The "brane theory": Big Bang's new rival?
1.9 How old are you?
1.10 The static concept of the world as formulated in the opening account of creation in the book of Genesis

Essentials

1.1 "*Remember that you are dust and to dust you shall return*"

The author of the opening account of the book of Genesis in the Bible managed to summarize in only a few splendidly written pages his society's view of the origin of the universe, the earth, life and the hierarchy among living beings. This is in essence the '*static concept* of the world', as experienced by the first urban communities in the Middle East where the origin of western civilization was located, and which remained largely unchallenged until the formulation of the theory of evolution by Darwin and Lamarck. God's crowning work was the creation of man:

> *Let us make man in our image, after our likeness...*
> *So God created man in his own image,*
> *in the image of God he created him;*
> *male and female he created them.*

If you read this text, which you find printed at the end of this chapter, note that the author systematically repeats that all living creatures were created "*each according to its kind.*" This clearly illustrates that the author had not even considered the possibility that species could change in the course of time and could evolve one out of another. Neither did he consider the possibility that being male or female was a rather late 'invention' in the course of evolution (Chapter 13). This is very understandable because it corresponds with what we actually see around us: species do not visibly change during a man's lifetime, neither do we experience that all of a sudden new species arise out of nothing. There is no such thing as *generatio spontanea* (spontaneous generation), a fact which Louis Pasteur unequivocally demonstrated. To the contrary, each living creature is the progeny of a parent organism. Thus a static concept of the universe suffices to explain what we see in nature, at least if we content ourselves with a superficial analysis.

The relationship between matter and living beings is also well known in all cultures: "*Memento Homo quia pulvis es and in pulverem reverteris*", says the priest in Christian churches on Ash Wednesday: "*Remember that you are dust and to dust shall you return*". But what kind of dust? In today's language it would read: "*Remember that your body consists of nothing but fossil stardust that is recycled again and again: your specificity does not reside in your body, but in your soul, which does not disappear when your body stops living.*"

This view reflects another concept of the universe and the place of living beings therein: the '*dynamic concept* of the world', in which everything continuously changes: "*panta rei*", as the Greek philosopher Heraklitus called it. In this concept, neither the universe nor the species that live on Earth came into being in the way we see them at this moment or as is depicted in Genesis.

Firstly, just as a living organism continuously changes from the moment it is in the fertilized egg until its death, so also the universe and all the stars and planets contained in it evolve from relative 'simplicity' to the present state of extreme complexity. Secondly, the millions of different species that inhabit the planet Earth at the current time were not created one by one in their present unchanging form; rather, all are the progeny of a common bacterium-like ancestral cell that came into existence nearly 4 billion years ago. In this point of view, *Homo sapiens* takes on a more humble position in the scheme of things than he does when he proclaims himself 'the measure of all things'. In the light of their enormous success in nature, one might better argue that the bacteria and the insects are the 'crown of creation', rather than man and all his mammalian cousins.

1.2 The origin of the Universe according to the standard cosmological model, also known as the "*Big Bang*" theory. Inflation

We live in a universe consisting for nearly 98 percent of two simple elements: hydrogen (^{1}H: 72%) and helium (^{4}He: 26%). It is also cold by human standards, its average temperature being not more than -270 °C, which is only 3 °C above absolute zero (-273 °C). The nearly 100 other elements of Mendeleyev's periodic table of the elements make up no more than 2 percent of the whole universe.

According to the 'Big Bang' theory, the visible universe in which we live began sometime between 9 and 16 billion years ago (Sneden, 2001) as a fireball of extraordinarily dense and hot matter. It has been suggested that its size was not bigger than a dime. To calculate the age of the Universe, scientists use a parameter known as the 'Hubble constant' (*Ho*). There is as yet no consensus about its exact value: the literature mentions figures ranging from about $Ho = 45$ up to about $Ho = 87$ km sec^{-1} Mpc^{-1} (km = kilometer; sec = second; 1 parsec = 3.08×10^{18} cm; 1 Megaparsec (Mpc) = 1 million parsec) (Pierce *et al.*, 1994; Freedman *et al.*, 1994). The higher the value of the Hubble constant, the younger the Universe. The main problem that astronomers face when trying to determine the value of *Ho* is how to measure distances in the universe as accurately as possible. Differences in the estimation of distances result in different values of *Ho*. With the issue of the exact value of the Hubble constant still unresolved, the age of the universe is commonly accepted to be around 12 billion years.

A more recent method, called radioactive cosmochronometry, is based on the quantification of the abundance of radioactive thorium in stars. Cayrell et al (2001) used this method to calculate the age of the universe as being roughly 12.5 ± 3 billion years. This agrees well with former estimations.

The original idea of an expanding universe comes from Georges Lemaître (1894-1966), who was a professor at the Université Catholique in Louvain, Belgium. Some specialists in the field have doubts about some aspects of this theory, and some even question its validity. One of the issues in this dispute is whether time has always been, or whether it came into being at some point not earlier than the moment of the creation of the universe. Nobody has as yet succeeded in fully explaining the nature of 'time'. I will give it a try in Chapter 11. The term 'Big Bang' was not introduced by Lemaître but by the astrophysicist Alfred Hoyle, who found the idea of an expanding universe ridiculous. In any case, it was the term 'Big Bang' that stuck – a term that is now commonly used to denote the onset of space and time (= 'space-time') in the universe by those who assume that before the big bang time did not exist. Whether this was the only big bang, rather than just the creation of our local universe in some 'eternal' superuniverse in which a succession of big bangs occurs, remains a matter of dispute - and perhaps of endless speculation.

Many questions remain to be answered. How can a universe evolve out of 'nothing' or, in the terminology of physicists, from a 'singularity'? Another question concerns what created the initial inhomogeneities in the universe that resulted in galaxies and other large-scale structures? Measurements of the cosmic microwave background radiation suggest that these inhomogeneities could have resulted from tiny quantum fluctuations generated in the early universe during a period of amazingly violent, rapid (faster than light) but very short (not more than 10^{-32} second) expansion called inflation (Gangui, 2001). Inflation theory makes it possible to explain why the universe is 'flat' (a technical term describing the large-scale curvature of space) and why it has roughly the same properties everywhere (Seife, 2001). Inflation theory is getting a rival, namely the 'brane theory', as will be explained later in this chapter (1.8).

Astrophysicists have now gained a pretty good idea of what happened from about 10^{-12} second after time zero, the nominal moment of perfect symmetry and infinite temperature and density (if it existed at all), up till today. The formation of the 100 or so naturally occurring chemical elements, which make up everything around us and of which about 30 are essential to life, is especially well understood. The expansion was accompanied by a rapid decrease in temperature.

The major issue is that the big bang created only the simplest elements, ^{1}Hydrogen and ^{4}Helium, and trace amounts of ^{2}Hydrogen, ^{3}Helium, ^{7}Lithium and the very unstable ^{8}Beryllium, and it filled almost 98 percent of the whole universe with them. The formation of all elements with a mass (or atomic number in chemical terms) exceeding 8 required high temperatures sustained over a much longer period of time than was possible under big bang conditions. But such conditions of high temperatures sustained over long periods of time came into existence not earlier than the time at which stars of varying sizes were born and evolved. This happened for the first time long after the big bang. Furthermore, it took several successive generations of stars being born, aging and exploding to return the elements to the space between the stars. This occurred before all the elements that were needed for planet Earth to support living beings finally ended up in our solar system. These elements were thus created in the most inhospitable of all environments, such as those prevailing in a sort of thermonuclear reactor.

1.3 Sequence of events leading to the formation of atoms

On the basis of data provided in papers by Whitrow (1959), Cox (1990), Horgan (1991), Riordan (1993), Weinberg (1994), James et al. (1994), and Kirschner (1994), I have attempted to summarize the sequence of events as follows.

Formation of subatomic particles

It is not possible to describe what happened between time zero and zero plus 10^{-12} second, when the temperature had already dropped to some 16 million degrees. The 'laws of nature', as we currently understand them, cannot be applied at temperatures above 16 million degrees.

At a temperature of about 10^{16} °C, a first phase transition took place, something like the freezing of water, in which the known particles for the first time acquired mass.
For reasons that are as yet unknown, *asymmetry*, without which there would neither have been a universe or life, was introduced in the system. At a temperature of 10^{15} °C, the universe must have been filled with a very dense gas consisting of all subatomic particles

that on the basis of modern high-energy nuclear physics are now known to exist, together with their antiparticles. In their collisions, all these particles were continuously being annihilated and created.

The temperature continued to decrease rapidly as the expansion continued. A crucial event for the development of the universe then took place: a very small (only one part in 10^{10}) excess of quarks over antiquarks and electrons over antielectrons originated. These quarks and electrons served as the building blocks for the first light atomic nuclei that were formed in the next few minutes.

Formation of the naked nuclei of ^{1}Hydrogen, ^{2}Deuterium and ^{4}Helium

When the universe was 1 second old, its temperature had further dropped to around 10 billion degrees, a million times hotter than the surface of the Sun. It was filled uniformly with subatomic particles – the electrons, protons, and neutrons that make up ordinary matter, plus particles of light called 'photons'. There were also large numbers of 'positrons', the antimatter opposites of electrons, and vast hordes of extremely light particles called 'neutrinos'. The density of the universe at that moment – i.e. the total amount of matter and energy packed into a given unit volume – is estimated to have been about 10 kilograms per cubic centimeter. Thus at this very early age, the universe was a thousand times denser than lead.

The temperature continued to drop very quickly in fractions of a second, thereby creating the right conditions for the formation of the nucleus of the basic chemical building block of the whole universe, hydrogen. Another name for the nucleus of hydrogen is 'proton', the single particle of which it is composed.

The possibility existed for neutrons and protons to combine, forming deuterium (^{2}H), a heavy form of hydrogen. At the high temperatures that were then prevalent, however, the deuterium nuclei reacted rapidly with more protons, and the ultimate product was the stable nucleus of helium (^{4}He), containing two protons and two neutrons. No further nuclear fusion reactions to produce heavier elements from helium could happen to any appreciable degree because the temperature and density had become too low by the time the helium was made.

By the time the universe was a few minutes old, all the neutrons had combined with protons to make heavier nuclei such as helium, and the positrons had been annihilated along with a fraction of the electrons to generate still more photons (light particles). A sprinkling of protons and electrons still permeated this fiery plasma. The temperature had fallen to several hundred million degrees, still much hotter than the core of the sun, and the density had dropped to about a hundredth of a gram per cubic centimeter (heavier than air but lighter than Styrofoam).

Physicists can calculate in which proportions hydrogen and helium can be formed under 'big bang' conditions. The agreement between the calculated values and the actually observed values is very impressive, being one of the strongest pieces of evidence that ideas about the big bang are correct: no other theory can explain the existence of hydrogen and helium in their observed proportions.

Naked nuclei capture electrons, thereby forming the atoms hydrogen and helium.

For the next 100 thousand years or so, there was not much change because it was still too hot for atoms to be formed through the combination of electrons with the nuclei of hydrogen and helium. The universe continued its steady expansion, gradually cooling off until its temperature had dropped to a few thousand degrees. At that point, its density was already far less than that of the best vacuum attainable in terrestrial laboratories. At this time the electrons permeating the universe finally slowed down to the point where they could be captured by the much more ponderous nuclei, thereby forming a gaseous mixture of atoms of hydrogen, helium, and a trace of ^{7}Lithium. Prior to this moment, the universe had been an incandescent plasma similar to the one existing inside neon lighting fixtures. Photons had been trapped within this plasma because they interacted readily with all the charged particles in it. When these charged particles became bound in electrically neutral atoms, the photons suddenly could break free and roam off to infinity in all directions, like light passing through thin air. The universe became transparent not earlier than some 300,000 years after the big bang.

^{4}Helium as a building block for other elements. The formation of atoms up to ^{16}Oxygen in small stars.

If after the big bang the expansion of the universe had been perfectly uniform all the time without any imperfection, then there would have been no stars and no life, only hydrogen, helium, trace amounts of lithium and beryllium and light. But for reasons we do not yet understand, *points of asymmetry* arose where gradual condensation of gases took place, leading to the formation of galaxies and of first-generation stars formed from hydrogen and helium, which 'cooked' some of these elements into heavier elements. Which elements resulted from this 'cooking' depended upon the size of the star.

A star is born when a large mass of gas contracts under its own gravity. This contraction produces heat and, at a given moment, the temperature rises high enough – some 10 million degrees – for fusion reactions to take place. First, the hydrogen atoms are stripped of their electrons, which means that they are ionized. The naked nuclei, which buzz furiously in the dense and very hot interior of the star, collide so vigorously that both strong and weak nuclear forces can come into play, forcing nuclei to fuse and thereby yielding great quantities of energy. This is called the 'burning' of hydrogen. A star like the sun that burns hydrogen has enough such fuel for 10 billion years. The fusion product of the hydrogen nuclei is helium.

As a star ages and nears extinction, its core contracts, which makes the temperature rise so high that two helium nuclei can fuse to form ^{8}Beryllium (4 protons and 4 neutrons) instead of the usual ^{9}Beryllium. This ^{8}Beryllium disintegrates almost instantly. However, specialists in the field think that within this short time span before fissioning, a ^{8}Beryllium atom absorbs another helium nucleus so that an excited state ^{12}Carbon is formed, the most vital element of living matter. In this excited state, the ^{12}Carbon nucleus emits a photon and decays into the stable state of lowest energy. Thus, carbon is the result of the fusion of three basic building blocks, namely 3 helium. At the present temperature, ^{12}C could again fuse with helium, thereby yielding ^{16}Oxygen. Carbon and oxygen are the most abundant elements heavier than helium that are present in stars. In stars of low and average mass (lower than 8 solar masses) no elements heavier than oxygen are formed. *This means that all elements present on Earth with a mass higher than 16-18 cannot have been formed in our solar system or in any other small star: they must therefore have originated from somewhere else in the universe.*

A star becomes a *red giant* when it nears its death. This phase lasts several hundred million years in small stars such as our sun. Such a red giant pushes off its outer layers, called a planetary nebula. Eventually the fuel supply gets exhausted and the inner core of the small red giant congeals into a *white dwarf*, in which oxygen and carbon are locked. This is the final end of such a star unless a too generous neighboring star supplies a stream of gas onto a white dwarf, provoking it into a sudden synthesis of new elements, typical of a *type I supernova*.

Formation of elements with mass exceeding that of ^{16}Oxygen but not higher than ^{56}Iron in stars of more than 8 solar masses.

Stars with more than 8 solar masses form the same elements as smaller stars. However, because of the prevailing conditions in their center at some period in their existence, big stars necessarily produce heavier and heavier elements. By the same principle as cited previously for the formation of carbon and oxygen in small stars, a whole array of lighter elements are formed with a mass which is always a multiple of 4 and with an equal number of neutrons and protons: ^{12}Carbon, ^{16}Oxygen, ^{20}Neon ^{24}Magnesium, ^{28}Silicon, ^{32}Sulfur, ^{40}Calcium, all relatively abundant elements. There are also other reactions in which other isotopes and also nitrogen are formed. ^{28}Silicon is finally converted to ^{56}Iron, in only one day when the star nears its end. ^{56}Iron is the normal end product of the fusion reactions in the core of a big star because it is the most stable of all elements. No more energy will be produced by fusion reactions. On the contrary, these reactions will consume energy.

Formation of elements with a mass exceeding 56.

In red giants, there is also some formation of elements heavier than iron in the layers around the core when certain reactions produce neutrons, which are captured by other nuclei. This process, which is called the s-process (meaning 'slow'), can yield elements up to ^{83}Bismuth.

At the end of its life, a big star also passes through a red giant stage before it is finally destroyed by a sort of inverse big bang lasting about one second. Under the influence of gravitation, the iron core suddenly collapses into a dense mass of solid neutrons to become a neutron star or black hole, raising the temperature to billions of degrees. The collapsed core of a neutron star is as dense as a nucleus. The outer layers of the star fall in, and then 'bounce back' on this hard core like a tennis ball hitting a wall. As the result of a sonic boom reaching the surface, the star suddenly becomes very bright and explodes, thereby spewing its contents out into space in a *supernova II explosion*. In this explosion it belches out the helium that was formed from hydrogen. It also launches carbon, oxygen, sulfur and silicon into the gas in its neighborhood.

The explosion itself generates more heavy elements, because it produces a flood of neutrons that are absorbed by the existing nuclei. This process is called the *r*-process (for 'rapid') as opposed to the *s*-process. In the latter, neutrons add one by one. In the *r*-process, so many neutrons are available that several of them attach to a nucleus all at the same time. Gold is formed by the bombardment of iron nuclei by neutrons. Gold is transformed into lead, which then gets transformed all the way up to uranium. Thus an exploding supernova ejects not only large quantities of light elements into a neighboring gas cloud, but also heavy elements. In this manner, second-generation stars are born containing hydrogen, helium and heavy elements, and during their evolution they 'cook' these into more heavy elements. This fusion process ends again in supernova explosions, which in their turn eject heavy elements into neighboring gas clouds. From this gas cloud containing hydrogen and helium plus more heavy elements, third generation stars are formed. These contain hydrogen, helium and enough heavy elements to form solid planets orbiting the stars. This is followed again by supernova explosions that eject even more heavy elements into a neighboring gas cloud.

Thus successive cycles of star formation and destruction enrich the interstellar medium with more and more heavy elements. This is a process that continues cyclically. At this moment in our story, we have only reached the stage where about 2 percent of the baryon mass of the universe consists of elements other than hydrogen and helium.

1.4 The formation of our solar system and planet Earth. The formation of planets is quite different from that of stars

Our solar system, the Sun with its nine planets, came into being some 4.55 billion years ago, which is many generations of exploded stars after the big bang. The formation of the Sun was not different from that of any other small star, involving the condensation of a gas cloud already enriched in heavy elements, a rise in temperature, the burning of hydrogen, etc.

According to Allègre and Schneider (1994) and Kirschner (1994), planets are formed in the following manner. When a cloud of gas enriched with heavy elements originating from exploded supernovae coalesces into a star, it first forms a rotating disk of gas and dust in which the initial temperature is very low. Cosmic dust aggregates to form particulates, the particulates become gravel, and this gravel gathers to form small balls; these balls grow by further accretion into large balls, and then into tiny planets or planetesimals, and, finally, the original dust aggregate reaches the size of the moon. This process of increasing in size by *accretion* has significant thermal consequences. As long as only tiny particles fall on the surface of a planet, the effects are minute. However, when sufficiently large bodies slam into a planet, such an immense heat can be produced in its interior that the cosmic dust melts into what is called a magma ocean.

The Earth was formed some 4.55 billion years ago from the debris of meteorites. It continued to grow through the bombardment of planetesimals until some 120-150 million years later. Early on, the temperature of the Earth must have been so high that a deep layer of molten rock covered the planet. Only after the Earth was no longer being hit by these planetesimals could the temperature drop enough so that finally the surface became solid. The core, together with the mantle, drives the geothermal cycle, including volcanism.

The relative abundance of the hundred or so elements present on Earth varies enormously. Oxygen, silicon and iron are common, while many elements such as helium, gold and platinum are millions of times as rare. Because of gravity, iron was concentrated in the core of the Earth while the surrounding layers of our planet, including the crust, are composed of silicon and oxygen (white sand is almost pure SiO_2), together with many other elements in smaller proportions.

1.5 The nascent atmosphere of the Earth. The origin of water and free oxygen gas

The gases emerging from the interior of the Earth by a process called 'outgassing' gave rise to the primordial atmosphere. Volcanism is an example of this. The gas composition of the primordial atmosphere was most probably dominated by carbon dioxide, with nitrogen as the second most abundant gas. Trace amounts of methane (CH_4), ammonia (NH_3), sulfur dioxide (SO_2) and hydrochloric acid (HCl) were also present. Free oxygen is supposed to have been very rare in early times. Such a gas mixture has what chemists call 'reducing' properties. Most of the hydrogen and helium escaped because the Earth's gravity was not strong enough to retain them.

Life is not possible without water. The explanation for this statement will be given in Chapter 9, which deals with biological electricity. This type of electricity is based on watery solutions of inorganic ions, by way of contrast to the electricity we normally think of that is carried by electrons through copper wires. The presence of water is not at all restricted to planet Earth. At a sufficiently high temperature, oxygen and hydrogen interact, with the formation water as a result. Great quantities of water-ice are present in the discs rotating around young stars, at least at a sufficiently large distance from their surface. At great distances, comets are formed, which can contain large amounts of water. The prevailing view now is that most of the water that is present on Earth was delivered by ice-comets colliding with the Earth over a long period of time. Because the hydrogen isotopic composition of water on Earth differs widely from that of the primitive sun, a large portion of all the water on Earth must have come from a distant source. It is thought that it was added by a few late giant impacts of carbonaceous meteorites originating from the cold region of the asteroid belts (Robert, 2001). When the Earth's crust cooled down, the rapid outgassing of the planet liberated voluminous quantities of water from the mantle, thus creating the oceans and the hydrological cycle.

The oxygen gas which now makes up about 20 percent of the Earth's atmosphere started being produced much later in evolution, namely not earlier than when the first green photosynthesizing organisms (blue-green algae or Cyanobacteria) appeared. Such organisms were able to carry out a chemical reaction whereby, with the help of light energy from the sun, CO_2 and water are absorbed and converted into the sugar glucose and O_2, thereby storing the light energy from the sun in the form of chemical energy. All of the major photosynthetic groups of bacteria arose prior to 2.8 billion years ago, and perhaps even much earlier (Des Marais, 2000).

The chemical reaction involved in the process of photosynthesis as carried out by contemporary plants is as follows:

$$\text{Light}$$
$$6\ CO_2 + 12\ H_2O \rightarrow C_6H_{12}O_6\ (= \text{glucose}) + 6\ O_2 + 6\ H_2O$$

By some 2 billion years ago, vast amounts of O_2, the first large-scale toxic pollutant gas on Earth, had accumulated in the Earth's atmosphere. It turned into a major catastrophe for all organisms that could not cope with its oxidative properties, but it created a niche for those that could. Such organisms were doomed to disappear and the ones that could stand high concentrations of O_2 survived and became the ancestors of all present-day dominant life forms on Earth. However, this mass extinction did not happen right from the moment the first O_2- producing organisms appeared. The reason for this is that O_2 is very reactive and it could be inactivated by reacting with the many reduced minerals before it could reach the atmosphere. Even in our atmosphere, reacting with some of the gases present would have inactivated it. The supply of reduced compounds was exhausted some 2 billion years ago, and from then on O_2 started accumulating in the atmosphere, reaching current levels roughly one billion years ago. An atmosphere rich in O_2 filters out the harmful ultraviolet radiation. This is not so important for aquatic organisms but it is beneficial for organisms that live on or above the Earth's surface. Furthermore, O_2 can be transformed into ozone (O_3), a gas which also absorbs ultraviolet radiation.

1.6 The formation of continents and oceans. Plate tectonics

Around the close of the 19th century, the Austrian geologist Eduard Suess postulated that the present-day continents of the Southern Hemisphere once formed a single giant continent called Gondwanaland. A few decades later, the German meteorologist Alfred Wegener postulated a supercontinent that he called Pangaea, that began to break up about 200 million years ago into the continents as we know them today, with oceans filling the widening gap between them (Press and Siever, 2001). 'Pangaea' floated on top of the hot liquid core of the Earth and broke up in parts as the result of convection streams of hot lava underneath. The different parts drifted away from one another. Accordingly, the North and South American continents were originally parts of Europe and Africa that detached and drifted away. This large-scale movement of continents over the globe is called 'plate tectonics'. The major mountain chains on Earth are places where continents have been pushed against one another. When interpreting data from paleontological findings, one must keep in mind that the continents were not always in the same locations as they are now. For this and numerous other reasons, the climatic conditions changed along with the drift of the continents.

1.7 The oldest fossils

Living matter is composed of about 30 different elements of the periodic table of elements. The most abundant ones are oxygen (^{16}O), carbon (^{12}C), hydrogen (^{1}H), nitrogen (^{14}N), and phosphorus (^{31}P). Other common elements in living matter are potassium (^{39}K), sodium (^{23}Na), chlorine (^{35}Cl), calcium (^{40}Ca), magnesium (^{24}Mg) and sulfur (^{32}S). Most of the other elements are metals that are required in low amounts for the proper functioning of certain proteinaceous molecules (e.g. iron in hemoglobin for the binding of oxygen), enzymes, etc. The elements incorporated in living matter are no different from those present in non-living matter. The major difference between non-living and living entities is therefore not at the level of their constituent atoms, but rather at the level of certain classes of molecules made from these atoms.

Life as we know it on Earth is not possible without water. The vast amounts of water that entered the atmosphere by the degassing of the crust of the Earth could not accumulate on the Earth's surface as long as it was too hot. The farther away from the surface, the more the temperature decreases. At a certain altitude the temperature becomes so low that water changes from the gas phase into the liquid phase, resulting in the formation of fog and clouds: meteorologists would say that under these conditions the 'dew point' has been reached. Because of the heat at the surface of the Earth, the first clouds must have been formed at much higher altitudes than today. As the crust of the Earth continued to cool down, the turbulent clouds with their many thunderstorms descended closer and closer to the Earth's surface. Finally, the falling rain drops no longer evaporated before reaching the ground: the Earth's surface became wet. In the course of time, more and more water accumulated in ponds, and later in seas, etc. The temperature in the first pools was probably near the boiling point of water, which on the face of it would not seem an ideal environment for living beings.

But there are bacteria that can tolerate – and some that even require – such high temperatures, such as those that live in the hot water springs in Yellowstone Park. Molecular biologists are currently making extensive use of enzymes that originate from such hot water bacteria (e.g. *Thermophilus aquaticus*) with a very powerful technique called the polymerase chain reaction (PCR), for which a Nobel Prize was awarded in 1993. It is not because the overwhelming majority of all present day living organisms cannot tolerate such extreme conditions that life could not have originated in nearly boiling water. Some authors have suggested other possible environments where life could have originated, such as around deep sea hydrothermal vents (the submarine fissures in the Earth's crust through which intensely hot gases are cycled), in ice, on the surface of pyrite crystals, *etc*. (For an overview of different scenarios and ideas about the evolution of life on Earth, see for example Horgan, 1991; De Duve, 1991, 1995; Orgel, 1994; Gould, 1994.)

Wherever life first appeared, the available conditions would have included:
- The Earth with its warm surface, all elements of the periodic table of elements present at the time, and a reducing atmosphere;
- A range of different environments with warm or hot water containing salts such as NaCl and KCl, certain Ca, Mg and Fe salts, and a variety of organic molecules.

There is a great deal of uncertainty not only about the origin of organic molecules on Earth but also about the order in which they were formed (Miller, 1987; De Duve, 1995). There are organic molecules (such as certain amino acids, which are the building blocks of proteins) present in the interstellar dust that traverses the universe. Substantial amounts of it may have fallen on Earth and may have been concentrated in water pools. On Earth, there may have been synthesis of this and other types of organic molecules out of the gases present in the reducing primordial atmosphere. Stanley Miller elegantly demonstrated this possibility by laboratory experiments a few decades ago (Miller and Orgel, 1973). De Duve (1991, 1995) points to the possibilities of thioester chemistry. Wächtershauser (1988 a, b) thinks that essential chemical reactions, such as the formation of the peptide bond required to link together amino acids into proteins, could have happened on the surface of pyrite crystals.

Another major unsolved problem concerns the nature of the first information-carrying molecules in living matter. What came first? Proteins, DNA (deoxyribonucleic acid, the present day universal carrier molecule for the genetic code), RNA (ribonucleic acid), etc? The discovery that RNA has autocatalyzing properties and can therefore perform certain specific enzymatic reactions itself (see Chapter 12), favors the view that the RNA world preceded the DNA world, but the chemistry which could have led to the formation of RNA in primordial conditions is still obscure. More recently, Lee *et al.* (1996) discovered a 32-residue peptide with self-replicating properties. If such molecules were present in "primordial soup conditions", perhaps a peptide world could have preceded all other ones.

Some investigators in this field think that the origin of 'life' coincided with the formation of the first information carrying molecules (DNA, RNA or perhaps special proteins). Others think that it coincided with the origin of self-replicating systems. I leave the formulation of hypotheses about the origin of the organic molecules that were required to finally produce the living state to chemists specialized in this domain.

My view, which will be expounded in the following chapters, is that 'life' came into existence much later, after a long period of prebiotic evolution in which many organic molecules were formed and spontaneously assembled into more or less ordered aggregates by processes of chemical self-organization such as protein-protein-, protein-nucleic acid-, and hydrophobic interactions. The major problem is that as long as we do not adequately define what life is, it is not possible to determine the exact moment of its appearance. The only thing we know with rather good certainty is that the earliest fossils of undisputed age, which were found in South Africa and Australia, are relics of blue-green algae and are about 3.5 billion years old. This suggests that organic matter was already present as long as 3.8 billion and, according to some specialists, perhaps even 4.2 billion years ago. The living state should thus have come into existence rather shortly after the Earth became a hospitable place to live in for a number of creatures requiring a bitter minimum of comfort.

1.8 Will the universe expand forever? The 'brane theory': Big Bang's new rival?

The big bang can be regarded as a phase transition, leading from an unknown phase into the ones we observe around us. It is impossible to imagine what that pre-big bang phase would have looked like. The following example may help to illustrate the problem. An observer looking through the glass of a champagne bottle for the first time before uncorking it would not be able to imagine the presence and nature of any other phase than the solid phase of the glass which he sees and feels. Upon uncorking, he sees gas bubbles forming, apparently out of nothing. An unknown phase becomes apparent. All the more so, since with the upward movement of the gas bubbles the observer can make the deduction that inside the glass yet another phase must be present, different from both the solid phase of the glass and the gas phase of the bubbles. Perhaps the universe is only one of the many structures that emerge as the result of a phase transition in an immense superstructure lacking outer boundaries. The problem is that our brain cannot possibly grasp the meaning of infinity.

At this moment the universe is still expanding, even faster and faster, despite the action of gravity. According to some astrophysicists this anomaly can be explained by assuming that there is an as yet unknown form of energy, which has been named 'dark energy' and which might constitute up to 70 percent of all the energy in the universe.

Not long ago, some specialists in the field thought that perhaps the expansion would not continue forever, but that it would rather stop and switch to contraction: the 'rising loaf of raisin bread', a comparison often applied to the universe, would then become a collapsing loaf. Time would be reversed, and the clusters of galaxies (the raisins) would move closer and closer together, touching and overlapping. The temperature would rise and rise because the total amount of energy would be being packed into a smaller and smaller volume. As the volume would decrease, the density of matter would increase. This would go on up to the moment where the volume has become so small and the density and temperature so high, that matter breaks up into a seething broth of subatomic particles - a plasma. In such conditions, nuclear and elementary-particle physics take over. Thus this story is exactly the opposite of the big bang. However attractive such a scenario may be, it is not supported by any observations. Therefore it is improbable. Stephen Hawking withdrew his original view that in a collapsing universe, the arrow of time would reverse: it does not (Hawking and Penrose, 1996).

Assuming that a large meteor does not hit the Earth, the nearest cosmological catastrophe will be the transformation of our sun into a red giant. This stage is still a few billion years away because there remains plenty of hydrogen to burn. A much greater threat to life on Earth are the many different forms of man-made pollution, including nuclear weapons. Where man comes, nature must retreat, but there are limits that should not be transgressed.

As in all sciences, in the field of cosmology new ideas are put forward on a regular basis. Not all astronomers and physicists agree that the widely accepted inflation theory adequately explains the formation and properties of the universe. Furthermore, in the 'standard model' it remains difficult to incorporate the force of gravity on an equal footing with the three other fundamental forces of nature (electromagnetism, and the strong and weak nuclear forces). In particular, the observed weakness of gravity on distances equal to or exceeding 1 mm is a problem (Arkani-Hamed et al., 1998; Antoniadis et al., 1998). The newest rival theory is called the 'brane theory', an abbreviation of 'membrane' (see Gibbons, 2000; Pease, 2001). This theory requires quite some imagination from the non-specialist -- perhaps too much to qualify as being a good theory. Normal mortals have no problem in accepting that space-time has four dimensions, namely the three dimensions of space and one dimension of time. In brane theory, the existence of not less than 11 dimensions is invoked. Six are rolled up (and can be roughly neglected), leaving 5 effective dimensions in which two perfectly flat four-dimensional membranes float, like sheets on parallel clotheslines. Seife (2001) attempted to visualize the outcome of the complicated physics and mathematics of the theory as follows. He says that our own universe corresponds to one of the sheets, and the other is a 'hidden' parallel universe. From the latter, a membrane is spontaneously shed. It floats slowly toward our universe, thereby flattening out. The floating membrane speeds up and splats into our universe. The point then is that some of the collision energy becomes the matter and energy that make up our cosmos. In this 'expyrotic model' (of P. Steinhardt, Princeton), our universe was not formed in a point-like big bang. Rather, it is a post-collision universe that was formed in a plate-like splash. Thus in brane theory, the universe is only a very tiny part of a truly gigantic superstructure with alternative dimensions beyond our experience.

1.9   How old are you?

All matter in the universe is composed of some 25 different elementary particles such as quarks, gluons, leptons, etc., which are combined in a number of ways via the Standard Model. It has taken billions of years and several generations of stars being born, aging and exploding in the most inhospitable of all environments, namely that in which nuclear fusion reactions take place, for the 100 or so elements which are found in the universe and on Earth to be formed. About 30 of these elements are present in living matter. The Earth came into being some 4.5 billion years ago, thus roughly 8 billion years after the big bang. The first organic molecules may have started accumulating on the surface of the Earth as early as 300-500 million years after the planet was formed. It probably took another few hundred million years before the first living entity appeared.

As far as its constituent atoms are concerned, living matter consists of nothing else than fossil stardust. All this stardust is composed of hydrogen that was formed shortly after the big bang took place approximately 12 billion years ago. This means that the answer to the question "How old are you?" actually is: "My hardware is 12 billion years old, but my software started to develop not earlier than the moment of my conception".

When dealing with the origin of the living state, it is convenient to make the distinction between the *prebiotic era* in which the organic molecules, which are needed to make up a living entity, were formed, and the era of *biotic evolution*. The latter started at the moment that a compartmentalized aggregate of organic molecules started to live. This moment will be defined in Chapter 12.

1.10 The static concept of the world as formulated in the Opening Account of Creation in the Book of Genesis

It reads:

In the beginning God created the heavens and the earth. The earth was without form and void, and darkness was upon the face of the deep; and the Spirit of God was moving over the face of the waters.
And God said, "Let there be light"; and there was light. And God saw that the light was good; and God separated the light from the darkness. God called the light Day, and the darkness he called Night. And there was evening and there was morning, one day.
And God said, "Let there be a firmament in the midst of the waters, and let it separate the waters from the waters." And God made the firmament and separated the waters, which were under the firmament from the waters, which were above the firmament. And it was so. And God called the firmament Heaven. And there was evening and there was morning, a second day.
And God said, "Let the waters under the heavens be gathered together into one place, and let the dry land appear." And it was so. God called the dry land earth, and the waters that were gathered together he called seas. And God saw that it was good. And God said, "Let the earth put forth vegetation, plants yielding seed, and fruit trees bearing fruit in which is their seed, each according to its kind, upon the earth." And it was so. The earth brought forth vegetation, plants yielding seed according to their own kinds, and trees bearing fruit in which is their seed, each according to its kind. And God saw that it was good. And there was evening and there was morning, a third day.
And God said, "Let there be lights in the firmament of the heavens to separate the day from the night; and let them be signs and for seasons and for days and years, and let them be lights in the firmament of the heavens to give light upon the earth." And it was so. And God made the two great lights, the greater light to rule the day, and the lesser light to rule the night; he made the stars also. And God set them in the firmament of the heavens to give light upon the earth, to rule over the day and over the night, and to separate the light from the darkness. And God saw that it was good. And there was evening and there was morning, a fourth day.
And God said, "Let the waters bring forth swarms of living creatures, and let birds fly above the earth across the firmament of the heavens." So God created the great sea monsters and every living creature that moves, with which the waters swarm, according to their kinds, and every winged bird according to its kind. And God saw that it was good. And God blessed them, saying, "Be fruitful and multiply and fill the waters in the seas, and let birds multiply on the Earth." And there was evening and there was morning, a fifth day.
And God said, "Let the earth bring forth living creatures according to their kinds: cattle and creeping things and beasts of the earth according to their kinds." And it was so. And God made the beasts of the earth according to their kinds, and the cattle according to their kinds, and everything that creeps upon the ground according to its kind. And God saw that it was good.
Then God said: "Let us make man in our image, after our likeness; and let them have dominion over the fish of the sea, and over the birds of the air, and over the cattle, and over all the earth, and over every creeping thing that creeps upon the earth." So God created man in his own image, in the image of God he created him; male and female

he created them. And God blessed them, and God said to them, "Be fruitful and multiply, and fill the earth and subdue it; and have dominion over the fish of the sea and over the birds of the air and over every living thing that moves upon the earth." And God said: "Behold, I have given you every plant yielding seed which is upon the face of all the earth, and every tree with seed in its fruit; you shall have them for food. And to every beast of the earth, and to every bird of the air, and to everything that creeps on the earth, everything that has the breath of life, I have given every green plant for food." And it was so. And God saw everything that he had made, and behold, it was very good. And there was evening and there was morning, a sixth day.

Thus the heavens and the earth were finished, and all the host of them. And on the seventh day God finished his work which he had done, and he rested on the seventh day from all his work which he had done. So God blessed the seventh day and hallowed it, because on it God rested from all his work, which he had done in creation.

These were the generations of the heavens and the earth when they were created.

Most importantly, the opening account of creation, like other such accounts in other religions, deals with the relations between humans and God. It is not a lecture in astrophysics. Humans say that God created man in His own image. Perhaps it would be more correct to say that religious humans created their God(s) in the image of man.

Contrary to what some people seem to think, there is no fundamental contradiction between the philosophy behind the story in the book Genesis and what astrophysicists tell us now. They both say that there was a beginning for the universe and that the creation of matter required ingenuity beyond our imagination. Despite lots of hard evidence to the contrary, creationists think that God created the universe and all living things piece by piece. Some astrophysicists say that God bestowed upon matter such ingenious properties that right from the beginning all laws necessary to generate life, when the conditions would allow this to happen, were present in matter. Others think that universes are being created and are collapsing all the time, in a cycle without beginning or end, and that there is no need for a creating God.

The text of Genesis quoted here is from the Revised Standard Version of the Holy Bible (Oxford University Press, 1989, with permission). Other stories of creation in other religions have been collected in a book by Vlaar (1996).

ESSENTIALS

THE UNIVERSE

1. The universe is thought to have come into existence some 12 billion years ago.
2. Theories explaining its origin:
   - Big bang theory: The universe was formed out of a "singularity". Its expansion is still accelerating.
   - Brane theory: Where 'branes' meet, multiple universes can come into existence.
   - The big bang is only a phase transition, not a unique beginning.
3. The basic building block of the whole universe: the hydrogen atom (H).
4. Formation of all other elements of the periodic table of the elements (Mendeleyev): By fusion of H nuclei and further fusion of the resulting fusion products, for example:

    In stars (= burning clouds of gas) of up to 8 solar masses:

    The fusion of a proton + a neutron yielded a heavy form of H, Deuterium or ^2H;

    Deuterium plus more protons yielded the stable element Helium (^4He);

    $^4Helium + {}^4Helium$ yielded $^8Beryllium$;

    $^8Beryllium + {}^4Helium \to {}^{12}Carbon$;

    $^{12}Carbon + {}^4Helium \to {}^{16}Oxygen$.

    In stars bigger than 8 solar masses, fusion reactions yielded additional elements with atomic number up the 56, which is $^{56}Iron$, the most stable of all elements, and normally the end product of all fusion reactions:

    $^{27}Silicon + {}^{27}Silicon \to {}^{56}Iron (= {}^{56}Fe)$

    During successive explosions of dying stars (= in supernovae explosions), all elements with atomic number higher than 56 (e.g. gold, lead, uranium) were formed.
5. Origin of our solar system (the sun and its 9 planets): ± 4.5 billion years ago.
6. Planets are accretions of fossil stardust originating from exploded supernovae. The impact heat of a sustained rain of meteorites caused the forming planet earth to become liquid.
7. Water (H_2O) is omnipresent in the universe. A few giant ice meteorites from outside the solar system probably delivered most of the water on earth.
8. The primitive atmosphere contained mainly CO_2 and N_2, with some trace amounts of methane (CH_4), ammonia (NH_3), sulfur dioxide (SO_2) and hydrochloric acid (HCl).
9. Life probably originated in extreme conditions some 3.7 billion years ago, or perhaps even earlier.

10. CO_2 is food for some types of photosynthesizing organisms, such as plants. Oxygen (O_2) in our atmosphere is an end product of photosynthesis by early organisms:

$$CO_2 + H_2O \xrightarrow{\text{light energy}} \text{glucose } (C_6H_{12}O_6) + O_2$$

O_2 was the first major pollutant on earth, killing numerous anaerobic species.

11. Plate tectonics caused movements in continents over the liquid core of the earth.

CHAPTER 2

HARDWARE BIODIVERSITY AFTER 3.7 BILLION YEARS OF MICRO-, MACRO-, AND MEGA-EVOLUTION

Genes, proteins, and the (first) central dogma

God said, "Let matter organize itself as communicating compartments with different forms, sizes, colors and languages, and let them all evolve in the course of time."

Contents

2.1 Introduction
2.2 The five-Kingdom classification system of Whittaker and Margulis (1978). More Kingdoms?
2.3 The origin of the eukaryotic cell type according to Lynn Margulis: unmatched natural 'genetic' engineering of cells
2.4 Elementary chemistry of living matter
2.5 Genes and heredity
2.6 A gene invariably codes for a chain of amino acids, and never for a (poly)saccharide or for a lipid
2.7 "One gene, one protein" is too simplistic. Introns and exons
2.8 How can genetic information change? Mutation versus modification
2.9 Micro-, Macro- en Mega-evolution. Evolution as a process of alienation resulting in a huge Babel-like confusion of tongues
2.10 Complexity. The human genome as compared to that of other organisms
2.11 Degree of kinship: man with other organisms
2.12 The descent of man: Paleontology versus the 'Eve -theory'
2.13 Philosophical consequences of the common descent view. *Homo* and his soul

Essentials

2.1 Introduction

To date, no one can make an accurate guess about the number of species that inhabit the earth. The estimates vary from a few million to several dozens of millions of species. The smaller the size of the organisms and the more inaccessible the places where they live (habitats), the higher the margin of error in the estimations. Even when using the most modern research tools, it often remains difficult to distinguish between species and varieties.

When we look at the world around us, it seems at first glance as if the plants (1.3 million recorded species) and the animals (several times more species than the plants) are the dominant forms of life on earth. However, microscopically small organisms, such as bacteria, nematodes and other worms and other invertebrates that live in the soil, in water or on the bottom of oceans, are not readily visible. When we take these creatures into account, we cannot escape the conclusion that the bacteria represent the dominant form of life on earth.

Over the centuries several attempts have been undertaken to design classification systems for ordering the multitude of different species. One of the these early systems grouped organisms according to their habitat: e.g. salt water, fresh water, the ground or the air (flying creatures). Another system was based on the distinction between animals, plants and fungi, without taking into account their habitats.

Carl von Linné, better known as Carolus Linnaeus, made history when he wrote '*Systema naturae*', the doctoral thesis he defended in 1758 at the University of Leiden in the Netherlands. It was only 11 folio pages thick. Apparently, such a thin volume sufficed to classify all the plants, animals and even the minerals that were known at that time. Linnaeus' system was truly ingenious because it grouped in a hierarchical manner all the known organisms on the basis of similarities in their anatomy/morphology (= classification based on homology) and not upon similarity in function of certain organs (= classification based on analogy). In Linnaeus' homology-based system birds, insects and bats are classified in distinct groups, while in the analogy-based system they would be classified together because they all have wings. In the analogy-based system, it does not matter that the wings originate in quite different ways during embryonic development, whereas in the homology-system it does.

Most probably Linnaeus did not realize that the classification system he had designed was in fact based on the degree of evolutionary kinship. Indeed, the theory of evolution according to Darwin and Wallace dates from much later. Charles Darwin published his '*On the Origin of Species by means of Natural Selection*' in 1859. Nowadays, evolution may be defined as "descent with modification [Darwin], closely related species resembling one another because of their common inheritance, and differing from one another because of the hereditary differences accumulated during the separation from their ancestors." Another definition says that evolution is the process by which related populations diverge from one another, giving rise to new species (or higher groups) (Dodson and Dodson, 1976). Thus, Linneaus had intuitively grouped organisms on the basis of their descent and on the time span since the different species had started to diverge from a parent species. The shorter the period of time that has passed since a given species diverged from a parental one, the more it still resembles it.

Although these conclusions may seem self-evident at the present time, it was not until the second half of the 20th century that a substantial number of biologists had accepted Darwin's theory as being sufficiently valid to incorporate its central ideas into their own research. In any case, the theory of evolution could not have acquired a solid basis before the laws of heredity had been elaborated down to the molecular level.

The milestones in this journey of scientific thought include the application of the work of Gregor Mendel to the genetics of the fruit fly *Drosophila*, the double-helix model of deoxyribonucleic acid (DNA, the carrier of the genetic information), and the establishment of the (first) 'central dogma' on the relationship between DNA and protein synthesis (DNA → RNA → Proteins, see later). Furthermore, the increasingly rapid methods that are continually being developed to read the nucleotide sequence of DNA have already resulted in the complete or nearly complete sequencing of the entire genome of a number of model organisms (see below). Many other organisms will be subject to genome sequencing in the near future.

2.2   The five-Kingdom classification system of Whittaker and Margulis (1978). More Kingdoms?

This system takes into account the architecture of cells (prokaryotic or eukaryotic, unicellular or multicellular), as well as a number of features relating to embryonic development (in plants and animals). It delineates five major *Kingdoms* (Fig. 2.1) One *Kingdom*, the Monera, comprises all prokaryotes. The other four *Kingdoms*, the Pro(toct)ista, the Fungi, the Plantae and the Animalia together form the Super-Kingdom of the Eukaryota.

The Prokaryota: The Monera with the Archaea and the 'modern' Eubacteria

In Greek 'karyon' means 'nucleus' and 'pro' means 'before'. The prokaryotic cell type (Fig. 2.2) has neither a nucleus delineated by an envelope consisting of a double membrane or any other internal membrane structures. The majority of the prokaryotes are usually very small cells that can only be seen under the microscope. In common parlance they are called bacteria. The blue-green bacteria, formerly called the blue-green algae, and the mycoplasmas also belong to the Monera. The very first cell that appeared on earth, the Progenote, was most probably a bacterium. All other organisms are descendants of this Progenote. Some authors subdivide the Monera into the Archaea or Archaebacteria and the (more modern) Bacteria or Eubacteria. Many Archaea live in extreme environments and are called 'extremophiles'.

Kingdoms with the eukaryotic cell type: Protista, Fungi, Plantae, and Animalia

Eukaryotic cells (Fig. 2.2) have a nucleus that is bounded by a double membrane (nuclear envelope). In their cytoplasm they also have several other membrane structures such as the rough endoplasmic reticulum (for protein synthesis), the smooth endoplasmic reticulum (e.g. for steroid synthesis), the Golgi apparatus (for packing of newly synthesized proteins into granules to be transported out of the cell), lysosomes (for intracellular digestion), mitochondria (for energy production) and, in some cases, chloroplasts (for photosynthesis)

Protista or Protoctista: This heterogeneous Kingdom comprises all eukaryotic organisms that do not possess the typical features of fungi, plants or animals. Some are unicellular, others are multicellular. The majority of Protista were formerly classified as 'Protozoa'. It was to be expected that, along with progress in (mitochondrial) gene sequencing in ever more members of this Kingdom, being so ill-defined, the Kingdom would in time be subdivided into more Kingdoms. A more refined system has already been proposed (see Raven and Johnson, 1996), comprising six eukaryotic Kingdoms: the Archezoa, the Protozoa, the Chromista, the Fungi, the Plantae, and the Animalia. This is probably not yet the end of the increase in the number of Kingdoms.

Fungi: These are the molds, the yeasts and the mushrooms. The body plan is mycelial or secondarily unicellular (in yeasts). They always utilize absorptive nutrition, and never photosynthesis.

Plants: These are the multicellular organisms with plastids in the cytoplasm of their cells (e.g. chloroplasts for photosynthesis). A seed of a plant contains a well-developed embryo.

Animals: An animal is an organism that in an early stage of its embryonic development organizes itself into a closed epithelium that is called a blastula (De Loof, 1992). Later in this book I will elaborate further on this definition.

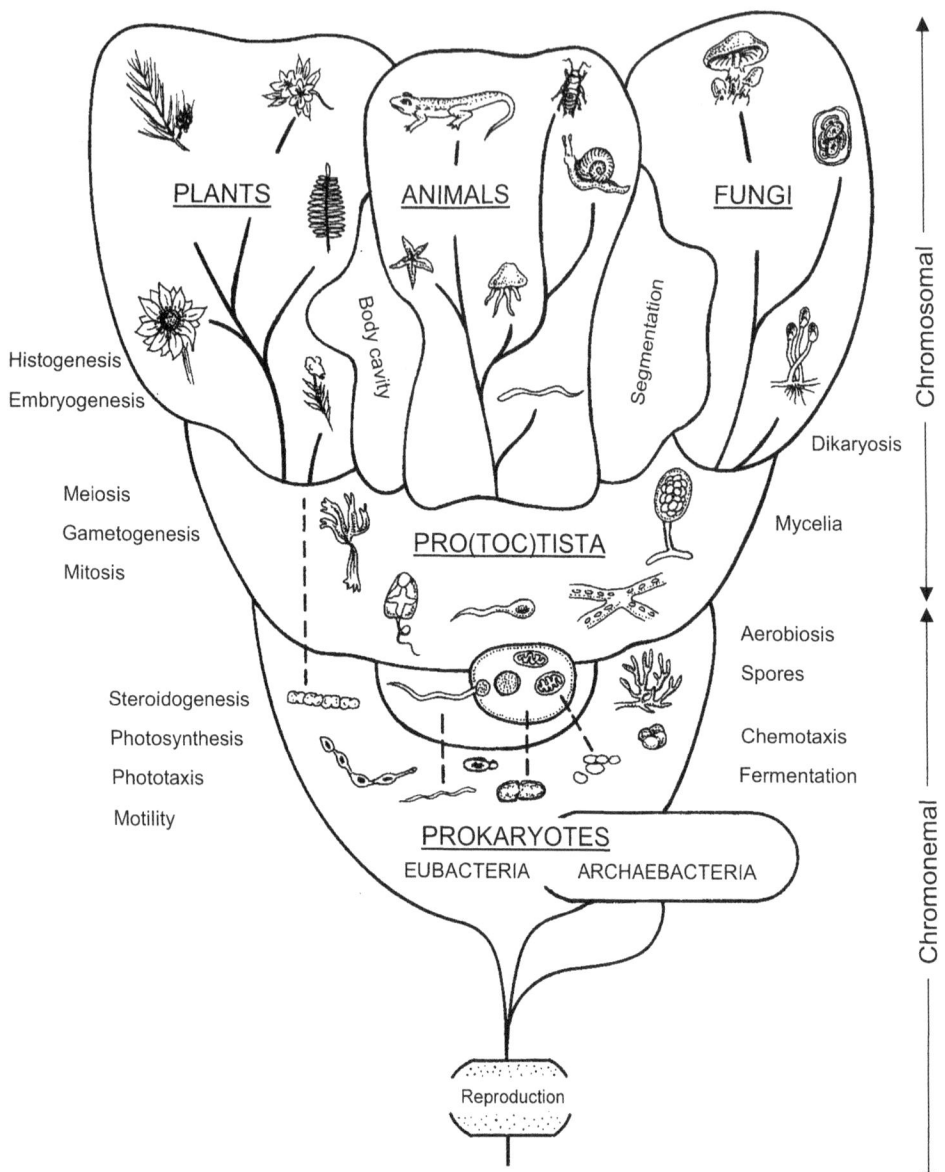

Figure 2.1 The 5-Kingdom classification system.

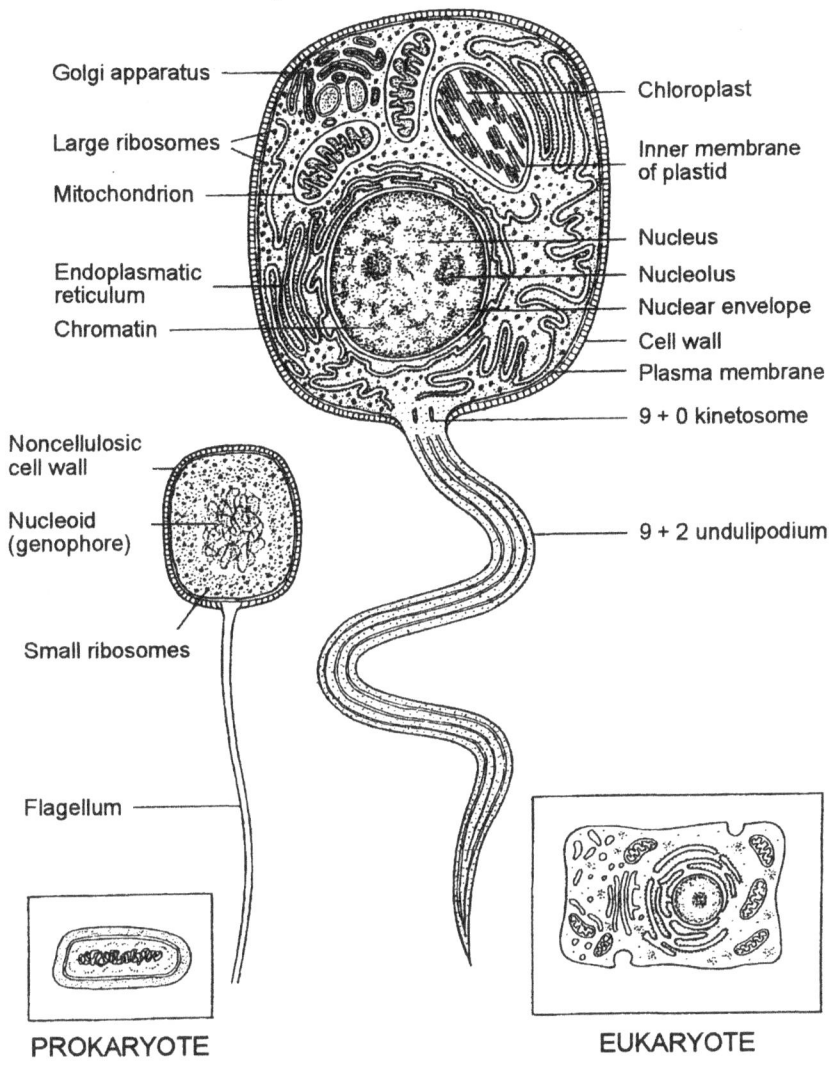

Figure 2.2 Morphology of the prokaryotic- and the eukaryotic cell type. All eukaryotic cells have a nucleus limited by a nuclear envelope as well as other membrane structures in their cytoplasm. The structural proteins present in a flagellum of a prokaryote differ from those in an undulipodium (redrawn after several authors, in particular after Margulis).

2.3 The origin of the eukaryotic cell type according to Lynn Margulis: unmatched natural 'genetic engineering' of cells

According to Lynn Margulis (1978, 1998), a prominent evolutionary biologist, the eukaryotic cell type originated when a few smaller bacterial species invaded a larger bacterium and established a permanent symbiotic relationship that was beneficial for all the organisms involved (Fig. 2.3). In the course of time, some evolved to mitochondria, others to cloroplasts. Perhaps the centriole system also goes back to still another type of bacterium. One of the arguments in favor of this splendid theory is that mitochondria and chloroplasts still have their own bacterial-type DNA. There are other even more valid arguments.

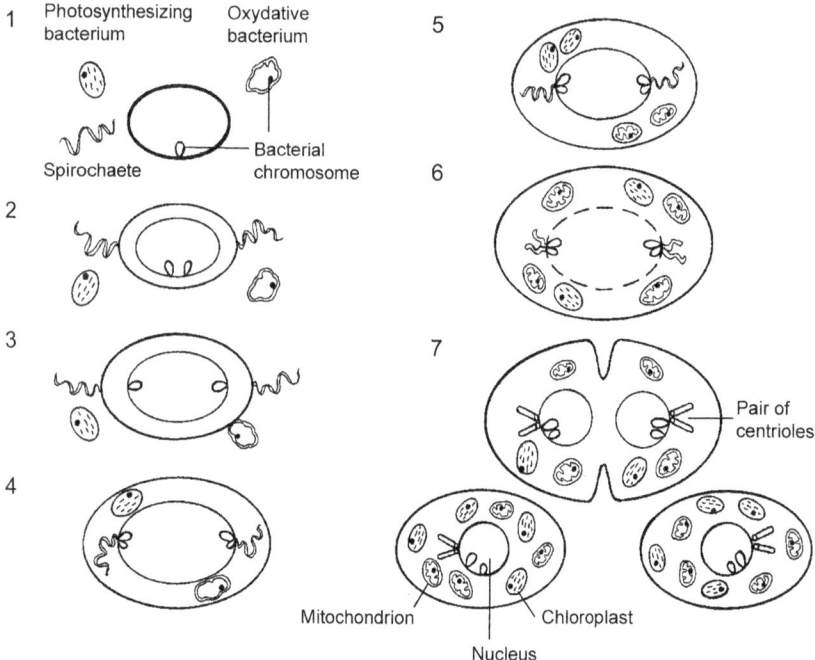

Figure 2.3 The origin of the eukayotic cell type according to the symbiont theory of Lynn Margulis.

One consequence of this theory is that, in fact, there is only one basic form of life on earth, namely the prokaryotic (bacterial) form. Indeed, eukaryotes are a derivative of this form. Another consequence is that all eukaryotes, man inclusive, are collections of descendants of urbacteria. We humans each regard ourselves as being only one single organism, although each of our mitochondria could also regard itself as being an individual organism. This hierarchical organization of the living state will be dealt with in the following chapters, in particular when a definition for 'death' will be deduced (Chapter 6). One of the questions will be: Can we be considered to be dead when some of our mitochondria are still alive?

I will come back repeatedly to this attractive 'symbiotic theory'.

2.4 Elementary chemistry of living matter

Elements

Of the roughly one hundred elements of Mendeleyev's periodic table of the elements, the following ones are preferentially used for the synthesis of organic molecules in living matter: carbon (C), hydrogen (H), oxygen (O), nitrogen (N), sulfur (S), and phosphorus (P). The elements Na, K, Mg, Ca, Fe, Mn, Cu, Co, and Zn also play an important role in certain biochemical processes, too diverse to be dealt with here.

Inorganic ions. Cellular electricity and signal transduction

Throughout the book, 5 types of inorganic ions will show up frequently. Na^+, K^+, H^+ and Cl^- are very important because of their role in biological electricity without which life cannot exist. Ca^{2+} and H^+ are key players in signal transduction.

Water

Below its boiling point, water (H_2O) is a polymer, which explains its properties in the liquid and solid (ice) states. The majority of biochemical reactions are only possible in an aqueous environment. Furthermore, biological electricity requires water. Without electricity, there is no life, as will be explained in Chapter 9.

Other small molecules

In addition to water, numerous other small inorganic and organic molecules occur in living matter. They are not relevant for the further reasoning in this chapter.

Lipids: essential for membrane formation

These are larger molecules that are poorly soluble in water. The common lipids, called glycerides, consist of a backbone of glycerol to which one (= monoglyceride), two (= diglyceride) or three (= triglyceride) fatty acids are linked. Some fatty acids comprise only a few carbon atoms, while others have over a dozen. Some fatty acids, in particular those present in animals, do not have double bonds and are therefore said to be saturated (with hydrogen); others (in plant oils) have one or even several such bonds (and are therefore called unsaturated). The plasma membrane of the cell, as well as other subcellular membranes, consists primarily of lipids of the phospholipid type. Lipid droplets inside cells function as energy storage reservoirs. Because lipids contribute little to biodiversity, further elaboration on this topic is omitted.

Polymers

The term 'macromolecule' is frequently used in biochemistry. Although this is not fully correct, the term is often used as a synonym for a polymer of elementary building blocks. The major polymers of living matter are:

> *Polysaccharides*: chains of simple sugars, covalently linked by glycoside bonds. Examples: starch from plants (amylum), 'animal starch' or glycogen, and cellulose (in plant cell walls). These are all polymers of the same monosaccharide, namely glucose. The very different properties of these three polymers result from the differences in the way the glucose moieties are linked.

Nucleic acids: chains of nucleotides.

Two types of nucleic acids occur: deoxyribonucleic acids (abbreviated as DNA) and ribonucleic acids (abbreviated as RNA). The building blocks of DNA and RNA are called nucleotides. Each nucleotide-monomer consists of a phosphate group ($-PO_4$), a sugar residue (ribose in RNA and 2'-deoxyribose in DNA) and a nitrogenous base residue. The phosphate group causes the acidic properties. In both DNA and RNA, four different nucleotides occur. They only differ in their nitrogenous bases.

Three out of the four bases of DNA are identical in RNA, namely adenine (A), guanine (G) and cytosine (C). The fourth base is different; being thymidine (T) in DNA and uracil in RNA.

The four nucleotides of which DNA is composed are:
    adenine-deoxyribose-phosphate   (abbreviation: A)
    cytosine-deoxyribose-phosphate   (abbreviation: C)
    guanine-deoxyribose-phosphate   (abbreviation: G)
    thymidine-deoxyribose-phosphate (abbreviation: T)

The four nucleotides present in RNA are:
    adenine-ribose-phosphate or adenosinemonophosphate (AMP)
    cytosine-ribose-phosphate or cytidinemonophosphate (CMP)
    guanine-ribose-phosphate or guanosinemonophosphate (GMP)
    uracil-ribose-phosphate or uridinemonophosphate (UMP)

Conventionally, the repeating monomers of DNA or RNA are represented by the single letters A, T, G, C or U. Furthermore, sequences are always written from 'left to right', or, in the appropriate jargon, from the 5' end to the 3' end.

DNA, the universal carrier of genetic information, has the form of a double helix. It can be very long, consisting of millions of nucleotides. The two strands of the helix are complementary, being held together by hydrogen bridges that can be formed only between complementary bases:
    Adenine (A) is complementary to thymine (T)
    Guanine (G) is complementary to cytosine (C)

It is for this reason that in drawings of the DNA double helix A is always situated opposite to T, and G is always situated opposite to C. This complementarity is also essential for the transcription of DNA into RNA (here uracil (U) is complementary to adenine (A)).

RNA only occurs as single strands and is never very long. There are three different forms of RNA, each with a specific function:
    - messenger RNA (mRNA);
    - ribosomal RNA (rRNA);
    - transfer RNA (tRNA).
This will be dealt with later.

Proteins: chains of amino acids, covalently linked by peptide bonds.

Apart from proline, all of the 20 'classical' amino acids found in proteins have a common structure (Fig. 2.4) in which a carbon atom (the α-carbon) is linked to a carboxyl group, a primary amino group, a proton and a side chain, which is different in each amino acid. Only the L-isomers of all these amino acids are routinely used for making peptides and proteins. Two special amino acids, selenocysteine and pyrrolysine, can also be incorporated into proteins in some organisms. Short chains of amino acids are called 'peptides'; long ones are called 'proteins'. As a result of the interactions among the different constituent amino acids, longer chains usually fold in a very specific way, thereby acquiring a specific 3-dimensional structure. This spatial configuration is usually essential for the proper functioning of the protein. Sometimes a change in only one amino acid of the chain, the result of a mutation, is sufficient to cause the loss of biological activity.

The functions of proteins are:
- enzymes: they strongly enhance the rate of biochemical reactions;
- transport and storage: for example, hemoglobin carries oxygen;
- structure (e.g. actin in the cytoskeleton) and development;
- nutrition, for example, milk (casein) and egg (ovalbumin) proteins;
- signaling, for example growth hormone;
- immunity: antibodies are proteins;
- regulation: by binding to DNA (e.g. transcription factors) or many other molecules.

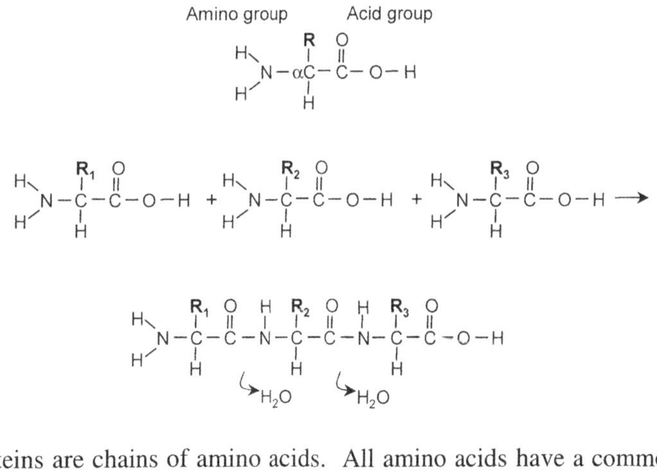

Figure 2.4 Proteins are chains of amino acids. All amino acids have a common basic structure in which a carbon atom (called the α-carbon) is linked to a carboxyl group (-COOH) a primary amino group (-NH$_2$), a proton (H) and a side chain (R from residue), which is different in each amino acid (R_1, R_2, ..., R_{22}). The reaction shows how amino acids are linked.

In the conceptual context of this chapter, we will focus only on the relationship between nucleic acids and proteins, because these are responsible for the bulk of biological variability that has emerged in the course of evolution.

2.5 Genes and heredity

Genetic information invariably resides in nucleic acids. The unit of genetic information is called the gene. Thus all cells of all organisms have genes. It is not the gene that does the work in cells. The gene only codes for a protein, and it is the protein itself that finally does the work. Following the explanation below of the principles of protein synthesis, a more detailed definition of a gene will be given.

It is not superfluous to emphasize again and again that the hereditary variability occurring in nature is primarily based on the variability occurring in proteins. The number of different proteins that can be made with the 20 different amino acids that can be incorporated into proteins is simply endless. For example, in producing a given protein, all of the 22 different amino acids can be used or only a few ones. The sequence in which the amino acids occur as well as the length of the chain can vary. After the chain of amino acids has been synthesized, some amino acids can undergo subsequent chemical modifications.

2.6 A gene invariably codes for a chain of amino acids, never for a (poly)saccharide or a lipid

Chains of amino acids, and thus of peptides and proteins, can only exert their function(s) properly when each of their constituting amino acids is correctly positioned in the chain. Therefore, each cell must have a highly reliable coding system to ensure that each amino acid is correctly incorporated in the right position.
When James Watson and Francis Crick described the double-helix structure of DNA for the first time (1953), they were aware of the possibilities of such a structure as a carrier of genetic information. At that time however, it was not yet clear how DNA, with only four different nucleotides, could possibly code for positioning 20 different amino acids in any protein.

After about 10 years of additional research in several laboratories, the explanation was finally found. It is now known as the (first) 'central dogma' (of biological hardware) and it reads:

$$DNA \rightarrow RNA \rightarrow Protein(s)$$

The formulation of this dogma required that solutions be put forward for several different problems.

First problem

How can only four different nucleotides (A,T,G and C) in DNA suffice for having 22 different amino acids incorporated in the proper sequence in whatever protein?
Solution
> Three adjacent nucleotides constitute a unit known as the *codon*, which codes for an amino acid. For example, the sequence AUG is a codon that specifies the amino acid methionine.'

With four nucleotides (letters), one can make 4x4x4 = 64 different triplets, while in principle 22 such triplets should suffice, one for each type of amino acid. These 64 triplets are all used. A few function as 'start codons' (AUG en GUG), while others (UAA, UAG en UGA) function as 'stop codons'. Some amino acids can be coded for by more than one triplet: in such cases the code is said to have degenerated. For the amino acids arginine and leucine, there are even six different synonyms, as illustrated in the following table (2.1.). Some modifications of the genetic code are found in certain bacterial, mitochondrial and protozoon species (Turner et al., 2000). The amino acids selenocysteine and pyrrolysine were recently discovered.

Table 2.1 The twenty two different amino acids that can be incorporated into proteins and their corresponding codons (Turner et al., 2000).

Alanine	GCU	GCC	GCA	GCG		
Arginine	CGU	CGC	CGA	CGG	AGA	AGG
Asparagine	AAU	AAC				
Aspartic acid	GAU	GAC				
Cysteine	UGU	UGC				
Glutamine	CAA	CAG				
Glutamic acid	GAA	GAG				
Glycine	GGU	GGC	GGA	GGG		
Histidine	CAU	CAC				
Isoleucine	AUU	AUC	AUA			
Leucine	UUA	UUG	CUU	CUC	CUA	CUG
Lysine	AAA	AAG				
Methionine	AUG					
Phenylalanine	UUU	UUC				
Proline	CCU	CCC	CCA	CCG		
Pyrrolysine	Uncommon amino acid: special coding					
Serine	UCU	UCC	UCA	UCG	AGU	AGC
Selenocysteine	Uncommon amino acid: special coding					
Threonine	ACU	ACC	ACA	ACG		
Tryptophan	UGG					
Tyrosine	UAU	UAC				
Valine	GUU	GUC	GUA	GUG		

STOP — UAA UGA UAG (= no amino acids except in exceptional cases).

START — AUG, the codon for methionine, is commonly used as a start codon. This means that the first amino acid of any protein should be methionine, at least in principle. This methionine, however, is often cleaved off from the nascent protein. The second possible start codon, GUG, is used only exceptionally.

Second problem

In eukaryotic cells DNA resides inside the nucleus but the incorporation of amino acids into proteins takes place in the cytoplasm. How does the required information reach the site of protein synthesis, namely the ribosomes? In other words, how is the genetic information transported from the nucleus to the cytoplasm? Is there a molecular transporter?

Solution

The discovery of a second type of nucleic acid, namely ribonucleic acid or RNA, that is always single-stranded, triggered the solving of this problem. RNAs are always formed in the nucleus. They can migrate out of the nucleus into the cytoplasm. The RNAs that carry the code for protein synthesis are called messenger RNAs (= mRNA).

The system functions as follows. DNA serves as the matrix for the synthesis of RNA. The two strands of DNA separate. Against one of these, a complementary strand of RNA can be formed with the help of the enzyme RNA polymerase. Opposite to thymine in DNA fits adenine (A) in RNA, opposite to guanine (G) fits cytosine (C) in RNA, and opposite to cytosine (C) fits guanine (G) in RNA. Opposite to adenine (A) in DNA, however, we find not thymine but rather uracil (U) in RNA.

The term 'transcription' concerns the transformation of the information residing in DNA (A,T,G,C) into information contained in the RNA language (A,U,G,C). Transcription cannot start just anywhere: the RNA polymerase enzymes need a recognition sequence at the beginning of a gene. The normal situation is that a specific region located upstream from that part of the gene that codes for the protein plays a role in the initiation of transcription. This is called the 'promoter region'. There is also a signal for initiation of transcription. The polymerases also need information for stopping the incorporation of nucleotides.

In 2001, Matthias Hentzle reported that some proteins can be synthesized inside the nucleus, thus without the help of ribosomes in the cytoplasm.

Third problem

How is the complementary code contained in mRNA translated into a chain of amino acids by the ribosomes?

Solution

Two additional types of RNA are required for this, namely ribosomal RNA (rRNA) and transfer RNA (tRNA).

rRNA is organized, together with a number of proteins, into more or less spherical structures that are called ribosomes. A ribosome consists of two subunits, a larger one and a smaller one, which, depending upon the conditions, can either associate or dissociate. In between the two subunits, there is a groove that can hold an mRNA molecule.

Transfer RNAs have the structure of a knot or cloverleaf-like structure. At one of their ends, they can carry a specific amino acid. There is at least one tRNA for each of the 20 different amino acids. Each tRNA has a specific 'anticodon' that is complementary to a codon present in the mRNA. This anticodon can bind at the complementary codon site in the mRNA, but only if the mRNA is 'imprisoned' in the groove between the 2 ribosomal units.

The linking together of the amino acids is done in the following way (Fig. 2.5).

- First, the 5' untranslated region of the mRNA binds to the small subunit of a ribosome.
- A first tRNA binds to the start codon. It usually carries the amino acid methionine.
- This results in a small conformational change of the small subunit that allows the larger subunit to bind to the smaller one. The mRNA strand is thus imprisoned.
- Next, a second tRNA binds to the second codon in the mRNA.
- The first amino acid, methionine, is enzymatically linked to the second one, which is still attached to its tRNA. This way, a dipeptide is formed.
- The first tRNA is detached and will pick up another methionine in the cytoplasm.
- The mRNA is displaced over the length of one codon so that the third codon on the mRNA is exposed.
- The complementary tRNA binds to it, and the already formed dipeptide is transferred to the third amino acid with the resulting formation of a tripeptide.
- This click movement will continue until a stop codon is encountered.
- The formed chain of amino acids detaches, and the peptide or protein has been formed. It will adopt its typical 3-dimensional structure and some of its amino acids will perhaps also undergo additional modifications. The protein will be transported towards its final destination, either inside or outside the cell. Protein synthesis is completed.

Although this is not at all common practice, I refer to this dogma as to the *first* dogma, thereby suggesting that there must be a second one. The reason is that this dogma deals only with the genetic information that is necessary for the continuation of the hardware. However, living communication systems also have a software aspect. The type of memory required for the functioning of this software is different from that which required for the hardware, and it is still poorly understood. I believe that a second central dogma is needed to deal with this non-genetic memory. I will elaborate on this in later chapters.

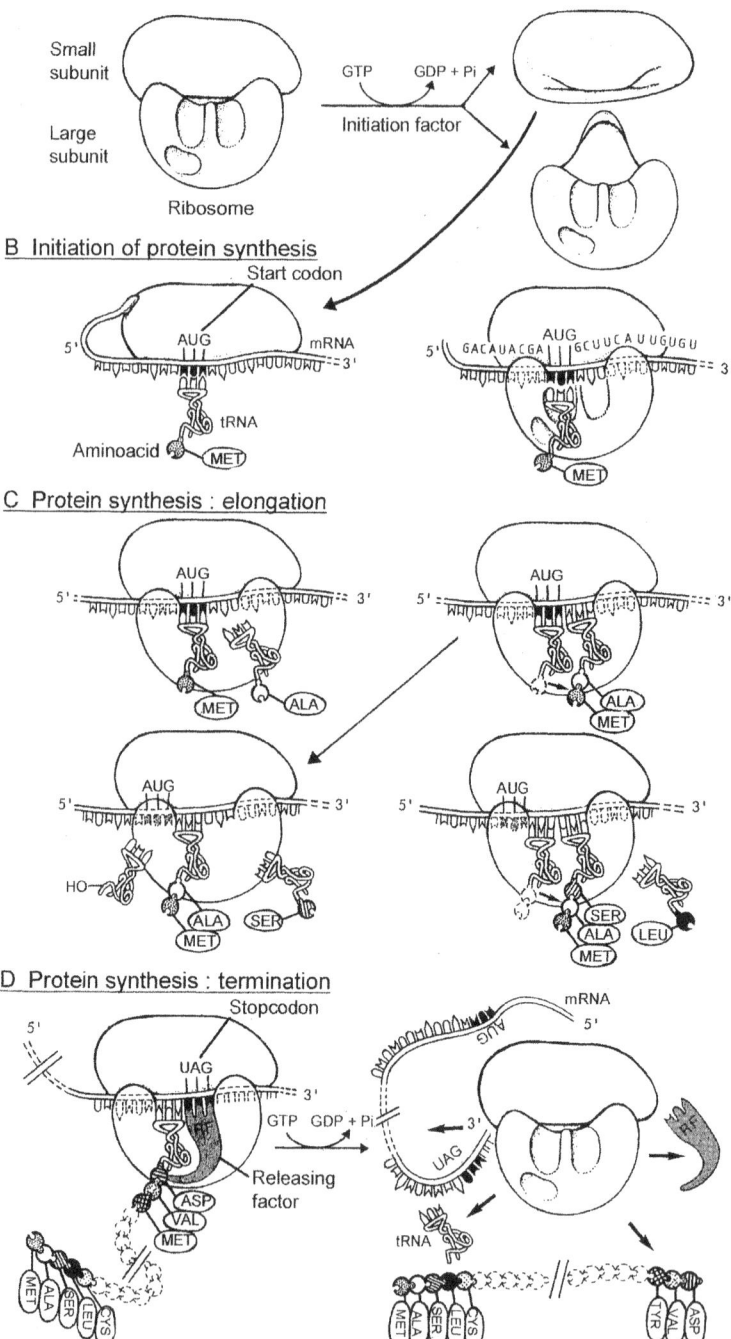

Figure 2.5 Schematic representation of the translation of the information contained in a messenger RNA (mRNA) molecule into a corresponding chain of amino acids. The major players are: the large and small ribosomal subunits, the messenger RNA, the different transfer RNAs that each carry one of the 22 different amino acids. See text for more details. Redrawn after C. De Duve, The Living Cell.

2.7 "One gene, one protein" is too simplistic. Introns and exons

According to the mechanism that has been outlined in 2.6, a given organism should not contain more proteins than it has genes. This has been found to be incorrect. It is possible to detect more proteins than there are genes, in particular in eukaryotic cells. The estimation is that man has three times as many proteins (~90,000) as he has genes (~35,000: see later in this chapter), without considering proteolytic processing or post-translational modifications such as phosphorylation and sulfatation (Galas, 2001).

There are several explanations for this anomaly. One possibility is that after their synthesis in the form of a precursor, some proteins can be cleaved into two or sometimes even more parts. Each part can have a distinct function. A second possibility, only valid in eukaryotes, is based upon the occurrence in some genes of exons and introns. To explain what this means, a more detailed description of the structure of a gene is required. In addition, some differences between a prokaryotic gene and an eukaryotic one need to be highlighted (Fig. 2.6).

Prokaryotes have only a single chromosome, which is compacted into a nucleoid and attached to the plasma membrane. It is circular. It consists of a double DNA helix, just as in eukaryotes. All codons situated in between the start codons and the stop codons code for an amino acid. In eukaryotes, however, the situation can be much more complex. Simply said, in between the start and stop codons of certain genes, there can be nucleotide sequences that are written in normal DNA language and that are normally translated into the corresponding mRNA, but which are finally not translated into amino acids. Such regions in the genes are called *introns*. The regions that do code for amino acids are called *exons*. A gene with one intron will have two exons. One with two introns will have three exons, etc. After the mRNA has been completely transcribed, the introns are enzymatically removed and degraded. The exons are linked together. The resulting RNA chain will be translated into a protein in the normal way.

Following example may serve to illustrate the mechanism. The fruit fly *Drosophila* has three sex-determining genes. One of them is called '*doublesex*'. In both males and females, the pre-messenger RNA of this gene contains five exons that are separated from one another by four introns. This mRNA is cleaved in a gender-specific way. In females, only the exons 1, 2, 3, and 4 are joined after the introns have been removed. In males, the exons 1, 2, 3, 5 and 6 are joined.

The following general concept emerges:

- A gene is materialized as a double stranded DNA fragment and is best defined as a functional transcriptional unit that includes regulatory sequences (gene promoter) in addition to exons and intron sequences (if present).
- The genetic code is universally valid for all organisms, not taking into account a few very special bacterial exceptions.
- Although the adage "one gene, one protein" can serve as a guiding principle, there are several possibilities for the synthesis of more than one protein from a single gene. The fact that an organism can produce more proteins than it has genes does not contradict the (first) central dogma.

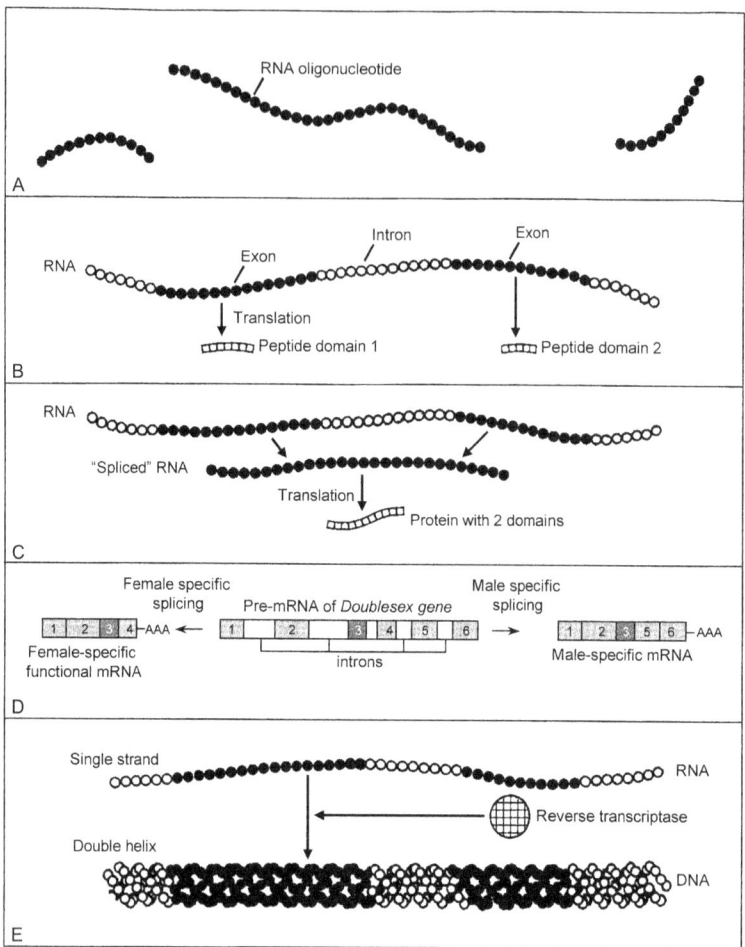

Figure 2.6 According to the 'Tomas Cech' hypothesis: (A) life came into being long ago in an 'RNA world', which preceded the 'DNA world' as we know it today. In this early 'world', RNA was the carrier of genetic information. (B) In the course of time, it was not the full nucleotide chain of certain eukaryotic genes but only parts of this chain that came to code for the corresponding protein. A coding region is called an 'exon'; a non-coding region is called an 'intron'. Prokaryotes do not have introns. (C) The messenger RNA that is transcribed from such a gene contains both the intron and the exon regions. The introns have to be removed before translation takes place in the ribosomes. This enzymatic process is called 'splicing'. (D) Sometimes, differential processing of messenger RNAs containing several exons and introns is used for controlling certain physiological processes. For example, in the fruit fly *Drosophila melanogaster* such differential processing is instrumental in determining either the male or the female phenotype. (E) Certain viruses called 'retroviruses' contain a gene that codes for an enzyme that can transcribe messenger RNAs in reverse order into the corresponding DNA sequence. Such enzymes, which are called 'reverse transcriptases', have become prime tools in molecular biological laboratories.

- As a consequence, a given organism that has more genes than another organism is not necessarily more complex, nor is it more advanced in evolutionary terms (see below). The number of proteins and the number of possible interactions with other proteins and macromolecules are more important than the number of genes.
- The idea that a single gene codes for a single given trait, (which grew out of certain clear-cut examples of Mendelian genetics such as the inheritance of flower color in peas), is often too simplistic, as well. Many physiological functions are controlled by more than one gene. Thus, not that many genes are required to generate substantial variation.
- The switching on and off of genes is done inside the cell, a fact which evidently does not exclude the possibility that this switching can be a response to stimuli coming from the environment.

2.8 How can genetic information change? Mutation versus modification

One intrinsic property of the genetic information is that it can change. Such changes are called mutations. Mutations are responsible for the appearance of totally new possibilities/features/traits in the gene pool. The term 'modification', by contrast, denotes a non-hereditary change in the phenotype (= external appearance) induced by the environment, without any change in the DNA. The tanning of our skin upon exposure to sunlight represents a modification. For the evolution of the 'hardware' of organisms, only mutations are important.

A mutation is a permanent hereditary change in the nucleotide sequence of the DNA, in particular in the coding region of a gene. Numerous things can go wrong with DNA, but cells also have mechanisms for quickly and efficiently repairing the majority of these errors and damaged sequences. Thus, permanent changes (mutations) are rather rare. Furthermore, only mutations that occur in the germ cells of sexually reproducing organisms are important. Mutations in the somatic cells of an organism are lost upon death.

During the cell cycle, the DNA helix is duplicated. The enzymes involved can make mistakes. High-energy radiation originating from supernovae that long ago exploded somewhere in the universe is incessantly bombarding all organisms on earth. This sometimes causes irreparable damage to a gene. Some chemicals can also damage DNA: they are referred to as 'mutagenic agents'.

Point mutation: one type of nucleotide is replaced by another, (e.g. A by T). Some point mutations have no effect on the resulting amino acid, but others do (see Table 2.1.). When a codon changes in such a way that it no longer codes for an amino acid but rather for the signal 'stop' (= nonsense mutation), then protein synthesis will stop.

Deletion: one or more nucleotides disappear. When one or two nucleotides are omitted, the total reading frame shifts and the resulting chain of amino acids will totally change (see example). If one or more complete codons are omitted, the reading frame does not change but the resulting chain of amino acids will be shorter.

mRNA before mutation	A U G Methionine	U G G Tryptophan	G C U Alanine	U A A Stop	A-A-A-A-... Untranslated
After point mutation	A U G Methionine	U U G **Leucine**	G C U Alanine	U A A Stop	A-A-A-A-A Untranslated
After deletion (of U in 2nd codon)	A U G Methionine	U G G **Tryptophan**	C U U **Leucine**	A A A Lysine	A A A Lysine

Insertion: the addition of one or more nucleotides. The effects are similar to those of deletion.

Inversion: a stretch of DNA is inverted in the chromosome.

Duplication: one or more genes are duplicated. The duplicates sometimes migrate toward other locations in the genome.

Translocation: a gene, or even a piece of a chromosome, takes up another location in the genome. This can result in a change in function.

Recombination: exchange of homologous regions between two DNA molecules. This often happens during the meiotic cell division that functions in the production of sperm and egg cells in eukaryotic organisms that reproduce sexually.

Transposition: transposons are small DNA sequences that can move about freely in the cell. In principle, they can insert themselves at any location in the genome.

There are still other sources of genetic variation, but they will not be dealt with here.

This all means that there is no lack of possibilities for change. It is remarkable that, all in all, so few mistakes are made. Mutations are rather rare, apart from the very frequent ones in certain viruses (e.g. in the AIDS virus). The rare mistakes resulting in viable progeny that have accumulated in the course of 3.7 billion years of evolution have resulted in the genetic polymorphism and enormous biodiversity in hardware that we observe around us.

The possibility of duplicating genes or even whole genomes is a very powerful tool for evolutionary change. While one gene can continue to exert its normal function(s), the other can mutate without causing too much harm. If in the end this results in a novel function that makes an organism more fit, there is a possibility that such a gene will do well in the population.

2.9 Micro-, Macro- en Mega-evolution. Evolution as a process of alienation resulting in a huge Babel-like confusion of tongues

In the classical, genetics-based view, Evolution refers to the process by which new types of organisms originate out of existing ones through an accumulation of mutations over sufficiently long spans of time. The focus is on increasing complexity, a term that is difficult to define (see 2.10 and Chapter 15).

Micro-evolution concerns changes at the level of the species (or lower, i.e. of populations) resulting in the formation of a new species (Fig. 2.7). This is the major focus of (neo)-Darwinism. Its genetic mechanisms are well understood. The basic rule of micro-evolution, in its strictest neo-darwinian interpretation, might read: **substitute one allele (= variant of a gene) for another** to drive adaptation at the **population/species level** (after Kauffman, 1993). This does not at all mean that one single mutation suffices for bringing about a new species. Chapter 15 will explain how the concept of 'cultural evolution' (i.e. non-genetic evolution through teaching/learning) can be fit into a broader theory including both cultural evolution and micro-evolution.

Macro-evolution
According to Futuyama (1998) "Macro-evolution is a vague term for the evolution of great phenotypic changes, usually great enough to allocate the changed lineage and its descendants to a distinct genus or higher taxon." It encompasses the origin of novel designs (e.g. forelegs, wings and fins in the different classes of vertebrates (Fig. 2.7): Gilbert, 1997), evolutionary trends, new kinds of organisms penetrating new habitats, and major episodes of extinction (Raven and Johnston, 1996). Its mechanisms are not very clear. I think that the disagreement among paleontologists, developmental and evolutionary biologists whether the cited definition is the best possible one might be reduced if a consensus could be reached on a basic rule. My proposal is: macro-evolution is based upon substantial **changes in an existing signaling pathway** resulting in changes in physiology and morphology **beyond the species level**.

The best documented signaling pathway in macro-evolution is the so called homeobox genes signaling pathway, known as the *Hox* pathway. Changes in this pathway are instrumental to the big changes in the anatomy of the different classes of both arthropods (insects, crustaceans etc.: Ronshaugen et al., 2002) and vertebrates (fishes, amphibians, reptiles etc.: Gilbert, 1997). Not everyone is convinced that there is need for 'macro-evolution'. Indeed, Carroll (2001) proposes that macro- = micro- = evolution. In other words, there is only one big picture of evolution. This is obviously true but that does not eliminate the need for an appropriate terminology to denote the level at which one is studying the mechanisms of evolution. I am in favor of maintaining the term 'macro-evolution'.

Mega-evolution
If one agrees to maintain the term 'macro-evolution' in its present interpretation, then there is need for still another term to denote the mechanisms of evolution at still higher levels than the ones micro- and macro-evolution are dealing with. I propose the term 'mega-evolution'. Mega-evolution operates at a still higher level than macro-evolution and it is not linked to given systematic entities. It deals with the formation of 'communicating compartments' of ever higher order, including at least 16 to date as will be outlined in Chapter 5. Basic rule: **add novel problem solving strategies**, either genetic or cognitive ones, so that **additional levels of compartmental organization** can start functioning.

The concept of mega-evolution was created in an attempt to explain the evolution of 'life', and hence it requires that 'life' first be defined (see Chapter 7). This level of evolution is approached from the principles of communication: a purely genetic approach does not suffice. It requires insight into the hierarchical organization of living matter and into the variety of ways in which cells, organisms, populations, etc. can interact. Since the introduction of neo-Darwinism, the genetic basis of the divergence of populations has been focussed on so heavily that little interest has remained in the consequences of evolution for the way biological entities communicate and live together.

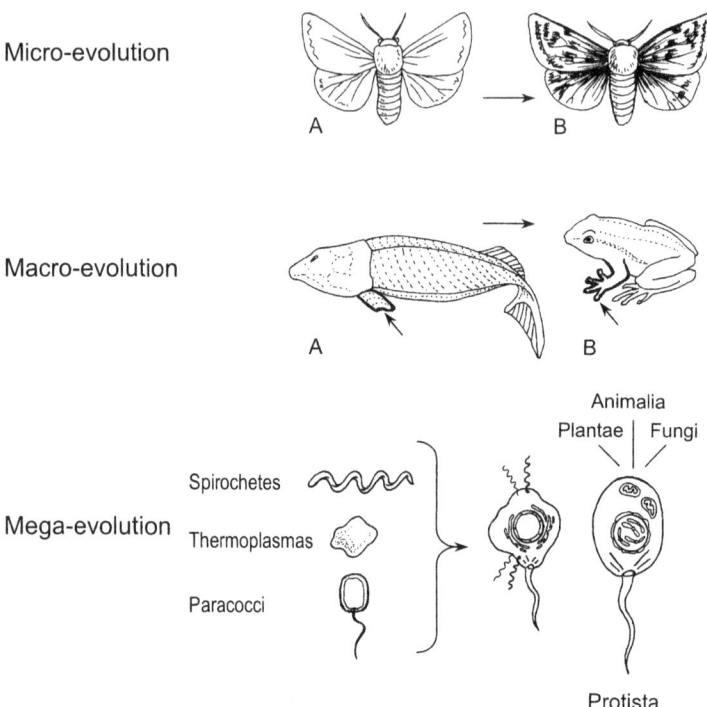

Figure 2.7 Schematic representation of the difference in scale of micro- macro-, and mega-evolution. Micro-evolution operates at the level of the population/species, and macro-evolution operates at the level of the genus or even higher. Mega-evolution, which crosses the borders of classical classification systems, deals with big changes in compartmental organization. One such big change was the formation of the eukaryotic cell type starting from several species of bacteria (the symbiont theory of Margulis). Many other examples of mega-evolution are given in Chapter 4.

A key idea in mega-evolution is genetic and communicational alienation. This is another way to look at evolution. Life on earth started with one single cell, the Progenote. As long as no mutations occurred in its progeny - thus as long as perfect clones were generated - the degree of kinship was 100 percent:. All these clones most probably used the same 'standard language' for their communication. From the first mutation onwards, however, the process of alienation started. The mutated organism was no longer 100 percent identical to its 'parent'. As time passed, more and more mutations accumulated. As a result, the diversity of phenotypes and communication systems increased. At a given moment the alienation had progressed so far that, parallel to the 'standard language', a variety of dialects had come into existence. In the end, some of the dialects achieved the status of a new language which was no longer understood by members of the principal population. The easiest way to avoid or minimize problems and conflict between the original and the mutant populations was for the two to separate and for one or both of them to colonize a new habitat. This process of alienation is going on all the time. To date, there are at least as many different 'dialects' as there are subpopulations in the different species that make up the biosphere.

Thus, evolution alienates both genetically and communicationally. One could say that evolution, by making use of mechanisms that result in genetic degeneration (mutations) and, where relevant, by utilizing processes of dialect formation and teaching-learning situations, has resulted in an enormous confusion of tongues among all the descendants of the Progenote. The inability of organisms to understand all possible languages has made it necessary to find means of cohabitation. The broad range of stategies for living together will be dealt with in the next chapter.

2.10 Complexity. The size of the human genome as compared to that of other organisms

It is not easy to define 'complexity' (Maynard Smith and Szathmary, 1995). A machine is considered to be more complex than another when it consists of more parts or/and when these parts are connected to each other in a more sophisticated way. According to this way of thinking, an organism with more genes should be more complex than one with lesser genes. Quite often an additional extrapolation is intuitively made by assuming that the larger the genome of a species, the more advanced the species is in evolutionary terms. However, this way of thinking is too simplistic and largely erroneous. This fact becomes apparent upon analysis of the following hard data resulting from recent intensive efforts in genome sequencing of an increasing number of species.

Organism	Approximate number of genes
Bacterium (*Hemophilus influenzae*)	~1,700
Fission yeast (*Schizosaccharomyces pombe*)	~4,800
Yeast (Bakers' yeast: *Saccharomyces cerevisiae*)	~6,150
Plant (Thale cress: *Arabidopsis thaliana*)	~25,700
Nematode worm (*Caenorhabditis elegans*):	~18,250
Insect (Fruit fly: *Drosophila melanogaster*) (Heidelberg Assembly, 2002)	21,396
Vertebrate (Man: *Homo sapiens*)	~32-35,000
Vertebrate (Mouse: *Mus musculus*)	estimation: ~25,000

Schizosaccharomyces pombe contains the as yet smallest number of protein-coding genes yet (= early 2002) recorded for a eukaryote (Wood et al., 2002). Until the end of the year 2000, it was thought that man would have in between 50,000 and 150,000 genes. The International Human Genome Sequencing Consortium reported in 2001 that *Homo* has 'only' ~32-35,000 genes (Nature, 2001, volume 409, several papers). Craig Venter et al. from the Celera consortium (2001) came to a similar conclusion. According to their data, *Homo* has 26,588 protein encoding transcripts and an additional 12,000 computationally derived genes with mouse matches or other weak supporting evidence. Only 1.1 percent of the human genomes are spanned by exons, whereas 24 percent are in introns, with 75 percent of the genome being intergenic DNA. Although some more sequencing remains to be done, it is thought that the total number of 35,000 genes will not change much any more.

Thus, with respect to his total number of genes, *Homo* is only five times as complex as a bacterium like *Pseudomonas aeruginosa* but less compex than rice (Goff et al., 2002; Yu et al., 2002). This is a serious blow for the feelings of superiority that *Homo* undeniably displays towards all other species. That there is no justification for such an attitude in terms of total number of genes becomes even more clear when additional data in terms of the size of the genome, most of it reported by Capy (2000), is taken into account.

First some terminology: one kilobase (Kb) = 1000 base pairs (bp) in DNA; one Megabase (Mb) = 1 million base pairs: one Gigabase (Gb) = 1 billion base pairs.

Virus
    Bacteriophage ΦX174 (11 genes):   5,386 bp

Unicellular organisms:
    Saccharomyces cerevisiae: 14 Mb
    Amoeba: > 200,000 Mb

Plants:
    Ferns: 307,000 Mb
Flowering plants (average):   50 Mb
    Arabidopsis: 125 Mb
    Rice: 430 Mb
    Maize: 3,000 Mb
    Wheat: 16,000 Mb

Invertebrates:
    Sponges: 49 Mb
    Caenorhabditis elegans (nematode worm):   100 Mb
    Drosophila melanogaster (insect):   165-180 Mb

Vertebrates:
    Fugu rubripes (Japanese pufferfish):   400 Mb
    (= the smallest vertebrate genome)
    Bony fishes (average):   139,000 Mb
    Mammals (average):   3,000 Mb
    Homo sapiens: 3,200 Mb

This table clearly illustrates that there is no good correlation between the size of a genome and our intuitive perception of organismal complexity (Szathmary et al., 2001). This and other data made Claverie (2001) conclude that neither the cellular DNA content (in mass) nor its gene content appears directly related to our intuitive perception of organismal complexity in terms of cell types, brain circuitry and cultural achievements.

For example, why should a 'simple' plant like wheat 'need' a genome that is much larger than that of *Arabidopsis* or even of *Homo*? In my opinion, the most spectacular figure in this list is that of the amoeba.

Claverie (2001) furthers suggests that it would be better to relate the complexity of an organism to the number of transcriptosomes (= universe of transcripts for a given genome) that a given genome can achieve. In this view, the human genome with its ~35,000 genes is not less than 10^{3000} times more complex than that of a nematode species with ~20,000 genes. This is evidently exaggerated as not all genes are essential, some theoretical transcripts are lethal, etc.

I think there is a much more satisfactory approach to measuring complexity, namely by estimating the different types of problems a given organism can solve. *Homo* will be orders of magnitude higher on this scale of complexity than any other organism. In Chapter 8 I will explain, with a flute as an example, how complexity in (communicational) output multiplies by introducing a very few additional mutations. It is the way in which combinations of gene products are made that matters, rather than the number of genes or gene products themselves.

A different and more positive view of these figures holds that evolution did a wonderful job with *Homo*: with a mere 35,000 parts a machine was made that can think and that can master all the known physical forces of nature.

Only 1.1 percent of all the DNA of *Homo* codes for proteins (Venter et al., 2001). It is far from clear what the significance can be of the remaining 98 percent of the DNA. This part is often referred to as 'junk DNA', the function of which, if any, remains to be elucidated. One view is that this junk DNA harbors genetic information from the past that is no longer in use and that awaits elimination from the genome. However, eukaryotes do not seem to have adequate mechanisms for eliminating DNA that is no longer in use. Another view is that this DNA serves as the factory where as the result of mutations, new genes can come into existence. My idea is that a long strand of DNA in the nucleus requires the presence of a similarly large chromosomal skeleton-complex. If this skeleton has some other function than just serving as an anchor for DNA, then the non-coding DNA might have an indirect function. In a later chapter, I will put forward the hypothesis that the actin present in the skeleton of the chromosomes as well as in the cytoplasm, at least in theory might be a suitable candidate for acting as a carrier of the non-genetic memory.

It is now known that the genetic difference among human individuals (the degree of heterozygoty) amounts to about 1 in 1,300 base pairs (Nature 2001, vol. 409, p. 911)

2.11   Degree of kinship: man with other organisms

The static worldview of the Bible tells us that God created the plants, the creatures of the sea, the birds of the air, the creatures of the land and, finally, man – all as separate entities. This story may be literally eloquent, but it is scientifically incorrect. The genomic analyses mentioned in 2.10 make it clear that less than 1 percent of all genes that *Homo* has are specific for *Homo*. The 99 percent remaining genes have homologues in other animals, and many of them even in 'lower' organisms. About 30 percent of all proteins of eukaryotes display substantial similarities in their amino acid sequence as well as in their 3-dimensional structure and their functions. This set of essential proteins is called the 'core proteome'. It must have been present in the common ancestor of all organisms that share this core proteome.

Some 3.7 billion years ago, when the earth's surface was still warm, the first form of life appeared on earth. Most probably it was a bacterium. This Progenote succeeded in dividing itself in such a way that the resulting daughter cells could regenerate, and divide again in an endless series that continues up to the present day in all organisms. The data obtained by molecular biological techniques clearly point in this direction and support the hypothesis that all organisms living on earth are descendants of the Progenote. In other words, all organisms that together constitute the whole biosphere form one huge family. The similarities in the genomes of the worm *C. elegans* or of the fruit fly *Drosophila* to those of *Homo* are striking and much greater than anyone had dared to predict a few years ago. Thus, even plants are our cousins, and bacteria are our many times-removed cousines. No matter how different the morphologies of all creatures are, genetically we are all family.

2.12 The descent of man: Paleontology versus the 'Eve theory'

Homo made its entry as a separate species on earth only a few million years ago, probably somewhere in central Africa. Skull 1470 of 'Lucy', found by the Leaky's in Olduvai Gorge in Tanzania, probably dates from 2.6 million years ago (Leaky and Lewin, 1977). According to the out-of-Africa theory, man first colonized the entire African continent, and later the entire earth. According to the relatively recent Eve theory, Eve, the woman of whom all presently living humans are the progeny, did not live some 3 million years ago (the view of paleontologists), but rather only about 140,000 years ago. This theory is based on molecular biological data derived from mapping mutations in mitochondria of humans of all races. If this theory were correct, it would mean that all the descendants of the populations of humans that inhabited the earth at the moment of the appearance of the Eve of 140,000 years ago were eradicated by the progeny of Eve. Paleontologists oppose this Eve theory because if it were true, then the fossils would display abrupt changes in the skeletons of fossilized humans corresponding to the displacement of the old races by the new race. Such abrupt changes are not seen: rather, only gradual changes have been found. Additional genetic analyses support the view of the paleontologists. According to Templeton (2002) there were several waves 'out of Africa'. The very first expansion of *Homo erectus* out of Africa into Europe and Asia took place some 1.7 million years ago. This was followed by two additional waves of modern man that in the mean time had developed in Africa as well. The first of the two waves took place some 600,000 years ago, and the second one some 100,000 years ago. Genome analysis suggests that some interbreeding took place between the 'archaic type' and the 'modern type', not complete replacement. This means that some of the genes (about ten percent) of Neanderthal man and/or of *Homo erectus* still persist in the contemporary European and Asian populations. Thus, Neanderthal man as a species was not completely different from modern *Homo*, and neither was he completely eradicated by Cro-Magnon man. Not all specialists agree with this view (Stringer and Davies, 2001).

2.13 Philosophical consequences of the common-descent view. *Homo* and his soul

All organisms inhabiting the earth are products of 3.7 billion years of evolution. They all go back to the Progenote. Thus, the argument "I was first", which is meant to claim privileges for *Homo*, is invalid. Likewise invalid is the argument "I have more genes and thus more rights". The veganists claim that it is wrong to eat animal proteins, but that vegetal or bacterial proteins in the diet are OK. This view thus holds that it is wrong to eat one's close relatives (in evolution), but that the eating of more distant relatives is OK. I do not understand the 'logic' behind this view.

One very touchy matter concerns the existence of the 'soul'. Charles Darwin expressed his views on this point. In some cultures and religions, only *Homo* is supposed to have a soul, as is stated in the Opening Account of the Book Genesis: "So God created man in his own image, in the image of God he created him; male and female he created them." In such a view, body and soul/mind are not one but rather represent two distinct entities. This view, which is deeply rooted in western societies, is reflected in our education system, which distinguishes between the positive sciences and the humanities. In medicine, as well, the distinction is made between somatic illnesses and mental illnesses. Either denying that *Homo* has a soul independent of his body or claiming that all organisms have a sort of soul sounds like blasphemy against God to many people. Why? Mainly because in not restricting the soul to *Homo*, man looses most of his claims to superiority over other species.

Psychology is the study of the 'psyche', which is Greek for what the Romans called the 'anima' or 'soul'. Hence we would expect textbooks on psychology to contain an adequate definition of 'soul'. This is seldom the case. In fact 'anima' refers to the Latin word 'animal'. 'Anima' refers to obvious features in terms of which animals differed from plants in the eyes of the ancient Romans. Scientific psychology is not the study of the 'soul', as defined by religions, but rather of the straightforward interaction between personality and environment (Cuyvers, 2000). Thus, psychology does not answer the question as to what 'soul' is, and neither does it answer the question as to whether 'soul' is restricted to *Homo*. Dictionaries usually give a definition of 'soul'. The one in Longmans' Dictionary of contemporary English reads: "Soul; 1. the spiritual part of a person that is believed to continue after they die; 2. the part of a person that contains their true character, where their deepest thoughts and feelings come from."

The common ancestor of all mammals, the class to which the species *Homo* belongs, lived about 125 million years ago. If one assumes that only *Homo* has a soul, then one must assume that God or some supernatural force performed a special act of creation some 3-4 million years ago when the first humans evolved out the ancestors that they share with other primates. If one thinks that this is not a plausible scenario, only two alternatives remain. Either the human soul is coded for by the 1 to 2 percent of the genes that make *Homo* differ from other primates, or 'life' and 'soul' are so intrinsically interconnected that body and mind form one indivisible unit, not only for *Homo*, but also for all organisms.

The human conception of 'soul' is intimately linked to the fact that man knows that he will die sooner or later. The idea of the soul functions to ensure that we will continue to exist after our physical dearth. I vividly remember what one of my teachers – a wise man – in our primary school, located in a rural village, told us about death. He told his pupils that he believed that, just as the soul of man goes to heaven after death, so also the soul of a horse goes to the heaven of horses, the soul of a dog to the heaven of dogs, etc. This was a children's version of what could be called the 'biosphere soul' in which all organisms are interconnected and in which the totality prevails over the individual interest. Urban societies in particular become alienated from the biosphere-as-a-whole concept. Biology makes us face the facts: the species *Homo sapiens* represents only a very tiny part of the biosphere. *Homo* differs less from other species that had been assumed until very recently. He is also a descendant from the Progenote, like all other organisms. *Homo* is no doubt ingenious in technical matters and in problem solving but he is also the most aggressive species on earth. More humbleness and respect for the biosphere as a whole are virtues to be valued.

In later chapters, when dealing with the software aspect of living systems, the immaterial dimension will be focused on again.

I have referred a couple of times to the 'central dogma' as the '*first* central dogma'. This dogma only deals with the genetic memory of living systems. However, communication requires not only hardware but also software. Software, in turn, can be used only if software memory is available. In my view, there needs to be a second central dogma dealing with the transmission and perpetuation of non-genetic information. This topic will be treated in Chapter 3.

ESSENTIALS

BIODIVERSITY

1. The very first ancestor of all living organisms on earth was most probably a bacterium. It was named 'the Progenote'.
2. Prokaryotes or bacteria do not have a nuclear envelope or internal membranes. Eukaryotes do have a nuclear envelope and other internal membrane systems such as the endoplasmic reticulum, Golgi apparatus, mitochondria etc.
3. The attractive symbiont theory of Lynn Margulis says that the eukaryotic cell type came into existence when a larger bacterium was invaded by some other smaller bacterial species, which coalesced to form an entirely new entity. This means that, in fact, there is only one basic form of life of life on earth, the bacterial form.
4. The number of different species on earth: dozens of millions.
5. Methods of Classification: there were several possibilities but the systems based on homology and genetic kinship finally prevailed.
6. The five kingdom classification system: Monera or Prokaryota (bacteria), Protista, Plantae, Animalia and Fungi. More and more additional kingdoms are likely to be introduced in the future.
7. Genes invariably code for proteins, never for lipids or for saccharides. One gene can code for more than one protein. The (first) central dogma concerns the genetic memory. It states: DNA → RNA → Protein(s).
8. Genes can undergo mutations, which can take the form of deletions, inversions, substitutions, etc. Mutations are at the very heart of the mechanism underlying the evolution of the 'hardware aspect' of species.
9. **Micro-evolution** operates at the level of populations/species. It is the explanation for how new species come into existence. Unit of change: the gene.
   Macro-evolution concerns higher order evolutionary changes, namely at the level of the Order, or even of the Class (e.g. from reptiles to birds). Unit of change: the signaling pathway.
   Mega-evolution deals with the evolution of 'life', more specifically with the revolutions underlying the coming into existence of ever higher levels of compartmental organization, all hierarchically organized. Its formulation requires that 'life' be adequately defined. Unit of change: the additional problem solving strategy.
10. Evolution is a process of alienation, both genetically and communicationally, resulting in a huge Babel-like confusion of tongues.
11. *Homo sapiens* has about 35,000 genes, the fruit fly *Drosophila* 13,300. Some amoebas and plants have a much larger genome than *Homo*.
12. The species *Homo sapiens* is thought to have originated in Africa and to have spread to the other continents in successive waves.

CHAPTER 3

THE NATURE AND PURPOSE OF COMMUNICATION, THE MAIN ACTIVITY OF ALL BIOLOGICAL SYSTEMS

SOFTWARE

UNCONSCIOUS AND CONSCIOUS PROBLEM SOLVING

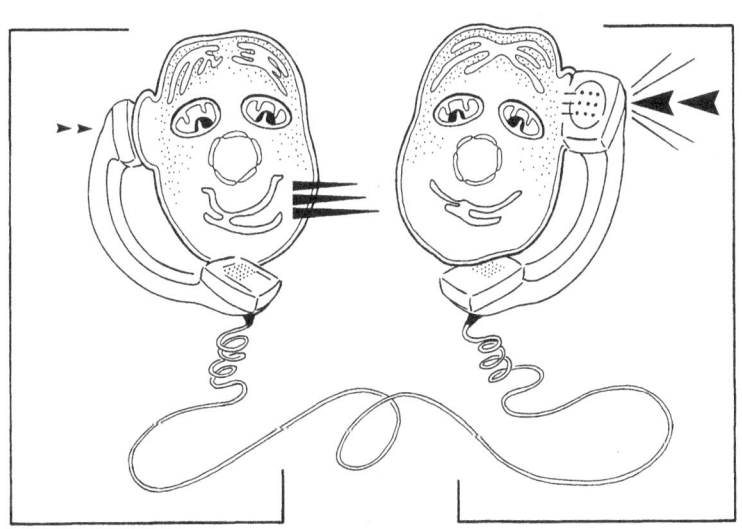

Communication is a complex process

Contents

3.1   Introduction
3.2   A few 'simple' questions
3.3   Definitions of communication, learning, memory, hardware and software
3.4   Basic architecture of a communication system. Why do we have a name?
3.5   The boundary of a communicating compartment must have 'functional holes'
3.6   Sources and sinks. The importance of gradient formation
3.7   Remembering. The need for memory capacity
3.8   Prerequisites for a reliable memory system
3.9   Some reflections and speculations on the molecular nature of the carrier of long-lasting memory
3.10 Characteristics and evolution of software
3.11 The energy question
3.12 Communication can master any force of physics
3.13 When is something information? A few definitions of information
3.14 The purpose of communication: Make others work for you in order to increase your own degree of contentment!
3.15 'Communication', a synonym for '(un)conscious problem solving'? Automation
3.16 The difference between communication and interaction. The (first) central dogma again
3.17 Communication and the anticipation of progressing time
3.18 Life as a double continuum. Two central dogmas?
3.19 Emotions, feelings, contentment, motivation
3.20 The basic architecture of a communication system in relation to the origin of the living state
3.21 Definitions of 'communicating compartment' and 'act of communication'

Essentials

3.1 Introduction

In the preceding chapter, biodiversity was dealt with from the point of view of a number of different species and their classification, as well as in relation to the size of the genome (in base pairs) and the number of genes. Later in this book we will concentrate more on the diversity of functions rather than of form. Hence, a relevant question is whether the chief activity of all living organisms, namely communication, likewise underwent a tremendous increase in complexity during geological time. In order to answer this and related questions, one has to clearly define what communication is and which goal(s) it serves. This is more difficult than one might think.

3.2 A few 'simple' questions

I used to tell my students that one can learn a lot from asking and answering simple questions about phenomena that are so perfectly common in our daily lives that we do not even consider the possibility that there might be a problem in understanding them. In Chapter 5 such a question will be asked, namely, '"When does an organism cease to be alive?"' Although we 'know' that a decapitated chicken is dead, the formulation of a plausible definition of death is not self-evident.

This chapter deals with the principles of communication. Again, as in the case of 'death', everybody thinks that he or she knows what the meaning of 'communication' and 'information' is, and how communication works, simply because we communicate all the time, from the very moment that we are born until we die. In general biology textbooks, the principles of communication, although essential in understanding 'life', are seldom explained.
Let us take this test: Here is a list of a few 'simple' questions. Try to formulate plausible answers.
a. The media bombard us daily with information about endless numbers of topics. But what is information? When does something become information?
b. Communication is essential to our lives. But what is it?
c. Why do we communicate at all? What is the purpose? What do we want to achieve by communicating?
d. What is the minimal architecture of the 'machinery' that is required to make biological communication possible? This question can be formulated in another way: "What are the essential elements of what is called a 'communication system'"?
e. Why do we need a name?
e. Communication is an activity and thus requires energy. Where does this energy come from and how is it liberated at the right moment?

I suppose that nearly all readers will have a hard time when trying to formulate answers, just as I had a few years ago when I tried to answer them myself. It cost me a lot an effort and I had to consult many colleagues from other disciplines before I could complete this chapter. I had not realized how complex communication is.

I had also not realized that messages, whatever their nature, are always in coded form, even if we think that they are not. When we hear a text in our mother tongue, we think that it is not coded. If the same message is transmitted in a language that the receiver does not master, he will think that the message is in coded form. To understand it, he must make a special effort to decrypt it. The decryption of coded messages and the generation of responses requiring energy are essential activities in communication.

3.3 Definitions of communication, learning, memory, hardware and software

Communication is a term that is commonly used in daily life. Everyone intuitively assumes that he/she understands its meaning. Therefore it may look strange when specialists in the field say that it is very difficult to formulate a definition that is universally acceptable for the variety of conditions under which communication plays a role. According to specialists in communication theory, over 150 definitions of communication have been formulated and there is still disagreement on which is the best one. One basic definition is: 'Communication is a transfer of information, and such transfer requires a functional communication system'.

In both living and mechanical (man-made) systems, communication invariably requires an instrument for communication (the hardware), programs for decoding-amplifying-responding (the software), memory capacity and energy.

Learned information is retained in a living system's memory. Learning is the acquisition of new information or knowledge. It involves the following sequence of events:
Consolidation
Sensory information → Short-term memory → Long-term memory

The following definitions with respect to the standard model of digital machines are taken from Bergstra and van Vlijmen (1999).

'The hardware is a mechanism containing a memory, in which it can somehow read and write. With great regularity (up to a few million times per second), the following cycle is executed:
- the introduction phase;
- the calculation phase;
- and the execution phase.'

The definition of software is more difficult. "The software is a special type of memory compartment. Such a compartment can be considered a program if circumstances prevail in which, during many subsequent executions of the central cycle, the regulation of how the memory has to be adjusted, in which channels the input must be read, and where a given output has to be delivered, causally depends upon the parameters that are stored in the compartment under consideration. This, together with other considerations, means that software has no sharp boundaries".

Although it is not yet common practice to do so, I will use the terms 'hardware' and 'software' in a biological context throughout this book. Young people, from their education and hobbies, are quite familiar with them. It is evident that the founding fathers of the theory of evolution did not think in terms of communication, which at the time did not yet play the dominating role that it does today.

To date, despite the tremendous increase in knowledge concerning the functioning of the brain in both vertebrates and invertebrates, little is as yet known about the processes that create our lifelong memories.

3.4   Basic architecture of a communication system. Why do we have a name?

In general, a basic communication system functions as follows: a *sender-encoder* produces a (coded) message, which is released into a *transmission channel*, through which it is transported to a *receiver-decoder-amplifier-responder*. This is represented in a simplified way in the diagrams in Fig. 3.1 and 3.2.

Because a message is usually sent to a particular receiver, this receiver should be identifiable. This is the essence of a name in the context of communication.

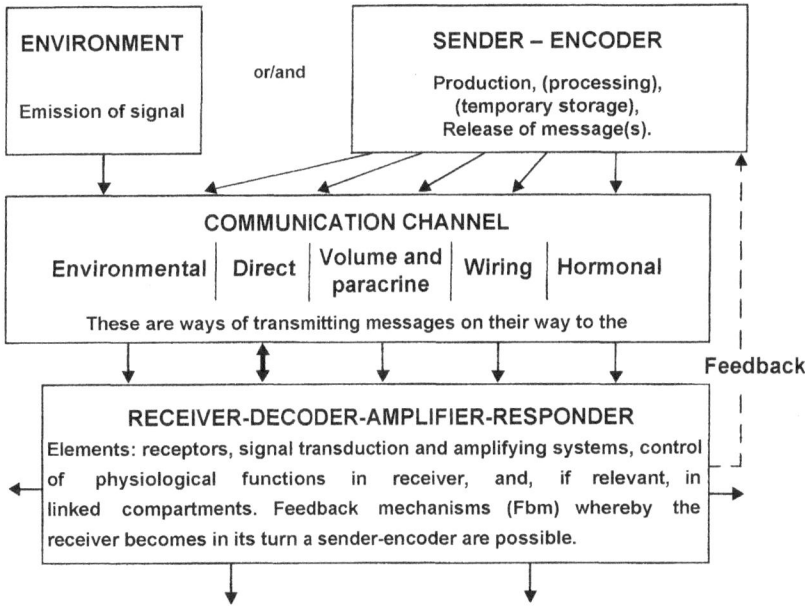

Figure 3.1 Schematic representation of the general architecture of a biological communication system.

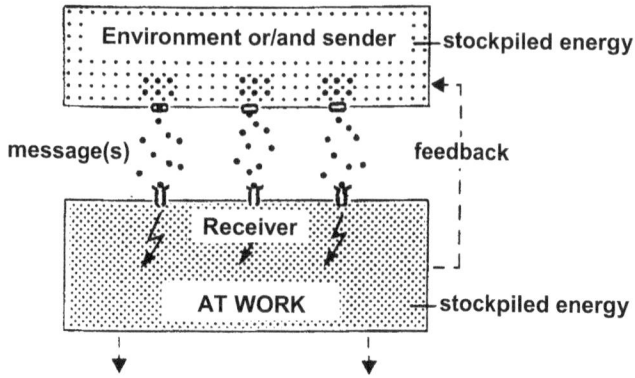

Figure 3.2 Schematic representation of a simple communication system, with emphasis on the need for stockpiled energy, gradients, and 'functional holes' in the limiting border of the sender and the receiver-decoder. The purpose of communication is to liberate at a given moment(s) specific forms of stockpiled energy in the receiver-responder to do some type of work at the appropriate moment.

3.5 The boundary of a communicating compartment must have 'functional holes'

The boundary of the sender-encoder must have the required properties for the message to leave the sender and somehow pass through the boundary of the receiver. Thus, *in the boundary of both the sender and the receiver there must be 'appropriate holes' that can open and close when there is a need to do so.* Senders and receivers with boundaries that are perfectly impermeable to any outgoing or incoming messages are completely useless for communication. At the cellular level, this means that a cell with a plasma membrane of glass or metal or any other impermeable material would not be compatible with communication and life. Thus, a moderately and selectively 'leaky' membrane is a prerequisite for communication at the cellular level. The boundary should not be too permeable, however, since then it would not be possible to maintain *gradients* over the boundary.

3.6 Sources and sinks. The importance of gradient formation

Another element that should not be overlooked is that communication is only possible if there is both a '*source*' and a '*sink*'. A messenger molecule will freely flow from one place to another in a unidirectional way only when there is a site where these molecules are present in a higher concentration (the sender) than in the target site (the receiver). This means that there must be a **gradient**, in this case a *chemical* gradient. Electricity can flow through a conductor only if there is a *potential* gradient. As I will discuss in Chapter 9, such potential gradients are essential to life. Wherever there is communication, there must be some sort of gradient: communication is impossible without gradients. In later chapters, in particular when we deal with the 'dissipative nature' of living systems (Chapter 5) and with the '*Gradient-Provoked Swelling and Self-Selection Principle*' (the 'GP-Triple S Principle') (Chapter 12), the importance of gradients will become more apparent. The same principle is also important for understanding that the gradients used in communication can act as a driving force of evolution (Chapter 15).

Building up and maintaining a gradient requires energy. Living systems must continually be investing energy (derived from their food, from photosynthesis, etc.) in the formation and maintenance of their gradients. If they do not do so, then the gradients will run down, and when a permanent equilibrium is achieved, communication is no longer possible. At that moment, the biological system is dead. Thus it follows that living systems can only be in continuous thermodynamic *disequilibrium*, since they depend upon chemical and electrical gradients - gradients which, by definition, are characterized by a lack of equilibrium. The 'magic' word to keep in mind throughout the following chapters is *gradients*, *gradients* and more *gradients*.

3.7 Remembering. The need for memory capacity

The process of remembering draws on a diverse array of cognitive processes to construct a representation that is experienced as a copy of the original past (cited after Nature Reviews Neuroscience 2, 624, Sept. 2001).

We rarely realize that it would be impossible to function properly if we did not have a 'brain memory' (in quotations because some authors think that memory does not exclusively reside in the brain). In essence, we would not even know what happened a second ago and we would not be able to learn anything. Communication would be impossible. A receiver-decoder can only function if it can store the program (the software) to decipher the incoming message. The function of memory is to store information for shorter or longer periods. Under proper conditions, this information can be retrieved and used for certain purposes.

An example: I ask you to raise your right arm and to keep it up for about 3 seconds and then to go back to the initial situation:

DO IT NOW!

Some 4 seconds later...

Your right arm is in its original position again.

If you think about it, raising your arm would not be possible if your brain memory had not contained information about every single item in the command: arm, raise, now, second, 3, hold, go back ...

This command was information for you if your brain managed to associate the different words and give a meaning to this association.

Memory also makes it possible to cope with progressing time. If you responded instantly, there would not seem to be a delay between the perception of the message and your reaction. But I wrote this chapter long before you read this command. Thus from my standpoint you reacted only after a long delay. Or, in other words, I managed to anticipate the progress of time by writing down a command that had to be executed later. This illustrates that some information/messages can anticipate the progress of time and exert effects long after they come into being. The phenomenon of 'time' will be discussed in Chapter 11.

3.8 Prerequisites for a reliable memory system

The biggest pending problem in contemporary biology concerns the deciphering of the mechanisms that govern the non-genetic memory, including the input of information, its storage in some carrier molecules, its role in the calculation phase, and its retrieval. For this non-genetic memory, I propose the term 'cognitive memory', by way of analogy with the term 'genetic memory'. Learning processes are of prime importance in the context of cognitive memory.

With respect to memory strength in humans, three terms are used. Short-term memory lasts for seconds to hours, long-term memory lasts for hours to months, and long-lasting memory stores information for months and even for the entire lifetime. The transition from one type of memory to the next does not seem to be a simple, continuous process (McCaugh, 2000).

In recent decades, very substantial progress has been realized in neurophysiology and in the cognitive sciences in general. Researchers have already succeeded in making an 'intelligent' mouse by using molecular biological techniques to modify the expression of the so-called 'NMDA receptor' in the nervous system. A lot of data has also been gathered on 'long-term potentiation', on consolidation, on the role of synaptic contacts between nerve cells and on many other aspects of the nervous system and learning processes. Some researchers contend that our cognitive memory is not restricted merely to our brains, but that our whole body is involved (the *Embodied mind* theory of Francisco Varela et al., 1992).

3.9 Some reflections and speculations on the molecular nature of the carrier of long-lasting memory

In the following paragraphs only a few points will be raised concerning one aspect of the difficult problem of the nature of the molecular carrier that is used by cells for storing information coming from the learning processes.

The concept of the 'genetic memory' of the cells being present in the DNA molecule illustrates well which properties are required for a good long-lasting memory system. The DNA strand present in the nucleus of each cell of our body is about 3 meters long. It is composed of 3.2 billion nucleotides Less than 2 percent of all this DNA codes for all proteins, whose synthesis is governed by about 35,000 different genes using an alphabet of only four letters. Despite its length, this thread can be packed reproducibly and in an orderly manner inside the nucleus.

The interesting properties of DNA are:

- it enables the storage of huge quantities of information in a very small volume;
- its double-helical structure allows identical replication during each cell cycle;
- it is composed of an extremely limited number of different building blocks (only 4 letters);
- by definition, it lasts as long as the cell itself;
- most damage and errors that occur can be repaired;

- DNA of multicellular organisms can be used differentially. This means that one cell type uses a certain portion of its DNA, another type uses another part, etc. Thus DNA can be used for sharing tasks in a network environment;
- DNA synthesis and repair is not very energy demanding.

There is no reason for being less demanding with respect to the properties of the cognitive memory. It would be a great boon for mankind if we did not forget essential things, either in the process of studying or in the process of aging. If DNA were the carrier of both the cognitive and the genetic memories, a most efficient system would prevail. Hypotheses pointing in this direction have already been forwarded but hard data in its favor are lacking. Some suggest that the ability of DNA to conduct photons, which are produced during certain metabolic processes in cells, might perhaps play a role. It has also been shown that DNA can conduct electricity.

If DNA plays no role in storing non-genetic information, another molecular carrier must be available. Small or medium-sized molecules such as saccharides, lipids and RNAs do not last for the lifetime of the cell. Very stable proteins are the most likely remaining candidates. I hypothesize that the proteins of the chromosomal skeleton of eukaryotes might be good candidates. Indeed, they have to last as long as the DNA that is attached to them. During each cell cycle the chromosomal skeleton has to be duplicated in a conservative way. How this is achieved is not yet understood.

In the silk moth *Samia cynthia*, Sauman and Berry (1994) found that *actin* is a key protein of the chromosomal skeleton. In eukaryotic cells actin is omnipresent as a major element of the cytoskeleton. This molecule has been very well conserved in the course of evolution. The actin of some mollusks is for 97 percent identical to human actin. This means that this molecule has been better conserved in evolution than DNA. This points to a very crucial role in cell biology. Actin filaments are formed out of actin monomers that are assembled in such a way as to form a double helix. The chromosomes in the nucleus are not just randomly distributed in the nucleus. On the contrary, they all have a well-defined and reproducible location (Agard and Sedat, 1983). Furthermore, the nuclear skeleton makes contact with the cytoplasmic cytoskeleton, which in turn makes contact with the plasma membrane skeleton (De Loof, 1986). Thus, the cytoskeleton, with actin as one of its major components, traverses each cell, running from the plasma membrane to inside the nucleus. In addition, there are also specialized parts of the actin-cytoskeleton that ensure tight contacts between neighboring cells, in particular in epithelial tissues. According to De La Cruz and Pollard (2001), during embryonic development the actin-based movements of nerve growth cones lay down a spectacular 1.6 million kilometers of nerve cell connections in our brains. Numerous other proteins, as well as some messenger RNAs, can associate with the cytoskeleton. An isoform of myosin I, a nonfilamentous member of the myosin superfamily of actin-based molecular motors, has also been identified inside the nucleus of some mammalian cell types (HeLa cells) (Pestic-Dragovich et al., 2000). In a later chapter, we will see that actin and myosin are fundamental to muscle contraction as well. Finally, actin can conduct electricity (carried by ions) in a very special way (Chapter 9) and it acts as a checkpoint in the cell cycle.

The hypothesis that actin might act as the molecular carrier of long-term memory cannot be valid if prokaryotes have no actin-like molecules. Indeed, because communication is not possible without memory, some memory system must have come into existence right away with the Progenote. Van den Ent et al. (2001) have recently solved the long-standing discussion as to whether or not prokaryotes do have actin. They found that, instead of actin, prokaryotes seem to use another protein, MreB protein, which has properties very similar to those of actin.

The idea of DNA as a carrier of genetic information and of proteins of the chromosomal skeleton as storers of cognitive information is also attractive from the point of view of co-evolution. In Chapter 2, I mentioned already that approximately 98 percent of human DNA does not code for proteins. This DNA is usually considered to be 'junk DNA'. However, it might be considered less 'junky' if it turns out that it contributes to making large chromosomal skeletons. If the hypothesis that a large chromosomal skeleton would be beneficial for creating a large cognitive memory, perhaps the superior cognitive capabilities of *Homo* versus other organisms might be due rather to his junk DNA that to his coding DNA. At the current time, however, all of this is no more than hypothetical speculation due to the lack of hard data.

I do not expect that the molecular mechanisms of the storage, coding, decoding and retrieval of the non-genetic information obtained through learning will be elucidated in detail in the near future. The tools that are needed for making substantial progress have probably not yet even been invented. We should not despair however. I remember the early 1980s when the sequencing of the whole human genome was suggested by a group of scientists with a clear vision, James Watson being one of the prominent members. Many people thought this was an absurd research proposal. It was argued that such a project would take at least a thousand years, even if all the biochemists in the world were involved. Another argument against it was that the project would be so expensive that no money would be left for other projects. Only 20 years later, a relatively small number of scientists have already completed the job. Progress in the sequencing technology and in computer science has made this possible. It can be expected that sooner or later similar progress will be made, particularly in the field of physics that will enable us to crack the code of the cognitive memory.

3.10 Characteristics and evolution of the software

When we think about the nature of software, we are confronted with a most remarkable phenomenon. Matter came into existence at the moment of the big bang. Part of it condensed into stars and planets. The building blocks of living matter are nothing else than fossil stardust. Thus far, no immaterial dimension is involved. At a given moment, however, some aggregate of molecules began to communicate. For the first time, messages written in coded form could be produced. The term 'information' took on a meaning. Along with information, the immaterial dimension of living matter came into being. Four billion years later, lumps of fossil stardust that compacted into compartments that we call *Homo* are able to reflect upon their own origin, properties, goals and future. Not only are they able to reflect, but they are also capable of acting and solving problems (and of creating a host of new ones, as well!).
Intuitively we are tempted to state that something that is immaterial does not exist. Nonetheless, information and software do exist and can undergo changes. If software were something material, then it would have mass. A word processing software program is not consumed little by little each time it is used. A memory compartment does not get

more mass as more information is stored in it. In principle, a software program lasts forever and never gets consumed.

It is the immaterial nature of software that makes it so difficult to study. Hardware can be studied *in vitro* by means of chemical and physical research methods. Software cannot be isolated in a test tube. One method for studying the general properties of biological software is to analyze which properties are attributed to the software of digital machines. By doing so, one assumes - hopefully correctly - that the same basic rules will apply (Bergstra and van Vlijmen, 1998).

1. Software programs are immaterial, yet they require a material carrier.
2. In principle, software makes it possible to govern the functioning of an automaton, whether non-living (machine) or living (organism). Software executes.
3. Biological hardware, like computers, cannot function without software. The possibilities of software cannot be visualized without suitable hardware. Hence, the hardware/software dichotomy is artificial, but it is maintained for practical reasons.
4. Multiple languages/coding systems are possible.
5. The software should make the hardware work fast, and it should be both user-friendly and flexible. The more problems a given software program can help to solve, the more valuable it is, provided it is user-friendly.
6. The software should not be more complex than the hardware can handle.
7. Memory compartments should last for the lifetime of the hardware. Programs can be silent for shorter or longer spans of time, and can then be reactivated when needed.
8. Software is subject to errors, usually due to malfunctioning of the hardware.
9. Viruses can infect programs.
10. Problem-solving and learning processes require software. This software of organisms, in particular of free-living ones, enables them to search for food, to protect themselves and other members of the species, to seek out sexual partners, to avoid danger, etc.
11. Software enables network formation. In biological systems this is important for all levels of compartmentalization (9 to 16 in Fig 4.1 of the next chapter) that comprise more than one individual (of the same species or of different species). Recognizing other individuals, which requires level 3 of consciousness (see Chapter 15) and participation in social activities, is not possible without suitable software.
12. Biological hardware always has a genetic memory that is used to ensure its continuity. Programs linked to the cognitive memory are not inheritable. Through teaching-learning, 'biological' software programs can be quickly multiplied. The propagation of information through such processes can proceed very quickly and massively, as illustrated by the worldwide broadcasting of certain television programs.
13. Reproduction of the hardware is a slow process, requiring a suitable environment. In the case of sexual reproduction, half of the genes of both parents are always lost in the first generation. Moreover, genes can only be transmitted 'vertically' to children, not 'laterally' (e.g. to the sexual partner). The transfer of non-genetic information can be done much more quickly and completely, and it can be done laterally, as well. In the case of correct communication, there is no loss of information when it is transferred from one individual to another. The skipping of generations is possible.

14. The development of hard- and software proceeds through very different channels. Hardware development is usually bound to a given environment because of the sophisticated infrastructure it requires. Software can be 'written' almost everywhere at any time by any competent person. This holds true for biological systems as well. This means that the evolution of software can be a very fast process.

3.11 The energy question

Task: Imagine that you teach physics and you want to test whether your students understand what you explained to them about the 'law of action and reaction' in physics.

Situation: You ask a student to raise his or her right arm and the obedient student does this instantly. Next you ask your class whether this is a genuine example of an action-reaction situation in which the energy contained in the command *"Raise your right arm!"* equals the energy that is required to make the student's arm conquer the force of gravity.

It is quite probable, as I know from my own teaching experience, that the majority of the students will prefer to remain silent because they are afraid of saying something wrong, and that the few audacious ones will not come up with the right answer.

Next, you can give them a few hints. One is that you now address the whole class and ask again: *"Raise your right arm!"* and, instead of only one student raising his arm, the arms of all obedient students conquer the force of gravity. The input of energy remained the same but the output multiplied: one does not expect this in a true action-reaction situation.

You can also give the following command: *"If I keep silent for 5 seconds, raise your right arm!"* Or, *"Raise your right arm tomorrow at 9 o'clock sharp!"*

It is very likely that the students will realize that the law of action and reaction in its physical sense does not apply in this communication situation. There is something that does not fit, but what?

Final hint: You take the hand of a student and you lift the student's arm yourself. Now you ask again whether this is good example of an action-reaction mechanism. The students are likely to answer that in this situation the law indeed applies.

Final question: *'In the situation in which a student raises his or her arm in response to the command, who invests the energy: the person who gives the command, or the student him or herself?'*

Solution: In both situations the final outcome is identical: the student's arm has conquered the force of gravity. However, the way this was achieved was fundamentally different. In the situation where the teacher lifts the arm of the student, it is energy delivered by the teacher himself that is used to make the arm conquer gravity. In the case of the command 'Raise your arm!', the command itself does not provide the energy but only causes the utilization of energy that was already stockpiled in the student's body to raise his or her arm.

The liberation of stockpiled energy can be illustrated with a mousetrap. In order to successfully catch a mouse with a trap, one has to do a few things. First, we will fix a piece of cheese or some other goodie to the trap so that the trap can emit a message to the mouse: in this way, the trap becomes a 'sender'. Next, we have to set the spring: this is an act of storing energy in advance that, later on when the mouse touches the trigger, will be liberated. Without stockpiled energy, no mouse will ever be caught. The energy 'invested' by the mouse in touching the trigger does not have to equal the energy, which is liberated when the spring is released.

3.12   Communication can master any force of physics

Communication itself is not at all a true force as defined in physics. Nevertheless, communication can mobilize substantial amounts of energy. Gravity is only a weak force compared to the electric forces in the atom and to the strong and weak nuclear forces. When, for example, we have to lift 50 kilograms, we do not experience gravity as being weak. Therefore, it may look like a great achievement if biological compartments can conquer this force. But a small magnet can lift a paper clip, despite the gravity exerted by the whole planet earth. Some physicists and engineers work with nuclear reactors, or with high-energy accelerators, in which the forces that keep atoms and subatomic particles together are broken and the elementary particles are set free. In view of this fact, we must conclude that communication - in this particular case, human (scientific) communication - can conquer any known force of physics under the proper conditions. Communication involves elements outside *the realm of physics and chemistry*.

3.13   When is something information? Some definitions of information

The definition of communication as 'the transfer of information' requires that 'information' be adequately defined. This is not so easy as one might think at first sight.

Some people who are gifted in mathematics think that information deals with the reduction of the degree of uncertainty in a given system and can most accurately be defined in mathematical language. But what if somebody tells a lie? Such an approach is beyond the scope of this book.

One possible non-mathematical definition, which was formulated by my colleague Dr. L. Van Poecke, reads: 'Information is any change within a communication system that makes a difference to any component of the system'.

Another definition is: "Information is everything that serves as an answer to a question." (Emmecke, 1994). In biology, what was the very first question? And the first answer?

I propose the following approach: 'A message contains information if, upon decoding by a competent receiver, this message can cause the mobilization of part of the energy that has previously been stockpiled in this receiver for the purpose of executing a specific task (or, in the terminology of physics, to do some sort of work). This will often result in feedback reactions.'

In short: *Information is anything that causes the mobilization of stockpiled energy in a competent receiver-decoder-responder for the purpose of doing some sort of work at a given moment.*

In certain circumstances, the absence of something, e.g. the girlfriend or boyfriend not showing up for a date, can itself be a form of information. This is an important element in the immaterial dimension of life.

3.14 The purpose of communication: Make others work for you in order to increase your own degree of contentment!

Because we communicate all the time, we seldom reflect on the question as to *why* we do it. When students are confronted with this question, they usually react with some embarrassment: "What a stupid question! If we didn't communicate, we would die!" Do we then communicate in order not to die, or is the answer simpler?

Making use of the definition of information, the answer to the question "What do we want to achieve by communicating?" is, in simple language: "We want the ones that receive our message(s) to do some sort of work – in the broad sense – for us." This is usually done to increase the degree of contentment in the sender.

In more polished language, this reads: The purpose of the release of a message from the sender is to liberate, at a specific moment(s), in a receiver a specific form of stockpiled energy for a specific purpose, such as for doing some kind of work, to prevent a system from breaking down (in physics this is called coping with entropy), etc.

Very often, the energy that is liberated will influence the behavior of the responding compartment. Behavior can be defined as the total sum of all movements made by an organism (or better a 'compartment'). Movement involves changes in the cytoskeleton of the cells that are involved in the movement. Thus there is an important link between communication and the cytoskeleton.

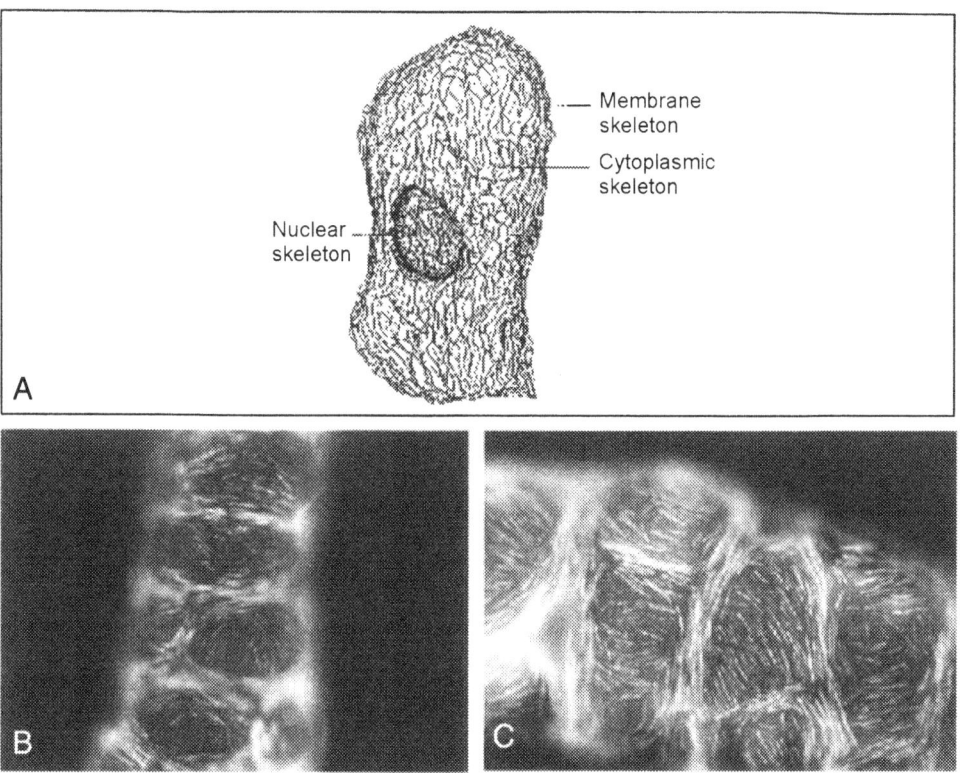

Figure 3.3 A. Schematic representation of the cytoskeleton, a network of proteinaceous fibres. The cytoskeleton comprises the membrane skeleton, the cytoplasmic skeleton and the nuclear skeleton, which are all interconnected (Luna and Hitt, 1992). Some cytoskeletal proteins are contractile, being responsible for active cellular motility. In addition to enabling movement, the cytoskeleton has many other functions. B and C are photos of the cytoskeleton of cells of the excretory system (= Malpighian tubules) of a flesh fly (from Meulemans and De Loof, 1992).

Up to now, in the literature that deals with the mechanisms resulting in the origin of life, little attention has been paid to this most important cellular structure. In Chapter 12 I will argue that in my opinion the cytoskeleton was just as important for the origin of the living state as were the nucleic acids. I think that 'life' is not possible without it. In Chapter 9 I will deal with the electricity conducting properties (cable-like properties) of actin. As mentioned earlier in this chapter, the cytoskeleton may also play a role in cognitive memory.

3.15 'Communication', a synonym for '(un)conscious problem solving'? Automation

In daily life, the term 'communication' is not used intuitively as an alternative for 'problem solving'. Yet, if one thinks about it, 'communication' largely overlaps with 'unconscious problem solving'. A sender invariably emits a message in coded form. The first task of a receiver is to solve the problem of how to decode the message. When humans talk to each other, they do not realize that their brain has to decode the sound waves into another form of message that can be interpreted by the brain. It is not because this seems to go automatically (*The zombie within (us)*, Koch and Crick, 2001), that it does not involve problem solving. In the course of evolution, organisms have developed means of simultaneously solving many different problems without having to think about them. Conscious problem solving represents only a very small percentage of the total problem solving activity going on in our body. The type of problem that can be solved goes hand in hand with the level of compartmental organization, as will be discussed in the next chapter.

Unconscious problem solving by biological systems is the equivalent of what we call 'automation' of the work packages done by all our machines. In the course of evolution, biological systems have managed to 'automate' numerous processes, at ever higher levels of compartmentalization (see next Chapter). In humans and other segmented animals, which will be classified as level 7 of compartmental organization in Chapter 4, all processes from level 1 (= cell organelles) to level 6 (= folded epithelia) are fully automated. Even in level 7 (= with coordinating brain function), nearly all brain processes are also automated. Very few processes remain to be handled consciously. The automation in level 8 (= segmented animals with their tools) steadily increases. It seems that once a given level of compartmentalization becomes nearly fully automated, then an additional level of compartmentalization can be added. The feedback aspect that is inherent in communication forms a basis for automation, and hence for the 'automaton nature' of living systems (Chapter 5).

3.16 The difference between communication and interaction. The (first) central dogma, once again

Having dealt with the different aspects of communication, we can now attempt to formulate the difference between communication and interaction. This is not a superfluous exercise, since my experience has taught me that the two terms are often erroneously used as synonyms. If they were synonyms, one could say that a single molecule is a communicating system because its constituent atoms interact with one another. This would hold true for numerous macrosystems as well, such as the sun and its planets. The final outcome would be that everything communicates with everything. The difference between living and non-living systems would disappear. I do not support this view. My view is that communication requires that a message be written in some coded form, transmitted through a communication channel and decoded by a competent receiver, and that an energy-requiring response should follow sooner or later. Interactions between atoms do not fulfill these criteria. Thus communication is a form of interaction, but interaction is not necessarily a form of communication. The central dogma relating to the hardware is an example of an act of communication between the DNA in the nucleus and the cytoplasm. The nucleus emits a message in coded form, namely as mRNA. This is transported through a communication channel, namely the liquid in between the DNA and the ribosomes. The ribosomes act as receivers that respond with an energy-requiring process, namely the synthesis of a protein.

3.17 Communication and the anticipation of progressing time

We have mentioned that some information can be passed down from generation to generation, either in its original form or in a modified form. Upon retrieval of this information later on, energy can be mobilized in a competent receiver-decoder-amplifier-responder.

The ability to retrieve information that had been stored beforehand in the memory system thus allows organisms to cope with progressing time. In some circumstances, the system allows you to arrange things ahead of time, for example in the execution of a will. It also makes it possible to recall information from former times (history).

The aspect of the anticipation of progressing time is interesting when one thinks about death. *Here, one has to* **make the clear distinction between the physical death of the compartment and the death of all the information this compartment produced during its lifetime.**

The physical instrument of communication can indeed completely die, thereby losing its ability to communicate at its highest level of compartmental organization as a 'receiver'. However, this does not necessarily mean that the instrument automatically and completely stops acting as a 'sender' in certain communication systems. By being there as a dead body it will probably cause movements in other organisms, which may come to eat it, or, in the case of humans, to bury it. The body's molecules disintegrate into small building blocks and energy for other organisms. The building blocks are continuously recycled and, in being so recycled, they influence communication activity in other compartments.

A much more important consideration here is that some acts of communication generated during the lifetime of a given compartment are stored as information in the memory of other compartments. Thus they can continue to influence the lives and communication of future generations, as we have already pointed out in the examples of founders of religions, etc. *As long as some information inherent to a given compartment remains somewhere in a memory, that compartment is not completely dead.*

3.18 Life as a double continuum. Two central dogmas?

If one rigorously pursues this line of thinking, one might come to the conclusion that 'absolute death' does not exist as long as there is a planetary level of compartmentalization of life (biosphere or level 16, Gaia).

Indeed, Life is a **double continuum** and only one of the continua ends at what we call death of the physical compartment.

> First, all living beings are *compartmentally* interlinked by being descendants of one and the same prokaryotic ancestral cell (the Progenote) (= *compartmental or genomic continuum*). The molecular carrier of the genetic memory is DNA.
>
> Second, all organisms which can learn (during life and over generations) are also interlinked by the continuous overflow from the past into the present and the future of non-genomic information (= *non-genomic information continuum*). In animals it is stored in the cognitive/brain memory. In which part of the cell the non-genomic information is stored remains unclear as already mentioned earlier. Very little is known about the non-genomic information continuum in organisms other than animals.

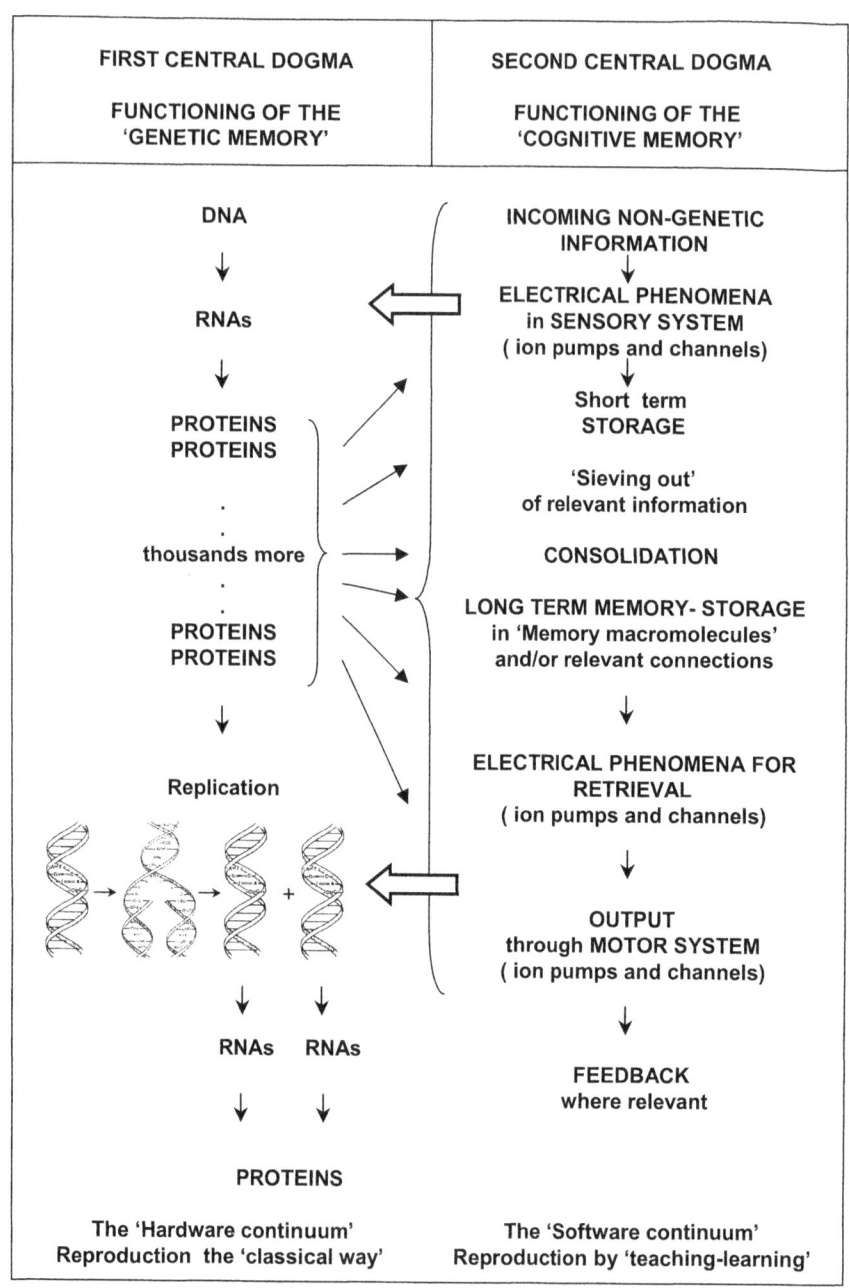

Figure 3.4 Tentative representation of the principles of the double central dogma system: the principles of the two types of memory and of the interactions between them.

The compartmental continuum is a chemical continuum, governed by the laws of genetics and biochemistry. The non-genomic information continuum is an immaterial continuum, because information does not contain units of matter, force or energy as these are defined in physics. Learning processes and the principles of communication govern the cognitive continuum. It does not reside in the genes. Both continua are evidently interdependent. They have their own distinct driving force of evolution. This approach necessitates the rethinking of neo-Darwinism, which attempts to explain evolution solely from the point of view of the chemical-genetic continuity of compartments. Neo-Darwinism is a *single continuum* theory: it focuses only on the 'hardware-aspect' of biological communication systems. I will further explain my '*hardware-software or double continuum theory of evolution*' in Chapter 15.

One consequence of the fact that there are two memory systems in cells and organisms – the genetic and the cognitive – is that there should be two central dogmas and not one, as is now generally held (Fig. 3.4).

First central dogma: the genetic memory and its role in the functioning and continuation of the hardware

Principles: replication of DNA and Protein synthesis (DNA → RNA → Proteins)

DNA has a double function. First, during the life of a cell it governs protein synthesis. Secondly, DNA can be identically duplicated. During mitosis, daughter cells are formed that contain an identical copy of the parental DNA. This ensures continuity over the generations.

Second central dogma: the cognitive memory and its role in the functioning and continuation of the software

Principles: cascades of signal transduction pathways are involved.

Stimuli trigger receptor molecules. Electrical processes may play a crucial role in transporting information from the cell membrane to the site of storage, as well as in the retrieval of this stored information later on. There is as yet no consensus as to the molecular nature of the structure in which non-genomic information may be stored. In this respect, I have pointed to the possibilities of cytoskeletal protein actin.
At the current time, only a preliminary and vague proposal can be made as to how the second dogma might eventually look.

3.19 Emotions, feelings, contentment, motivation

It is not common practice by authors of contemporary textbooks on General Biology to elaborate on feelings and emotions. This is left to the discipline of psychology. The usual argument is that such things are too private and subjective to be described in the usual terminology of the exact (biological) sciences. This attitude seems to be changing, mainly as the result of progress in the neurosciences (Damasio, 2001).

It certainly is difficult to obtain certainty about what different persons actually feel when they say they feel 'happy' or 'sad'. On the other hand, anybody who owns a dog or a horse can see the continuity of emotional phenomena from non-human species to humans, a fact that Darwin himself noted. Emotions and feelings must have very ancient biochemical roots, perhaps going back to the Progenote. In Chapter 15, which deals with the mechanisms of evolution, I will attribute an important role to feelings.

It seems to me that the following two definitions given by Damasio (2001) are really quite good.

'An emotion, be it happiness or sadness, embarrassment or pride, is a patterned collection of chemical and neural response that is produced by the brain when it detects the presence of an emotionally competent stimulus – an object or situation, for example.'

'Feelings are the mental representation of the physiological changes that characterize emotions.'

In Chapter 15 I will emphasize the importance of the feeling of 'contentment' with respect to adaptation and evolution. All organisms unconsciously but incessantly evaluate their over-all internal and environmental physiological status. They integrate their perception of temperature, luminosity, pressure, availability of food, etc. in an over-all feeling of 'contentment. If an organism feels contented, it can proceed as usual. However, when the over-all feeling turns into discontentment, the organism has to 'do something'; adaptation strategies have to be invoked. The feeling of contentment and the motivation to do something, in particular to solve a problem, are often closely linked.

According to Damasio (2001), feelings are amenable to scientific analysis, just like any other cognitive phenomena, provided that the appropriate tools are used.

3.20   The basic architecture of a communication system as it relates to the origin of the living state

When thinking about the origin of life, the principles of communication should be taken into consideration. In my opinion, the essential difference between living and non-living matter is that living matter can decode coded messages and provide an energy-consuming answer to them, whereas non-living matter cannot. Another way to say this is that living matter communicates, and non-living matter does not. There are evidently other features in which living systems differ from non-living ones, but none is so essential as communication. This will be dealt with in Chapters 6 and 7.

Communication requires a communicating system, which itself has both a hardware and a software component and which uses energy. Electricity is always involved. Biological systems make their own electricity (Chapter 9).

If communication is indeed essential to the living state, life could not have come into existence earlier than the time at which all the required essential elements were present and assembled properly in that first communication system. How this assembly may have happened and which type of gradient may be regarded as the original archetype will be dealt with in Chapters 9 and 12.

3.21 Definitions of 'communicating compartment' and 'act of communication'

In classical biology, the cell is considered to be the basic building block of living matter. Immediately a problem arises. Is it the prokaryotic cell type or the eukaryotic cell type? According to Margulis' theory, it should be the prokaryotic cell type, but that is not what textbooks on cell biology tend to say.

I think that there are good arguments in favor of introducing another basic building block of living matter, namely the 'communicating compartment'. The possibilities of this approach will become more clear in the following chapters. I will use the terms 'compartment' and 'communication' in a broader context than Mitchell (1979) did.

My definition of 'biological compartment' combines properties of an autopoietic system with those of a communication system. The reason why I systematically prefer to use the term 'compartment' over 'system' resides in what, to my mind, is the too general character of 'system'. The term 'system' is commonly used in physics but it remains ill defined.

The term 'autopoiesis' is not frequently used in classical biology. Hence, some explanation may not be out of place. According to Fleishaker (1988), an autopoietic system has the following properties.

> 'The unit should have an identifiable boundary and constitutive elements, and it should be a mechanistic system, i.e. the component interactions and transformations are determined by component properties. The boundary of the unity should be constituted by relations among its components, i.e. the boundary should be determined by component interaction, not imposed from outside. The system components and the boundary components should be produced by component interactions and transformations.'

A remark: one should realize that autopoiesis was tailored at the level of the cell and that there may be restrictions if it is applied at higher order systems (Cullen, 1996). It is beyond the scope of this book to deal with these restrictions.

The properties of a communication system are:
> In brief, such a system consists of a sender-encoder, a transmission channel, and a receiver-decoder-amplifier-responder. Furthermore, in order to be functional, a communication system should be able to stockpile energy, its boundaries should be moderately 'leaky' in order to allow messages to pass its boundaries, and there should be appropriate gradients.

The following definition of **'communicating compartment'** results:

> *'A biological communicating compartment, or simply "compartment", is a unit based on carbon chemistry and on electricity carried by inorganic ions. This unit*
> * - is limited by a moderately 'leaky' boundary with appropriate 'holes';*
> * - can stockpile the right form(s) and amounts of energy;*

- can generate gradients that can be used for communication for the purpose of enabling the compartment to function from its lowest to its highest level of compartmental organization.

More simply, a biological compartment is a unit that has all the necessary properties of a biological communication system. These properties will be described in more detail in Chapters 7 and 8.

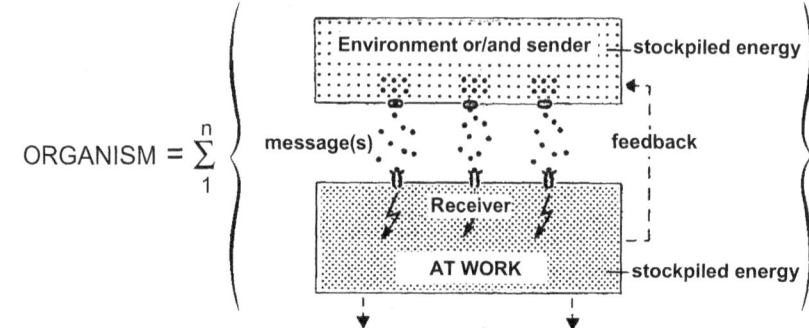

Figure 3.5. Schematic representation of the basic architecture of any living organism, considered from the point of view of communication systems. In a single cell such as a bacterium, in which there is only one communicating compartment, $n = 1$. In multicellular organisms, n represents the number of all communicating compartments, all different from each other, which are present at the successive levels of compartmentalization of the organism. This number can amount to many billions in large organisms. No two fully identical cells exist in nature. Hence, neither do two fully identical communicating compartments.

The importance of 'from its lowest to its highest level of compartmental organization' will become clear in the next chapter, which deals with the hierarchical organization of living matter, as well as in chapter 6, which deals with the definition of death.

The elementary unit of communication activity is the '**act of communication**'. This act is the cascade of events, including the sending of a coded message, its transport through a communication channel, and the energy-requiring response in a competent receiver.

Just as all cells can execute numerous biochemical reactions at the same time, so can complex biological communicating compartments simultaneously execute numerous acts of communication at all successive levels of their organization. These successive levels and the variety of languages that are used will be the subject of Chapters 4 and 8.

The basic idea of Fig 3.5 is that biological systems are communicating compartments. They can be influenced by the environment and by other compartments. In multicellular organisms, one cell can act as sender, while another acts as receiver. By means of feedback mechanisms, the roles can be inverted.

ESSENTIALS

COMMUNICATION

1. 'Life' and 'death' are not abstractions but rather concern properties of 'something', namely of a 'communicating compartment'. The long definition of the term 'compartment' reads: "A biological communicating compartment, or simply 'compartment', is a unit that is based on carbon chemistry and on electricity carried by inorganic ions. This unit is limited by a moderately 'leaky' boundary with appropriate 'holes'; it can stockpile the right form(s) and amounts of energy; and it can generate gradients that can be used for communication to enable the compartment to function from its lowest to its highest level of compartmental organization." Depending on the conditions, a compartment can act either as a sender or as a receiver, and feedback mechanisms are the rule.
2. Communication is the transfer of information.
3. The basic unit of communication activity is the *communication act*. This involves the production and release of a *coded* message by a sender, its transport through a communication channel, and decoding-amplifying-responding by the receiver (by mobilizing part of the stored energy).
4. By itself, information is usually immaterial, but it may need a material carrier for its transport. Something, or even the absence of something, can act as information if it is capable of mobilizing part of the stored energy in a competent receiver in order to carry out a given task.
5. The terms hardware and software, which are borrowed from computer science, will be routinely used in the biological context. Hardware and software are obviously interdependent.
6. Without a preexisting memory system in which 'software programs' can be stored, and without preloaded energy, communication is not possible. Memory allows the organism to cope with progressing time. Living systems have two memory systems: one is the genetic system (DNA) for the transfer of genetic information to the next generation and the other, which is called the 'cognitive memory', for the storage of non-genetic information. The cognitive memory is best documented in the species *Homo sapiens*, but it probably is as universal as the genetic memory. Each memory has its own set of rules. This view implies that in order to fully understand biology, one should rather think in terms of 'life as a double continuum' and of two central dogmas instead of only one, as is still common practice. The second central dogma should deal with the functioning of the cognitive memory. It is as yet only incompletely understood.
7. A number of arguments are given in support of the hypothesis that the cytoskeleton, in particular its actin component, might play a crucial role in storing cognitive information. Perhaps the actin part of the chromosomal skeleton, which by definition should last as long as the DNA it carries, is a good (theoretical) candidate for the long-term storage of (part of the) non-genetic information.
8. The transport of messages requires that there be 'higher-lower' situations, or, in other words, that there be *gradients*. The importance of gradients will be repeatedly emphasized in the next chapters.

9. Because a sender emits a message that will make the receiver mobilize part of its stored energy in order to do some sort of 'work', one of the goals of communication could be defined as: "Make competent receivers work for you!"
10. Because communication invariably uses messages that are written in coded form, communication invariably includes problem solving. One could even say that to a large extent the two terms are synonyms, although they are not perceived to be so in daily life. In daily life, we use 'communication' as another term for 'unconscious problem solving'. The term 'problem solving' is used when we face difficult problems that require deliberate thinking. In both cases, the same principles of communication are used, but at higher compartmental levels. The degree of difficulty of the problems that can be solved goes hand in hand with the complexity of the level of compartmental organization, as will be outlined in Chapter 4.
11. What is our motivation for engaging in problem solving? To increase our degree of 'contentment'. This is an important fact for understanding the mechanisms of evolution (Chapter 15).
12. Any force of physics can be mastered by communication activity. Thus, by inventing 'life', nature introduced a system to master itself. A remarkable achievement indeed.

CHAPTER 4

DIVERSITY AND HIERARCHICAL ORGANIZATION OF COMMUNICATING COMPARTMENTS

PRINCIPLES OF MEGA-EVOLUTION

REVOLUTIONS

Nature: one level, two levels, three levels,…
The final outcome of evolution: a Babel-like confusion of tongues
(after P. Bruegel)

Contents

4.1 Introduction: Why should Mega-evolution be introduced to complement Micro- and Macro-evolution?
4.2 Compartment. Complexity
4.3 Definition of Mega-evolution. Revolutions. Continuous versus discontinuous evolution
4.4 Why is a novel classification system needed that is not based on kinship? Bringing order to the multitude of communicating compartments. Principles of the 16-level classification system
4.5 Compartments restricted to one and the same individual organism: mono-organismal compartments
4.6 Compartments, which consist of more than one individual of the same species: polyorganismal-monospecies compartments
4.7 Compartments consisting of individuals belonging to different species: hetero-species compartments
4.8 The mechanisms instrumental to Mega-evolution. Some common denominators in the mechanisms for creating higher levels of compartmentalization
4.9 Newly formed compartments expand. Need for maintaining order. Installation of headquarters
4.10 Mechanisms for generating daughter compartments
4.11 General conclusions

Essentials

4.1 Introduction: Why should Mega-evolution be introduced to complement Micro- and Macro-evolution?

In biology textbooks the term "biodiversity" usually denotes the multitude of species that inhabit the earth. They came into existence as the result of mutations that accumulated in the course of evolution (Darwin: descent with modification). These genetic differences are responsible for variability in morphology and physiology. Micro-evolution applies to the formation of new species. Macro-evolution is the generation of morphological novelties in the evolution of new species and higher taxa. The transformation of the fish fin into the amphibian leg is an obvious example. In the same context, the molecular genetics of the differences in the basic body plan between birds and mammals, are already well understood, in particular with respect of the role of the so-called Hox genes (Gilbert, 1997). All of this has been briefly outlined in Chapter 2.

Although equally relevant and obvious, morphological and physiological differences between different cell types of the same organism are less readily recognized as good examples of biodiversity. The reason is that such cells – apart from a few rare exceptions – all have the same genome. However, with respect to the different modes of living together, this form of biodiversity is just as relevant as those involving two or more different species.

The different cells and cell types that constitute a multicellular organism, as well as the millions of species that make up the biosphere, are all interconnected by a variety of means. Such means cannot be deduced from genomic analyses. Indeed, in addition to the genes, a variety of other factors play a role. As a consequence, some aspects of evolution exceed the scope of both micro- and macro-evolution. This is illustrated by a few examples.

- According the theory of Margulis (1981), and as already mentioned in Chapter 3, the first eukaryotic cell came into existence as the result of symbiosis. A few bacteria of one species invaded a larger species and established a permanent relationship with it that was beneficial for both parties involved. The larger species became dependent upon the intruder species, and vice versa. This means that that the formation of the eukaryotic cell is not an example of the formation of a new species out of a single ancestral species. Hence it is not an example of micro-evolution. Neither is it an example of macro-evolution. Symbiosis evolution involves changes in the communication systems both of the host and of the guest(s).
- Heterosexuality. The emergence of gender, i.e. the generation of both males and females, had a much greater impact than the simple formation of a new species out of an existing one. Heterosexuality occurs in several Kingdoms. This process implies the well coordinated coevolution of several aspects of sex-related communication such as: hormones, pheromones and their respective receptors; meiosis in special organs; and behavior.
- Evolution through the utilization of tools. Although he is not the only species that uses tools, man is the most obvious example of the possibilities of using tools for evolution. Tools changed the fate of *Homo,* as well as that of several animal and plant species that are used by *Homo* for agriculture and companionship. The evolution in the use of tools did not follow from additional mutations. Some tools greatly facilitate software evolution. This will be dealt with more extensively in Chapter 15.

These few examples illustrate that truly drastic changes have taken place in the course of evolution and that changes in genetics do not suffice to account for all of them. To describe their effects, in particular on the way organisms live together and communicate, a new term will be introduced: Mega-evolution. This term applies to the evolution of 'life' from the point of view of communication.

4.2 Compartment. Complexity

In this chapter, the term "compartment" will be frequently used. In Chapter 3, the "communicating compartment" is defined as follows:

"A biological communicating compartment - or simply "compartment" - is a unit based on carbon chemistry and on electricity carried by inorganic ions. This unit is limited by a moderately "leaky" boundary with appropriate "holes"; it can stockpile the right form(s) and amounts of energy; and it can generate gradients that can be used for communication to enable the compartment to function from its lowest to its highest level of compartmental organization.

A simple definition reads: "A biological compartment - or simply "compartment" - is a unit that has all the necessary properties of a biological communication system.

One relevant question relates to which criteria can be used to state that compartment A is more complex than compartment B. My view is that a given compartment A is more complex than B when A needs more levels of subcompartmentalization than B does in order to be functional. The usual situation is that more complex compartments can simultaneously solve more problems and also more difficult problems.

The 'family tree' in levels of compartmentalization is here represented in a 2-dimensional way due to the limitations of printing methods. In reality, a number of levels can overlap or intertwine.

4.3 Definition of Mega-evolution. Revolutions. Continuous versus discontinuous evolution

Mega-evolution deals with the formation of communicating compartments of ever-higher order. At least 16 such compartments have been identified to date and are discussed in this chapter. All of them are hierarchically organized. Using a variety of specific languages, These compartments can communicate with a number of other (levels of) compartments, a process that results in novel ways of living together. Mega-evolution theory is an attempt to explain the evolution of 'life' from the point of view of communication. Such an attempt naturally requires that 'life' be adequately defined (see Chapter 7).

This sort of evolution is approached on the basis of the principles of communication: a pure genetic approach does not suffice. Mega-evolution crisscrosses the borders of the present day classification systems that are based on genetic relatedness. My classification system is not based on a distinct group of genes. Rather, it is based on sets of genes that allow the coming into existence of communicational connections, or – in more scientific terms – of novel signal transduction pathways. Such sets of genes will differ from level to level, as well as from association to association. In fact, Mega-evolution describes the solutions that nature has invented to allow an ever-greater number of individuals, species, populations, etc. to live together – all of which have become increasingly alienated from one another in genetic terms.

Mega-evolution involves major **'revolutions'**. A revolution is a key (physiological) innovation that has enabled the coming into existence of a new level of compartmentalization. Each successive revolution has caused a fundamental change in existing relationships, with far-reaching consequences for the rapid evolution of that compartment, and perhaps also for other compartments that are somehow linked to it.

In biology the discussion between the protagonists of the concept of continuous evolution and those who favor a theory of evolution in leaps and bounds has not yet been definitively settled (Chapter 15). One must realize that an additional level of compartmentalization can only become functional upon the acquisition of quite a lot of additional genetic information. This can happen through gene duplications, mutations, etc. (Chapter 2). It can take a long time before this happens. Only after the last link in the chain has become functional can the novel compartment suddenly arise, seemingly out of nothing.

This succession of events can be illustrated by the following example. A lot of preparatory work was done before the first automobile appeared on the street: all the different parts had to be designed and optimized. It was only after the final part was put in place that the contraption actually became an 'auto-mobile' that could function, provided it had fuel. Thus it was not apparent that a vehicle was under development until the 'working' model was made available to the public.

As soon as the car is moving, additional rounds of improvements occur to polish the interactions between the parts. In very few years after the combustion engine was invented and utilized in cars, the jet engine was invented – an invention that revolutionized air traffic. These examples illustrate that (1) there is no contradiction between continuous and discontinuous evolution, and (2) it is logical that revolutions seem to happen in ever-quicker succession.

4.4 Why is a novel classification system needed that is not based on kinship? Bringing order to the multitude of communicating compartments. Principles of the 16-level classification system

At this early stage of the book, the reader may be reluctant to accept that an alternative classification system should be introduced at all. For non-specialists, classification systems tend to be boring. The reason why I introduce this system is not that I love systematics as such or that I feel that my classification system is less boring than the others. It arose out of necessity because I had to find a way to solve a basic problem concerning the deduction of a plausible definition for 'death'.

Let me anticipate just a bit concerning this problem, which occurs in all biological systems that are more complex than bacteria. In Chapter 6 I will call this problem 'the duality of death'. It can be illustrated with the following series of questions and answers. Is a chicken dead when one of its wings is amputated? No. When both wings are amputated? No. When both wings and a leg are removed? The answer is still negative. How long can one continue removing additional parts? Actually for quite a long time.

But what if an intact chicken is decapitated? Then it is dead. Is it dead instantly, or only after all its constituting organs and cells are dead? If upon decapitation the heart is removed and properly incubated, it can continue to function for some time. Thus the heart is still alive, but we had concluded that the chicken itself was dead upon decapitation. But can the chicken be declared dead as long as some of its constituting parts are still alive? This situation, which is not at all restricted to complex animals, illustrates that "death" is linked to a hierarchical level of compartmental organization. In Chapter 6 we will see that 'death' invariable applies to the *highest level* of compartmental organization. The fate of the lower levels of organization - whether they continue to live or not - is irrelevant in the definition of death.

In later chapters I will demonstrate that 'being alive' likewise applies to a given level of compartmental organization. In the definition of 'life' that will be presented in this book, the levels of compartmental organization are explicitly mentioned.
It took me quite some time to determine how many different types of compartments nature has generated in the course of evolution and to design a logical classification system. I have attempted to bring a certain degree of order into the multitude of possibilities by categorizing them into:
1. **Compartments restricted to one and the same individual organism**: *mono-organismal compartments*. Example: an individual segmented animal with its organ systems, organs, tissues, cells and cell organelles.
2. **Compartments that consist of more than one individual of the same species:** *poly-organismal monospecies compartments*. Example: a colony of bees.
3. **Compartments of individuals belonging to different species:** *heterospecies compartments*. Example: a host-parasite compartment.

This system of classification, which is summarized in Fig. 4.1, is not the only possible one (see Buxbaum, 1995). It should be considered a rough scheme that is subject to further improvement. This illustration is meant to be approached from the bottom. On the left side, there are 16 different levels of compartmentalization. On the right, you see the mechanisms that the different levels use to generate daughter compartments (to ensure their perpetuation in the course of time). Levels 1-16 in this classification system are not meant to represent a linear series of increasing complexity. Some of these levels can still be subdivided or overlap with other ones. The simplest level - that of the bacterium-type organisms - is called level 1. The most complex one - that of all life on planet earth taken together - is level 16.

This chapter continues with the description of the major characteristics of the 16 levels of compartmental organization. When going through all these levels, one should always keep in mind the following questions:
- Which revolution made the coming into existence of this type of compartment possible?
- Among how many different levels of subcompartmentalization should there be communication for this particular level to be functional?

4.5 Compartments restricted to one and the same individual organism: mono-organismal compartments

Level 1 External membrane compartmentalization: the 'prokaryote type' or 'simple balloon type'.

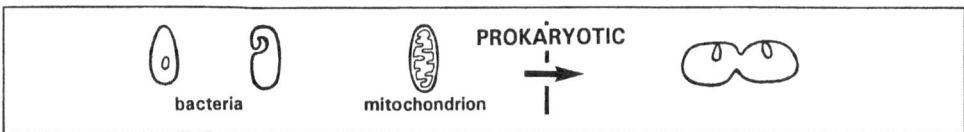

Problems: How to establish and maintain concentration gradients over a leaky lipidic plasmamembrane? How to establish cellular polarity? How to select out of several possibilities the principles of the genetic and non-genetic memory and how to select the ways in which these two types of memory are used?"

With respect to cell polarity, it is as yet poorly documented whether or not there are regional differences in the composition of the plasma membrane of prokaryotes. The chemotactic behavior in bacteria (e.g. when they direct themselves towards a food source) suggests preferential localization of certain types of membrane receptors in distinct areas of the limiting membrane. It is likely that the first cell contained a cytoskeleton that contributed to cellular polarity (Chapter 12).

Membrane folding with concomitant specialization is seen in cell organelles such as mitochondria and chloroplasts. The outer membrane and the inner membrane have different functions. The exact origin of the outer membrane of such organelles is still uncertain.

The simplest form of compartmentalization is encapsulation by a single, unfolded membrane. All cells are surrounded by a plasma membrane consisting of a *liquid lipid* bilayer. Some types of macromolecules with sufficiently large hydrophobic moieties can become trapped in the membrane and float freely in it like icebergs in an ocean unless anchored in some way (e.g. by the restricted fluidity of the membrane or by attachment to elements of the cytoskeleton). The lipid part of the membrane is not permeable to polar substances such as water and ions. Their transport across the membrane, which is absolutely essential for communication and other cellular activities, requires the presence of specialized proteinaceous macromolecules in the membrane, which can form appropriate 'pores' under proper conditions. This is thought to be the archetype of compartmentalization of the living state. It is still the basic form of compartmentalization present in prokaryotes.

Communication at the level of the plasma membrane became possible from the moment that this membrane became selectively permeable for certain ionic species (especially inorganic ones) by the incorporation of proteins with ion transporting properties (ion channels and ion pumps: see Fig. 9.5). It allows the building up over the plasma lemma of an ionic/electrical gradient, the archetype-gradient that is still being used for certain aspects of communication in all cells of all organisms (Chapter 9).

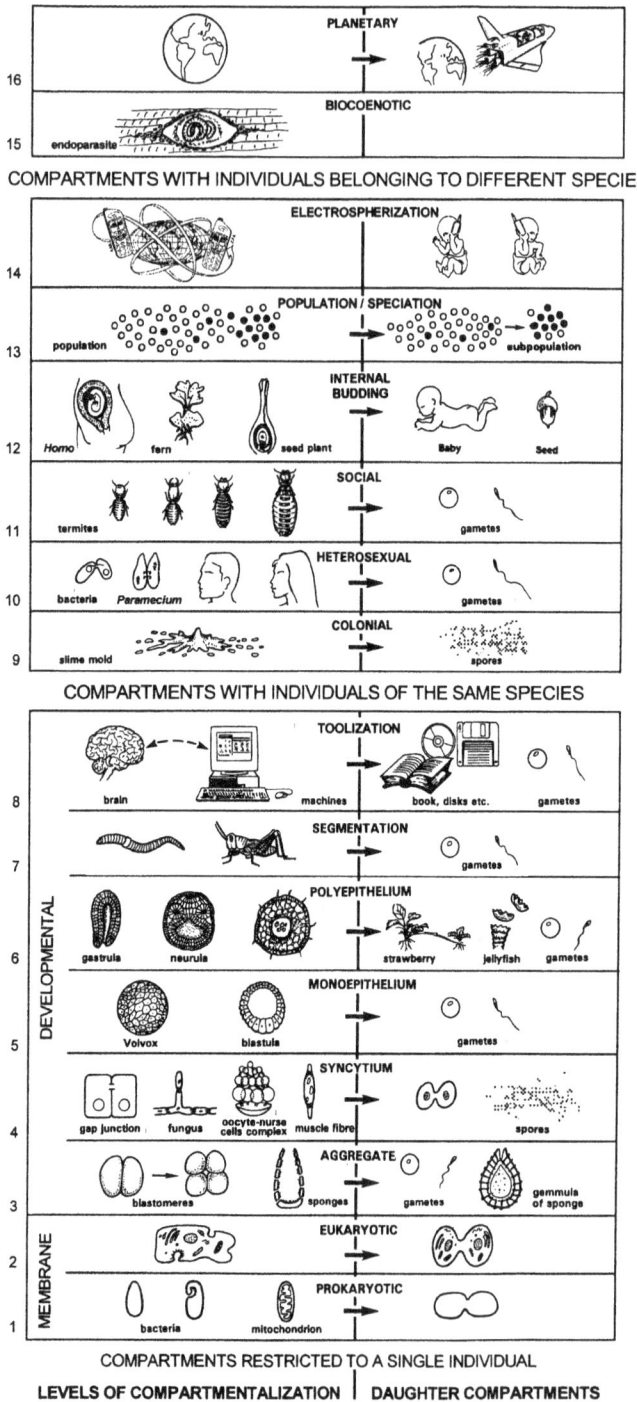

Figure 4.1 Successive levels of compartmentalization (left side) and means for generating daughter compartments (right side).

The simplest level of compartmentalization is the monomembrane level, to which the bacteria and membrane-limited cell organelles such as mitochondria and chloroplasts belong (1). The most complex is the planetary level (16). Different levels of compartmentalization are possible within one and the same organism (2-8), within a single species- (9-14) and within polyspecies compartments (15-16).
To generate additional levels of compartmentalization, several mechanisms can be used together.
One that is widely used involves the internalization of novel compartments within existing ones: a. membrane-limited compartments within an outer membrane (the eukaryotic type, 2); b. epithelium-limited compartments within an outer epithelium (6); c. organism(s) within an organism (11, 13); d. subpopulation(s) within a population (12); subelectrospheres within the master electrosphere (14).
Another system involves the aggregation of compartments (3,4,5,9 etc.). The communication between the constituent compartments can be intermittent, as for example through gap junctions between neighboring cells (4a) or through electromagnetic waves (14), or - more permanently - as in the case of cytoplasmic bridges, (e.g. in Fungi or in meroistic insect ovarian follicles or in myoblasts that have fused into a muscle fibre).
A third system is segmentation (7).
A fourth system is the spreading of the genes needed for reproduction over more than one individual (10,11).
For the splitting off of daughter compartments (reproduction), the variability in systems is rather limited: mitosis, meiosis, systems for asexual reproduction (e.g. 3,6), systems for speciation (13), etc.
In some systems it seems that several mechanisms can be simultaneously operational. Other approaches than the one used in this figure are possible for categorizing levels of compartmentalization.

Revolutions: the 'invention' of the principles of communication and of the voltage-gradient over the plasma membrane (Chapter 9), thereby establishing the electrical dimension of the living state. The cytoskeleton started to exert its different functions.

Level 2: External and internal membrane compartmentalization: the eukaryotic cell.

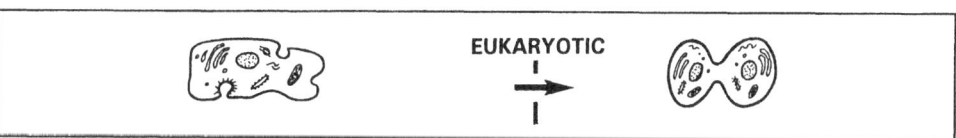

Problems: How to internalize membrane compartments and how to establish means to coordinate their activities?

The eukaryotic cell type represents the next level of compartmentalization. It is characterized by the presence in the cytoplasm of membrane-limited compartments: the nucleus, the endoplasmic reticulum, mitochondria, chloroplasts, vacuoles and lysosomes (Fig. 2.2).

According to Margulis' splendid symbiont theory, the eukaryotic cell originated with the invasion of a given prokaryote by other prokaryotes. Several of the common cell organelles (e.g. mitochondria and chloroplasts) are thought originally to have been bacteria that came to live inside another larger bacterium. The formation of the first eukaryotic cell represents the most sophisticated example of (natural) genetic engineering known in nature. Indeed, the entire genomes of different species were thus combined into a new functional unit. Without this event, neither plants, nor animals, nor Fungi, nor Protists would have come into existence. Only bacteria would have inhabited the earth. Some contemporary protesters against genetic modification (improvement) of plants and laboratory animals seem to be unaware of the fact that we humans are one of the end products of a natural genetic engineering experiment in which the eukaryotic cell was formed.

The origin of membrane-limited compartments such as the two types of endoplasmic reticulum, the nuclear envelope and the Golgi apparatus is still uncertain. According to some hypotheses, some of these structures are infoldings of the plasma membrane. The limiting double membrane of the nucleus is called the nuclear envelope. Whether the 'pores' which are always present in this envelope are always open, as in a sieve, or whether they can be closed under certain conditions is still a matter of debate.

By way of analogy to the balloon-type model of level one, level two corresponds to a large plastic bag filled with balloons of different sizes, forms and colors.

Revolutions: 1. Different species of bacteria become cell organelles and make use of each other's gene products; 2. The proliferation of internal membranes allows novel cellular functions.

Levels 3 and 4: Multicellular non-epithelial compartmentalization.

Level 3: The cell-aggregate type.

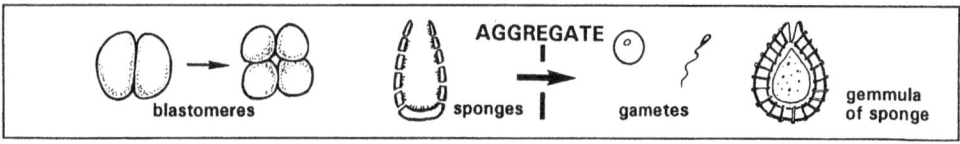

Problem: How to establish multicellularity?

The '*invention*' of *cell adhesion molecules,* which are present in the plasma membrane, and of means that allow the cytoskeletons of neighboring cells to make firm contacts (especially important in epithelia) made the formation of cell aggregates and colonies possible. Such mechanisms are still used in all 'higher' organisms for keeping cells together in tissues. In connective tissue, intercellular matrix molecules play a role in keeping the cells together. There are many types of cell adhesion molecules (CAMs). The ones that require Ca^{2+} for "sticking" are called CADCAMs. True multicellular animals are called metazoans. Müller (2001) has reviewed how the metazoan threshold was crossed. The 'gluing' together of specific cells is evidently in no way restricted to the Animal Kingdom.

Level 3.1: The morula type.

The cells that are formed when a fertilized egg starts cleaving are called blastomeres.

There are several possible mechanisms for keeping the blastomeres together. A morula, which resembles a mulberry, is a stage in animal development that precedes the blastula (for figures, see Chapter 14).

Level 3.2: The sponge type.

Sponges do not have true tissues. They are called 'cell aggregate' animals. Cell adhesion molecules are very important for maintaining their aggregated form.

Level 3.3: Germ cell line- and somatic cell line compartmentalization.

While unicellular eukaryotic organisms can form new compartments by mitotic cell division, animals and plants can generate special types of cells from which the gametes will eventually be formed by meiosis.

In animals, the germ cell line usually gets isolated very early in embryonic development (see Chapter 14). This is clearly the case in insects like *Drosophila*, where the prospective germ cell line is already set apart when there are only eight nuclei present in the egg. In all vertebrates the germ cell line is also segregated early. In the human embryo, the germ cell precursors can already be found in the third week of development, when the gonads have not even been formed. Next, they migrate towards the gonadal anlage and settle there. This special treatment of the germ cell line is the first step in the compartmentalization of cells whose progeny, the gametes, will eventually be expelled from the body to contribute to the formation of new compartments (fertilized eggs).

To continue with the balloon analogy: in level 3 the plastic bags of level 2 are now glued together.

Revolution: multicellularity due to the appearance of cell adhesion molecules.

Level 4: The syncytial compartment.

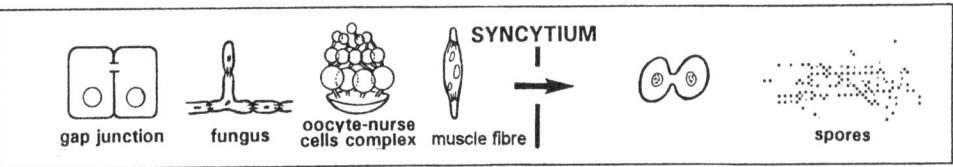

Problem: How to coordinate the activities of neighboring cells glued together by cell adhesion molecules?

Level 4.1: Cells interconnected by ***gap junctions***.

Another important event in evolution was the 'invention' of a way to enable cells, which stick together, to 'talk' to each other through 'holes in the hedge'.

The formation of cytoplasmic bridges was an important event. The smallest of these bridges (gap junctions), when open, allow free passage of ions and other small solutes up to a molecular mass of approximately 2200 in vertebrates and 2800 in some invertebrates. Gap junctions can open and close, depending on certain conditions in the cytoplasm (e.g. changes in Ca^{2+} concentration).

Level 4.2: Cells interconnected by ***true cytoplasmic bridges***.

In some other systems, such as the plasmodesmata in the phloem of plants, in *Volvox* and in some types of ovarian follicles of insects, the cytoplasmic bridges can be quite large.

Level 4.3: ***Fully fused cells***.

Cytoplasmic bridges can become so large that the cells completely fuse and become a new compartment with more than one nucleus. Such a unit is called a syncytium.

One very clear example of this is muscle fibers, each of which originates from the complete fusion of several individual myoblasts. This is the reason that striated muscle fibers have more than one nucleus. Smooth muscle myoblasts do not fuse, and therefore they have only one nucleus.

At this level of compartmental organization, the fact is often overlooked that the physiological unit of function is not 'the cell' but rather the group of cells that are interconnected by cytoplasmic bridges. This is because some signaling molecules can move throughout the syncytium, be it sometimes in a restricted or unidirectional way as will be discussed when we deal with self-electrophoresis (Chapter 8). The different nuclei present in a syncytium do not necessarily express the same set of genes.

Revolution: communication between adjacent cells by means of substances that can pass through cytoplasmic bridges.

Levels 5 and 6: Epithelial compartmentalization.

Level 5: external epithelium only: the blastula type.

Problem: How to arrange and connect cells so that they form a spherical structure with the properties of an epithelium?

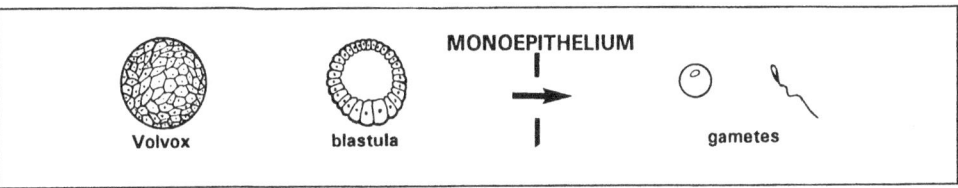

This next level of compartmentalization is achieved when cell clusters, already attached to each other by cell adhesion molecules or other means, make contact with each other's cytoskeleton by means of special contacts. This allows the epithelial compartmentalization form to become established.

Epithelial cells are typically interconnected by tight junctions, in such a way that water and solutes, when transported across the epithelial cells, are required to pass *through* the cells and not via intercellular spaces. This makes it possible for the composition of the environment enclosed by the epithelium to be different from that of the outside world. Transepithelial gradients become possible. Although this may appear simple, it took about 1 billion years of evolution to successfully achieve this level of organization (compartmentalization). It seems to have been a difficult process.

Animals necessarily pass through this stage. They all develop from a blastula, which is the stage when they become organized as an epithelium for the first time: this is the definition of an animal (De Loof, 1993).

Volvox is a Protist, and therefore neither an animal nor a plant. Cytoplasmic bridges interconnect the individual cells, though the structure as a whole looks superficially like an animal epithelium.

Revolutions: realizing multicellularity by means of specializations of the cytoskeleton linking adjacent cells so tightly together that water and small molecules have to pass through the cells and not in between them. This system greatly improves the efficiency of some physiological functions. Novel types of gradients.

Level 6: External and internal epithelial compartmentalization.

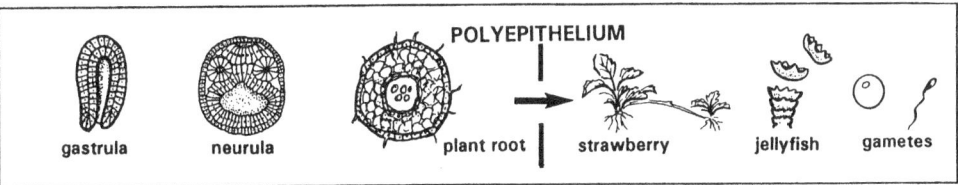

103

Problems: How to make such a structure (e.g. a blastula) invaginate? How to establish intercellular communication over longer distances? How to make cells that are genomically identical acquire different morphologies and engage in different activities?

There are very few organisms that arrest their development at the monoepithelium stage. All animals proceed at least one step further. The infolding of the epithelium of the blastula (blastoderm) yields the primitive gut or archenteron. This is the first degree of internal epithelial compartmentalization. Higher complexity is achieved by subsequent infolding. All animals are to a large extent organized as folded epithelia, this being their basic body plan. Some of the cells may generate descendants which are no longer epithelially organized (blood cells, muscle cells). The more epithelial compartments, the more complicated the internal communication systems that are required.

Revolution: the coming into existence of the basic body plan of animals.

Level 7: Segmentation

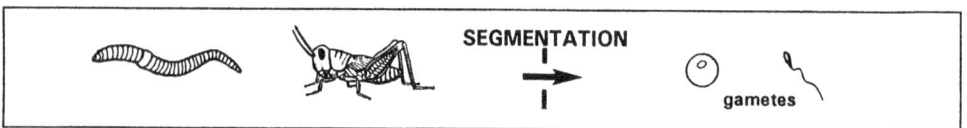

Problem: How to increase the body size of an animal without significantly increasing the complexity of its physiological systems?

Nearly all higher animals are segmented. The major representatives of segmented animals are the annelids, the arthropods, the vertebrates and some other groups. Segmentation is the multiplication of a basic unit resulting in an organism in which all units (segments) are required for its proper functioning. Animals such as tapeworms are not segmented in the true sense: they consist of proglottides that are formed by a sort of strobilation process. Strobilation is a process whereby duplicated units separate and are released from the parent. Some Coelenterates also undergo strobilation. This is schematically represented in the right half of level 6 in Fig. 4.1.

Revolution: increase in body size without the necessity of concomitant increase in complexity of physiological systems. Growing big has advantages over remaining small, e.g. with respect to finding food as a predator.

Level 8: Toolization (a new word: the contraction of 'Tool utilization').

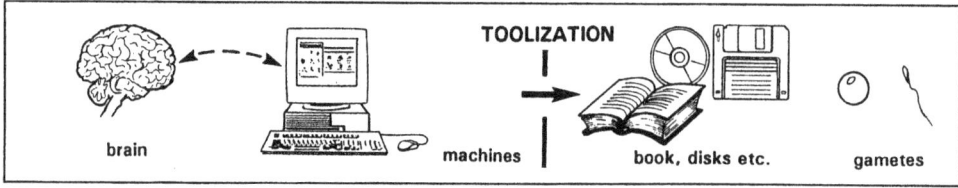

Problem: How to expand the cognitive and coordinating activities of the brain in a short span of time without having to increase the brain volume?

Some organisms have acquired the ability to use tools and, in doing so, have enlarged the size of their own compartment. Computers, for example, can be regarded as mechanical extensions of the human brain, based on silicon- and metal chemistry and on electricity carried by electrons (versus carbon chemistry and electric current carried by inorganic ions in the brain itself: see Chapter 9). They are not communicating compartments that can operate independently of man. This is an important consideration when addressing the question as to whether computers are alive or not (Chapter 9). In fact, computers are machines that are used by man largely for self-communication. Such tools, however, can become so refined in communication (e.g. a computer voice that sounds like a real voice; a touch screen that is extremely sensitive) that it may look as if the tools are part of the compartment that designed them (see Chapter 10). *Homo sapiens* is not the only species that uses tools.

Revolution: improving ones living conditions with less effort.

4.6 Compartments consisting of more than one individual of the same species: polyorganismal-monospecies compartments.

Level 9: Colonial compartmentalization.

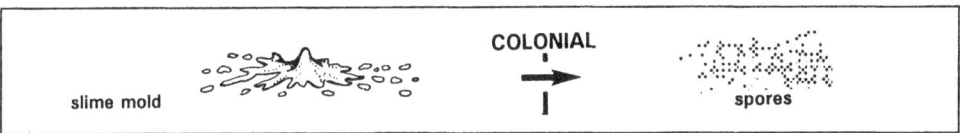

Problem: How to maintain genetic diversity in a population of individual organisms that normally reproduce asexually?

Under certain conditions a number of cells that can live freely under certain conditions can aggregate and form a colony type of organism. This aggregation requires that the individual cells adhere to each other in some way. One typical example is that of the slime mold *Dictyostelium*. In its simplest form, this organism looks and behaves like a solitary ameba. It lives in the soil and feeds upon organic materials. When there is a sufficient supply of food, the individual amebas are solitary. When there is a food shortage (*e.g.* after a heavy rain) some individuals secrete a signaling molecule, called cyclic adenosine monophosphate (cAMP). This molecule induces chemotaxis in other individuals in the neighborhood. The result is the aggregation of a number of individuals. Gradually, a new phenotype appears: the colony eventually comes to look like a small mushroom and starts forming spores. Corals are also colonies.

Revolution: colonies of individuals are able to achieve more complex goals and a greater genetic variability than the individuals are able to achieve on their own.

Level 10: Heterosexual compartmentalization

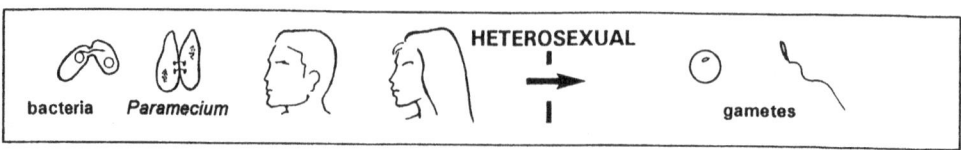

Problems: How to retain the full genome of a population over generations when the individual members no longer possess this full genome?

In the case of heterosexual reproduction, a single individual does not have the complete set of genes that is required for perpetuating its own genome. As a consequence, the individual is dependent upon another individual of the opposite sex, though of the same species, for securing the survival of its genome into the next generation. Depending upon the species, there are a variety of possibilities for generating sexual dimorphism (males and females). Hermaphrodites, if self-fertilizing, can reproduce as a single organismal compartment.

Revolutions: the introduction of meiosis allows much greater genetic variability in the resulting offspring than does mitosis, thereby facilitating the formation of new species. Copulation and gamete release elicit 'orgasmic' feelings.

Level 11: Social compartmentalization.

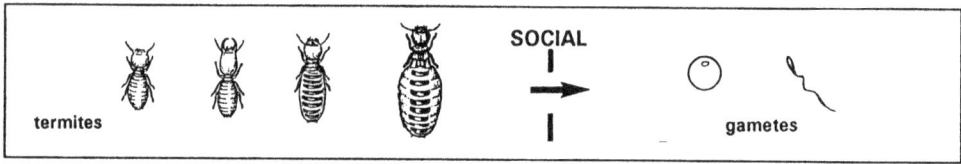

Problem: How to get individuals to work together on tasks that cannot be accomplished by any single individual working alone?

In some cases sexual reproduction gets compartmentalized to a few (or even two) individuals, who do almost nothing else than reproducing themselves, while being dependent on workers for food, shelter, protection, brood care, etc. Typical examples of this are the casts in social insects.
Social compartmentalization comes into play. Nature has been very inventive in generating all sorts of social compartments. Compared to social insects, *Homo sapiens* is still in a presocial stage of development. Nevertheless, man creates for himself an almost endless variety of social compartments, some overlapping with one another: the family, the neighborhood, the town, the state, the religious community, the school, the language community, the political party, the bridge-, tennis-, soccer-, football- clubs, the lovers of Bach or of heavy metal music, etc.

At the cellular level, the formation of the eukaryotic cell (level 2) could as well be regarded as an example of social compartmentalization, at least in the context of Margulis' theory.

Revolution: a group of organisms that organizes itself adequately by giving the right task to the right individual can achieve more than an equal number of non-socially organized individuals can achieve. The whole is more than the sum of its parts.

Level 12: Internal budding: baby-inside-mother compartmentalization.

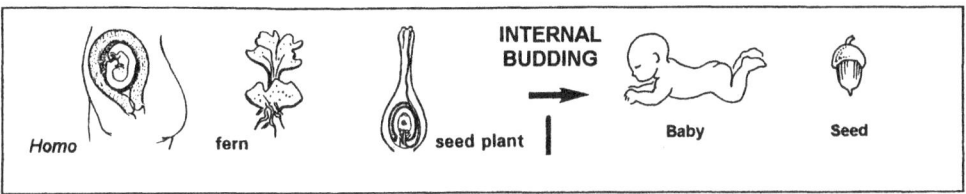

Problem: How to internalize whole organisms into existing organisms of the same species and how to coordinate the resulting complex?

In some organisms, embryonic development occurs inside the body of the female. After a period of time, the baby leaves the mother's body. Intrauterine development of mammals and some other viviparous species (e.g. some flesh fly species) are clear examples of this. The seeds of flowering plants are embryonic plants that develop inside the 'parental' plant. The sporophyte of ferns develops from within the gametophyte.

Revolution: females can remain mobile during the early developmental stages of their embryo(s), and the embryos are better protected.

Level 13: Population compartmentalization.

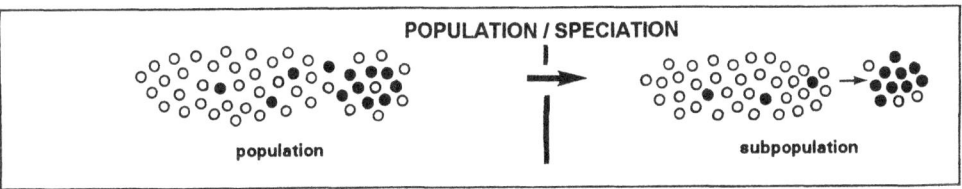

Problems: How to generate subpopulations in an existing population? How to expand the territory a population occupies?
A group of organisms of the same species that live together and form a reproductive community is a population. Early in its development, when a species still consists of a small number of individuals, the species forms a single population. The larger the population and the larger the territory it occupies, the more likely it is that subpopulations will form in due time. When reproductive interactions between the subpopulation and the parent population become less and less frequent, this populational compartmentalization may lead to the creation of a new species.

Revolution: this is the same as in social compartmentalization, but it occurs on a much grander scale and with emphasis on reproductive advantages (greater genetic variability).

Level 14: Electrospherization: connecting individuals by electromagnetic waves. Electrical/electromagnetic bridging.

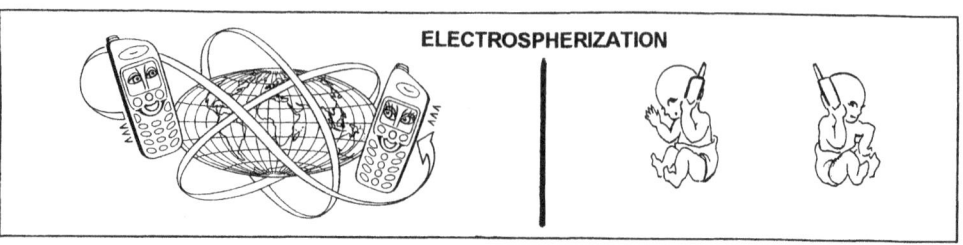

Problem: How to connect individuals, sometimes living at great distances from one another, by electromagnetic waves?

Some fish have electrical organs that produce electric currents. In some cases, these currents are used to paralyze prey. Weak self-produced electric currents are used in some species for communication among individuals of the same species. In such cases, the animals have electro-receptors.

Man has invented machines for producing and receiving electromagnetic waves for communication. These waves are used in radio, telephone, television, fax, Email and telex. This entire process is now culminating in the large-scale use of portable phones. If one thinks about it, mankind is in the process of realizing a new level of compartmentalization in which electromagnetic waves are the mechanical equivalent of the cytoplasmic bridges by which cells in a blastula can communicate with one another. A new type of sphere is being established in which individual organisms are interconnected by a great variety of electromagnetic bridges. To date, this process has culminated in the use of cell phones. I propose the term electrospherization for this phenomenon. While the computer is a mechanical extension of man as an individual, cell phones constitute a mechanical extension of man as a population.

Revolution: rapid exchange of information over distances ranging from short to very long enables individuals to act in concert. Improving one's living conditions with less effort.

4.7 Compartments consisting of individuals belonging to different species: heterospecies compartments

Level 15: Nutritive and protective compartmentalization.

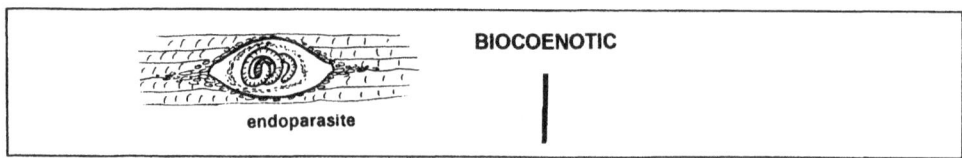

Problem: Food chains: How to enable different species to survive, some of whom are no longer able to produce their own food (e.g. by photosynthesis)?

Organisms that cannot synthesize all their required nutrients for themselves (such organisms are called heterotrophs) must ingest organic molecules originating from other organisms. The "heterotrophs" contrast with the "autrotrophs", which synthesize most of their own food, utilizing processes such as photosynthesis.

Many organisms depend on others for food. There are many forms of nutritive compartments: symbionts, commensals, parasites, predators, food chains in the broad sense, etc. Agriculture is relevant in this context. Domesticated farm animals are in fact tools for food production. Sometimes, different types of compartments are linked for reproduction. This is obvious in some parasites.

Although dependence on food is an important factor in generating life communities, it is not the only one. Some organisms live together for reasons other than nutrition, such as for shelter or for transport.

At the cellular level, the formation of the eukaryotic cell (level 2) could as well be regarded as an example of a heterospecies compartment, at least in the context of Margulis' theory.

Revolution: new habitats can be colonized and the unlimited reproduction of some species restrained, thereby facilitating the multiplication of others.

Level 16: Planetary compartmentalization: the Gaia compartment.

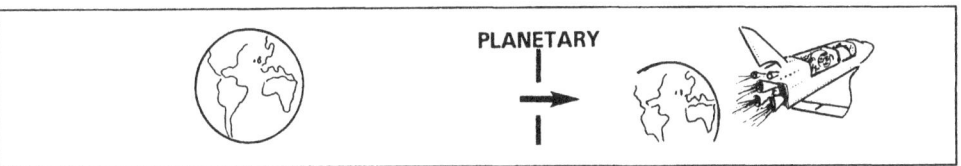

Problem: How to prepare for colonizing other planets?

The total sum of all biological compartments on the whole earth makes up the planetary biological compartment. This could be called the Gaia compartment. All living beings on earth are somehow interconnected by communication, but this does not mean that the earth itself can be considered an organism, as was proposed in Lovelock's (1995) Gaia hypothesis. Whether there is life on other planets in our Universe is as yet unknown.

Revolution: the integration of all lower (level 1-15) revolutions into a more or less harmonious entity of a higher order, thus making the planet Earth a sort of self-sustaining spacecraft cruising through space.

4.8 The mechanisms instrumental to Mega-evolution. Some common denominators in the mechanisms for creating higher levels of compartmentalization

It is clear that the acquisition of additional molecular tools for creating a novel communication system requires gene duplications and mutations. Mega-evolution involves simultaneous changes in (larger) sets of genes, sometimes in more than one species at a time. This will not yet be dealt with here. Later, in Chapter 15, I will present in more detail my 'double continuum' – or 'hardware-software' – theory of evolution.

Here, I will only indicate certain common denominators for major mechanisms used to create higher levels of compartmentalization with their inherent novel communication systems and expansion. The analysis of the 16 levels of compartmental organization shows that, on the whole, not so many different mechanisms are needed to create a broad range of variation. Apparently, a similar functional result, though at ever-higher levels, can be obtained by using different sets of genes.

These mechanisms are:
- *internalization* of a smaller compartment(s) into an already existing compartment of the same nature;
- *gluing together* compartments by a variety of means;
- *utilization of tools;*
- *sexuality and development;*
- *nutritive and protective deficiencies.*

Now I will regroup some levels of compartmentalization, as outlined in Fig. 4.1, to illustrate the cited mechanisms.

1. Some examples of 'internalization of compartments in an existing compartment'.

My first example is the internalization of membrane compartments within an existing membrane-bound compartment. This results in the eukaryotic cell. According to Margulis, some cell organelles such as mitochondria, are semi-independent organisms. From that perspective, a eukaryotic cell could also be considered a level-15 compartment.

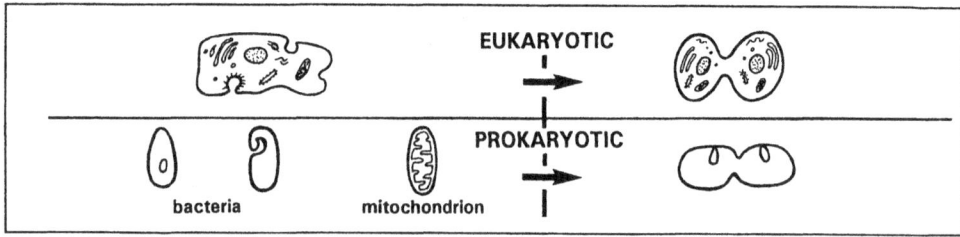

A second example is the internalization of epithelial compartments within an existing epithelial compartment: all animals are formed this way. First a closed, often more or less spherical epithelium is formed. This is called a blastula. Next, at a given point, an invagination is formed that results in the formation of the primitive gut. In 'higher' animals both this gut and the primitive skin continue to form infoldings. Many of them are severed and become separate internal compartments.

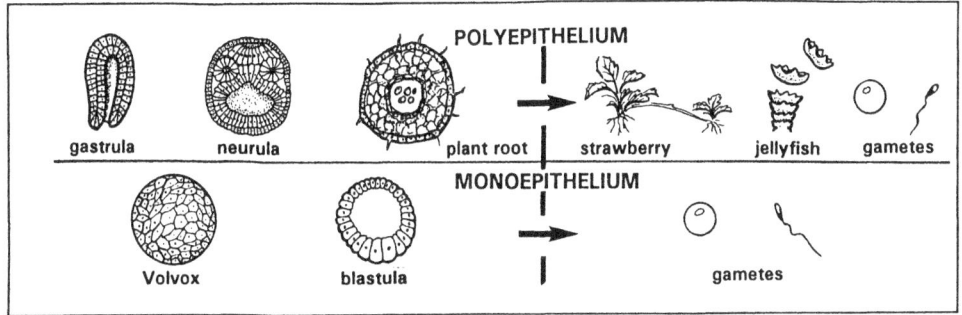

My third example is the internalization of organismal compartments inside an existing organism: e.g. intrauterine development in some animal species. The baby-inside-mother level of compartmentalization is due a genetic defect. In animals, the primary cause is the loss by the female of the ability to produce egg yolk proteins. One means of survival is the replacement of the yolk nutrients by parasitic intrauterine development (see Chapter 13).

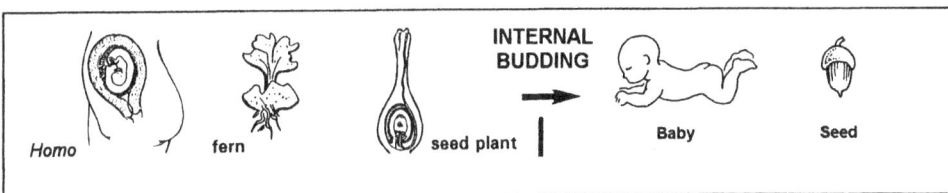

The next example is electrospherization. Different wavelengths and coding systems are used to enable the formation of subcompartments, such as different telephone-, radio- and television networks.

The last example is the internalization of a subpopulation(s) within an existing population.

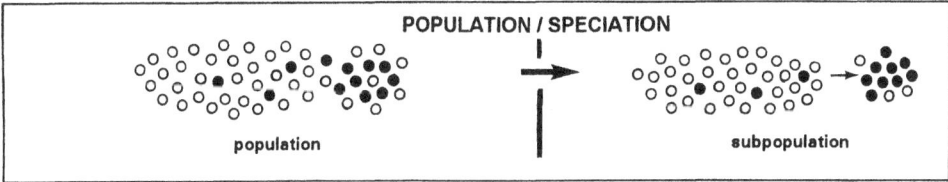

2. Some examples of 'gluing together' of similar compartments (e.g. levels 3 and 14).

- Cell adhesion molecules allow multicellularity.
- Specialization of the cytoskeleton enables the formation of epithelia.
- Connections by sounds and electromagnetic waves. Cellular phones as a recent example.

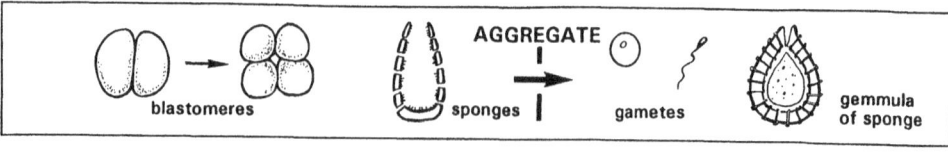

3. Utilization of tools.

This applies in particular to *Homo sapiens,* though some animals also use tools.

4. Heterosexuality and development.

Sexual reproduction with males and females came into existence rather late in the process of evolution. Asexual reproduction was established first. At a given moment a defect occurred in the genome of an asexual organism whereby the complete genome was no longer present within a single individual. As result, two individuals had to cooperate by means of sexual reproduction to ensure the survival of the whole genome. This will be dealt with in Chapter 13.
Embryonic development can also be regarded as a sort of compensation for the functional deficiency of a number of genes. All the cells in our body do have the same genome, but certain mechanisms render the somatic cells unable to use the complete genome at all times (see Chapter 14). The division of labor was the solution for this problem.

4. The introduction of defects in the nutritive self-support mechanisms necessitates the formation of obligatory or facultative nutritive compartments (symbiosis, parasitism, and food chains).

The first major bifurcation in ways of living together was probably caused by the loss in some bacteria of the ability to produce its own food by means of photo- or chemosynthesis. The deficient organisms became heterotrophs. They had to eat other organisms. Later in evolution, many additional defects showed up in heterotrophs. They resulted in the appearance of symbionts, parasites and predators. Ruminants, for example, cannot make a number of essential amino acids. To compensate for this defect, they rely on microorganisms living in their paunch.

5. Other defects, such as insufficient self-protection, may stimulate the development of other means of living together.

4.9   Newly formed compartments expand. Need for maintaining order. Installation of headquarters

The common situation is that as novel compartments and compartmental levels come into operation, they expand both in terms of numbers and in terms of the territory they occupy. Gradients play a crucial role in this expansion. Such expanding compartments are confronted with the problems of transporting messages over longer and longer distances with the consequent weakening of the signal. As the total number of units increases, more and more individual units have to be informed at the same time, and the action the receivers are supposed to undertake must somehow be coordinated. This may necessitate both a means for central decision-making (installation of a headquarters) and a system for controlling the quality of the response to the rules (a police system).

One can easily deduce the mechanisms that must be used when a given compartment enlarges by analyzing what happens when a human population starts growing. The colonization of North America may serve as an example. When the first immigrants arrived, spoken communication was initially sufficient. As the distance between the individuals increased, they had to shout to still make themselves understood. As the distances among the individuals continued to increase, more and more sophisticated mechanical means for transporting messages had to be invented and put into use. Soon, a system for central decision making had to be installed: a government. But government rule requires a police force to ensure that there is obedience to the rule of the government.

The installation of a headquarters is not restricted to *Homo sapiens*. Other examples are the nucleus for a cellular compartment, the brain of animals, the leader of a herd, and the queen of a termite colony.

Yet, there are complex compartments that contain several levels of subcompartmentalization but that do not have a headquarters. Plants and Fungi have nothing that can be compared to the brain of animals. For communication at their highest level of compartmental organization, plants use a diffuse system of signal transduction. All cells are interconnected electrically. They use action potentials but these are slower than the ones in the nervous system of an animal. Such a diffuse system can be compared to our World Wide Web. It has the advantage of being less vulnerable than the concentrated nervous system of animals. In a school of fishes or in locust swarms, there is no leader either.

There are limits to the size a given compartment can grow. Such limitation is inherent in the nature of communication systems.

4.10 Mechanisms for generating daughter compartments

These are schematically represented in the right half of Fig. 4.1. Nature has not been very inventive in creating a wide variety of mechanisms for forming daughter compartments. Bacterial compartments (level 1) are multiplied by binary fission. Mitosis and meiosis, followed by cytokinesis, is by far the predominant mechanism in eukaryotes. Meiosis comes into play when gametes (sperm and egg cells) are to be formed. Several mechanisms of asexual reproduction exist (= without gametes), such as strobilation in jellyfish, the formation of proglottides in tapeworms, or of gemmulae in some sponges, or of stolons in strawberries, etc.

At level 16, the Gaia compartment started its first trials to spawn daughter compartments when it launched its spaceship programs. We are now watching the very first preparations ever made by a terrestrial species, *Homo sapiens*, in an attempt to colonize other planets. A 'manned' spaceship is equivalent, at the Gaia-level, of the baby at the level of a heterosexual couple, or of a gamete at the level of an individual. Like all stages before, this next step of compartmentalization will be difficult to realize but, once done, it will open up enormous possibilities.

In addition to the genetic continuum made possible by the duplication of DNA (deoxyribonucleic acid), there is also the possibility of the continuation of another type of information by teaching-learning in organisms that have the ability to do so. Teaching-learning is thus a form of reproduction. Its possible role in micro-evolution will be discussed later (Chapter 15).

4.11   General conclusions

Nature is a master in combining simple elements, again and again, to form an endless and increasingly complex array of living and non-living entities. The use of only four nucleotides, combined in triplet codons as a means for specifying the composition of all peptides and proteins synthesized by all organisms on earth, is the best-documented example of this fact. This chapter gives a glimpse into the possibilities for generating variability on the basis of a single building block, the communicating compartment of level 1. Several compartments of level 1 gave level 2, the eukaryotic cell. Through the making of associations of level 2 eukaryotes, level 3 appeared, etc. This story closely resembles the formation of the universe that started with only a single atomic building block: hydrogen. All other elements are fusion products of this elementary building block. All structures of a higher order, ranging to up to stars, planets and living systems, are in fact multiples of hydrogen.

There can be no doubt that the first dogma is of prime importance in understanding biodiversity. However, there are other universal principles that are just as important for the living state as the principle of the genetic code. One of them is compartmentalization, as outlined in this chapter. Another is communication. In general biology textbooks one finds that the basic unit of the living state is the cell. In Chapter 15 I will discuss the idea that this may be too restrictive a view. I think that it is better to take *the communicating compartment as the basic building block*, and to consider the prokaryotic cell type as being its *smallest* and least complex autonomous unit.

Evolution alienates, both genetically and communicationally. This chapter has illustrated some of the possibilities that can be obtained through the second aspect. The really major breakthroughs in evolution (Mega-evolution) were realized every time a higher level of compartmentalization was achieved. For this to be successful, novel means of communication had to be introduced. These will be dealt with in chapters 7 and 15.

Nature is hierarchically organized. The interest of a higher level always takes precedence over that of the lower levels. This is already apparent from the structure of the family tree of levels of communication as was outlined in Fig. 4.1. It will become even clearer when the definitions of 'death' and of 'life' will be dealt with in later chapters. Taking precedence at the highest level is an intrinsic property of 'life' itself.

ESSENTIALS

REVOLUTIONS IN MEGA-EVOLUTION

1. The mechanisms so far described governing micro- and macro-evolution are insufficient to explain certain major revolutions that have taken place in the course of geological time. These revolutions concern changes in the hierarchical organization of living matter.
2. Hierarchical organization automatically implies the participation of more than one entity. The evolution of these entities may imply changes either in some of the participating entities or in all of them. There are conditions in which the entities belong to different species.
3. The term 'communicating compartment' or simply 'compartment', the full definition of which was given in Chapter 3, will be used throughout the book to denote a level of hierarchical organization.
4. A 16-level classification system is presented which brings order to the multitude of communicating compartments. It crosses systematic borders.
5. Levels 1-8 represent compartments restricted to one and the same individual organism: mono-organismal compartments. Level 1 corresponds to the monomembrane- or prokaryotic cell type of compartments; level 8 corresponds to segmented animals that use tools.
6. Levels 9-14 correspond to compartments consisting of more than one individual of the same species: polyorganismal-monospecies compartments. Here mechanisms are operating that enable 'social life'. The heterosexual compartment (= the 'invention' of males and females) corresponds to level 10.
7. Levels 15-16 correspond to compartments consisting of individuals belonging to different species: heterospecies compartments. Host-parasite combinations and food chains fit in level 15. The total biosphere, or the Gaia compartment, is the highest one, namely level 16.
8. The mechanisms used to reach ever-higher levels of compartmentalization are few and simple. They have been repeatedly used in the course of evolution, though at ever-higher levels. The internalization of a given compartment into a preexisting larger one and the gluing together of compartments are examples of such mechanisms.
9. The interest of the highest compartmental level always prevails over that of any lower level. This is one of the explanations for the fact that in the perception of humans, "nature is red in tooth and claw". This rule should not be misinterpreted to justify certain political and economic systems (dictatorships, feudalism, fascism, extreme forms of capitalism and socialism) that make the interests of the population (compartmental level 13) subordinate to the interests of the leaders, who themselves belong to a lower compartmental level, namely level 8.
10. Newly formed compartments that do well - or, in other words, that are 'fit' - expand. This introduces the need for discipline and order, and in some instances the installation of headquarters as well. It also forms the basis for conflicts at borders.
11. Mechanisms for generating daughter compartments are rather few in number and have been well conserved in the course of evolution.

CHAPTER 5

TRADITIONAL ATTEMPTS TO DEFINE 'LIFE'

Literature survey

For a fish, understanding the meaning of 'being wet' is as challenging as grasping the 'very nature of life' is for the thinking *Homo*

Contents

5.1 The necessity of defining 'life'
5.2 'Life': of extraterrestrial origin?
5.3 The 'circular definition' approach
5.4 The approach based on listing the 'properties of living matter'
5.5 The approach based on the 'classical laws of thermodynamics'
5.6 The approach based on the far-from-equilibrium thermodynamics of Ilya Prigogine: "Living systems are 'dissipative systems', as is shown by their heat production"
5.7 "Living beings are characterized by the fact that they are continually self producing. Autopoiesis and cell suicide (apotopsis)."
5.8 "A system in which information carriers can duplicate themselves is alive"
5.9 "A self-correcting system is alive"
5.10 "Life is a machine"
5.11 Other approaches: "Feeling alive"
5.12 A list of traditional approaches to the definition of life
5.13 Still something missing? Where is the 'intellectual dimension' of life?

Essentials

5.1 The necessity of defining 'life'

The ultimate goal of any introductory course in biology should be that the students come to understand the very nature of 'life'. To my knowledge, no single contemporary textbook achieves this goal. This leaves both teachers and students with a feeling of dissatisfaction.

Any theory of evolution is an attempt to provide an answer as to how 'life' evolves in the course of geological time. The most logical approach is first to define 'life'. Next, its properties should be defined. Finally, based on these properties, the force(s) driving evolution should be searched for.

No matter how logical this approach may now appear, it was not the one taken by Darwin and the majority of later evolutionary biologists. There is a major problem. Indeed, none of the different approaches that have been taken in the past have resulted in a good definition of 'life'. In this chapter, a concise overview is given of the major approaches. In Chapters 6 and 7, another approach that makes use of the principles of communication will be proposed.

5.2 'Life': of extraterrestrial origin?

Although the simplicity of a static concept of the world (Chapter 1) may be comforting to many, it is no longer in agreement with the data provided by science. The creationist view, at least according to the verbatim interpretation of the book of Genesis, is no longer acceptable. The data provided by the recent genome projects all confirm Darwin's concept of 'descent with modification'.

The view that life did not originate on earth but somewhere else in the Universe and then was somehow brought to earth, as advocated by some renowned researchers, such as Alfred Hoyle, looks like an escape maneuver. It simply displaces the problem of what life is and how it originated instead of solving it.

Let us be practical and consider the possibility that life originated on earth as the result of a natural process of evolution. A major problem arises with this 'practical scenario'. If life on earth indeed came into being 'spontaneously', somehow and sometime in the course of evolution, how can we determine the moment that life began when we do not even know what 'life' is? Thus, before trying to answer the question when life began, we should first find out what is or are *the* distinguishing parameter(s) that demarcate the boundary between the living and the non-living state. In other words, we must find a good definition of 'life'.

5.3 The 'circular definition' approach

"Life is what all living organisms have in common."
"Life began when the pre-life period ended."

These are just two examples of 'circular' definitions or statements. They have hardly any informative value.

5.4 The approach based on listing the 'properties of living matter'

Biologists and physicists face a similar basic problem: neither of them has as yet been able to plausibly define the very nature of their study object, namely 'life' and 'matter', respectively.

They circumvent the problem by listing as many properties as possible of their study object. For example, physicists will say that matter consists of molecules, molecules of atoms, and atoms of subatomic particles such as electrons, protons and neutrons. There are even smaller particles such as (up and down) quarks. In addition, the forces operating in matter are gravity, electromagnetism, the strong nuclear force, the weak nuclear force, and, perhaps, the ultra-weak force.

In Henderson's *Dictionary of Biological Terms* (11th edition, 1995), 'life' is defined as follows.

> "Living organisms can be distinguished from other complex physico-chemical systems by their storage and transmission of molecular information in the form of nucleic acids, their possession of enzyme catalysts, their energy relations (e.g. photosynthesis, respiration and other enzyme-catalyzed metabolic activities), their ability to grow and reproduce, and their ability to respond to stimuli (irritability). Entities such as viruses, which satisfy only some of these criteria, are also generally considered to be part of the living world."

In Chapter 10 I shall explain that in my opinion viruses cannot be alive. In one of the textbooks on general biology (Postlethwait and Hopson, 1991), which uses a similar approach, five 'life traits' are listed that combat disorganization (order, adaptation, metabolism, movement, responsiveness), three that perpetuate a population despite the death of the individuals (development, reproduction, inheritable traits of information or genes) and one that is important for generating diversity (evolution).

Although all these traits are essential for maintaining the living state *over generations*, only some of them are necessary for a cell to be alive for a time span *shorter than its reproduction cycle*. This distinction is important. Indeed, differentiated cells (for example fully differentiated neurons) usually do not grow any more, and they neither divide nor evolve. Nevertheless they are alive and functioning. The enucleation of an amoeba, a mammalian red blood cell or a fertilized egg, for example, does not cause instant death. At first sight, the essential traits in the *short run* are those five that are responsible for the orderly state.

A first set of basic questions is then: How orderly must 'biological order' be, how many different possibilities and combinations of metabolic reactions must simultaneously be available, how much responsiveness or movement must be present for a state to earn the label 'living'? The only answers to such questions are quantitative and would result in endless discussions about numbers.

An additional basic question is whether or not there is a decisive trait that makes the crucial difference between the non-living and the living state (Margulis and Sagan, 1995). There is none among the five parameters of the short-term life: any of them can be present in non-living systems as well. Enzymes can carry out metabolic reactions in a test tube, crystals are highly ordered, etc. Thus, taking this approach we can only say that the living state requires the simultaneous presence of at least the five short-run parameters. *But their combination, 'life', is clearly more than the sum of all five traits.*

Although all listed traits are undeniably necessary for the living state to exist, they are not the traits that we intuitively feel as being absolutely needed to conclude that something is living. It is therefore necessary to seek alternative approaches.

5.5 The approach based on the 'classical laws of thermodynamics'

A third approach is to investigate whether some of the basic laws of physics are not also valid for the living state or whether living matter might possess certain 'forces' that are absent in non-living matter. Most attention has been paid to the two laws of thermodynamics.

First law: the various forms of energy (light, mechanical, chemical, electrical, radiation) can interconvert into one another, and when they do so, they do so quantitatively, that is, without any energy being created or destroyed in the process.

Living matter (cells) is particularly effective in performing energy transductions. Photosynthesizing plants convert light energy from the sun into chemical energy. The human eye converts light energy into chemical energy and then into electrical energy. Our muscles convert chemical energy into mechanical energy that results in movement. Thus, this law is obeyed as scrupulously by living matter as it is by the heat engines investigated by the early students of thermodynamics.

Second law: This law states that when a system performs work it runs down, not only because the free energy decreases but also because the entropy, its state of disorganization, increases.

The formal relationship among four essential properties of matter (G, H, T and S) is stated by the formula:
$$\Delta G = \Delta H - T\Delta S$$
where:

ΔG is the change in free energy (= energy that can be used to do work) measured in joules.

ΔH is the change in enthalpy, or total energy content, also measured in joules. Enthalpy is the energy given off or absorbed, and it can be measured directly in an isolated system by the change in the temperature that the system undergoes.
T is the absolute temperature, which is measured in degrees Kelvin.
ΔS is the change in entropy, which is measured in joules per degree Kelvin.

To state the second law a different way: in a given system the change in the energy capable of doing work is equal to the amount of heat given off or absorbed minus the product of the absolute temperature and the change in entropy (disorganization).

According to this law, all systems have a 'tendency' to decrease their ability to do work and/or to increase their state of disorganization. In other words, the normal end situation is maximal disorganization.

At first sight, living systems do not obey this law: they are able to create order, as evidenced in embryonic development. But living systems can only do so as long as energy is 'pumped' into the system. When the food supply stops and energy is no longer being provided, a living system will sooner or later die, and death means maximizing its degree of disorganization. The minerals in its body disperse, and its constituent complex organic molecules all break down into carbon dioxide, water, ammonia, etc.

Schrödinger, in his book "*What is life*" (1946), emphasized this second law and, according to him, "Living organisms stay alive by virtue of their ability to get rid of the entropy that is created by the processes by which the organisms live."

Schejter and Agassi (1994) attempted to correct some of the limitations of this approach. They summarize their view as follows:

"The thermodynamic history of a living system can be distinguished from all possible non-living systems, including non-living local instabilities. The distinguishing factor is the appearance of a permanent structural modification that is a difference between final state and initial state, through which the entropy of the system has decreased. This is because the temporary (dynamic) local instability - the living organism - has originated a permanent (static) modification in the environment. Only living systems are able to produce such a sequence of events".

As we will see later in this chapter and in the next, there is a much simpler way to circumscribe that 'permanent structural modification': it is nothing else than a 'compartment' or, in other words, life is always linked to a 'compartment' as this has been defined in Chapter 3.

The limitations of the approach based on the second law of thermodynamics will become clear in Chapter 6 where we will discuss the death of a small population of deer while the individual deer of this population nevertheless stay alive, grow, cope with entropy, etc.

Coping with entropy is evidently very important for organisms but we do not intuitively experience it as being the property that distinguishes the living from the non-living state. With all due respect, I do not think that Schrödinger's approach is the right one.

5.6 The approach based on the far-from-equilibrium thermodynamics of Ilya Prigogine: "Living systems are 'dissipative systems', as is shown by their heat production"

'Dissipare' in Latin means 'to distribute'. A steam engine cannot use all the energy present in the coal that is burned in its oven to do work. Less useful forms of energy - notably heat - are exported or dissipated away. Not all energy that we ingest with our food can be used for doing work. A substantial part is lost as heat.

A dissipative system is far from its thermodynamic equilibrium, and therefore unstable by definition. If left undisturbed, it will run down. However, dissipative systems display some degree of order. The generation and maintenance of this ordered state requires the import of useful energy. According to the pioneer in this domain, the Belgian Nobel laureate Ilya Prigogine and his team (Progogine and Nicolis, 1971; Prigogine,1980; Prigogine and Stengers, 1985), "Living organisms are far-from-equilibrium objects separated by instabilities from the world of equilibrium. They are necessarily 'large' macroscopic objects requiring a coherent state of matter in order to produce complex biomolecules that make perpetuation of life possible." This approach is very important for a better understanding of 'life', as we will discuss in later chapters, but it is not sufficient. In summary, a 'dissipative system' is: 'a far-from-its thermodynamic equilibrium system where, by importing useful forms of energy and by exporting less useful ones, an ordered state can be generated and/or maintained.'

5.7 "Living beings are characterized by the fact that they are continually self-producing. Autopoiesis and cell suicide (apoptosis)."

Maturana and Varela (1980) invented the term autopoiesis, which means 'self production', or 'making itself' to describe the criteria 'necessary and sufficient' to define life at the cellular level. In their view, cells are chemical machines that are continuously producing themselves. Furthermore, they are autonomous, which means that they can specify their own laws and what is proper to them. The criteria for autopoiesis are:

    i. identity,
    ii. mechanistic operation,
    iii. self- self-limiting,
    iv. compartment production by compartment transformation (for more details, see also Zeleny (1981), Fleishaker (1988), and Cullen (1996)).

In autopoiesis, the normal fate of a cell is to grow, divide, grow again, divide again etc. In other words, a cell functions as a self-sustaining machine. There is no role for death, as this terminates autopoiesis. But cells do die.

Research about the causes of cell death in healthy organisms has revealed that (some) cells can die in two different ways, either by necrosis or by a process which is called apoptosis (second 'p' is silent). Necrosis is the ultimate death of a cell following an accident, an action of a toxin, exposure to heat or to ice crystal formation in its cytoplasm, or through oxygen deficiency in cells that need this gas. Cells that die by necrosis swell and die when their plasma membrane bursts. It is a rapid process. Apoptosis also leads to cell death, but the mechanisms that it involves are totally different from those encountered in necrosis. It is widespread and *necessary* in certain developmental stages of healthy organisms.

Apoptosis is derived from the Greek and means 'dropping off', as of the leaves from trees. Apoptosis is *natural cell suicide* carried out by cells that are active and healthy at the moment that they switch on the genes that lead to self-destruction. Doomed cells shrink and repackage themselves into digestible parts so that neighboring cells can consume them. The whole process takes a few hours. Apparently, any normal cell type can undergo apoptosis. This means that all living cells are genetically programmed not only to grow and divide but also to kill themselves. The observation that completely isolated cells cannot survive and that in fact they commit suicide led a researcher in this field (Professor Martin Raff from London, cited in Cullen, 1996) to suggest that the normal state for a cell is death. In multicellular organisms cells literally need company and support in the form of stimulating chemicals (e.g. cytokines) in order to stay alive. If cells do not get this support, they initiate their own deliberate cascade of self-destruction. Thus, the more 'autonomous' such cells are allowed or forced to become, the more likely they are to commit suicide. A cell will become cancerous and thus pass into a mode of continuous self-(re)production only if it can ignore apoptotic signals (Cullen, 1996). Cellular compartments are continually balancing on the edge of decision between self-perpetuation and self-destruction. The degree of 'attention' they receive from their environment dictates which way they go. One argument against this approach is that the individual protistan does not fit well into the 'needing company' concept.

5.8 "A system in which information carriers can duplicate themselves is alive."

This approach implicitly assumes that reproduction is an essential - or even the key - property of the living 'state'. But there are flaws in this logic: sterile organisms, for example an ox, cannot be classified as non-living simply because they cannot reproduce. Reproduction is essential for ensuring the continuity of life over more than a single generation, but it is not an absolute prerequisite within the framework of an organism's lifespan.

The carrier of the genetic information of organisms is deoxyribonucleic acid (DNA), a molecule in the form of a double helix. When a cell divides, the helix unwinds and two new strands are formed to complement the two single parent strands, so that the original situation is duplicated.

The approach based on the principle that a cell is alive because it can duplicate its DNA - and therefore itself - does more or less fit the facts. Nevertheless, the viability of enucleated cells and the multiplication in the laboratory of nucleic acid strands by the so-called 'Polymerase Chain Reaction' (PCR) do present certain problems. 'When a cell such as an egg cell or an ameba is enucleated, it does not instantly die. Thus, DNA is not required for an organism to be alive in the short run. As for the PCR method, it does not require the presence of living cells, but only of test tubes, the proper enzymes, other reagents and the proper protocol for multiplying nucleic acids. No molecular biologist would ever think that he or she was 'creating life' by doing PCR.

5.9 "A self-correcting system is alive."

In the view of general systems theorists, a self-correcting system is alive. Deoxyribonucleic acid (DNA), the carrier of genetic information, can be damaged by a variety of factors (certain chemicals, ultraviolet and cosmic radiation, etc.). Most cells have adequate repair mechanisms to correct 'mistakes' in the code. Thus, because of its ability to correct mistakes, a cell is considered to be living. Certain servomechanisms, such as a word processor equipped with a spelling correction program, can also automatically correct mistakes. No biologist would ever consider such a system to be 'alive', however. The question as to whether a computer is alive or not will be dealt with in Chapter 9.

Somewhat along this same line of thinking is the notion that DNA is an essential characteristic of all living organisms, whereas non-living entities, structures, etc. never possess such an information carrier. Thus the presence of this molecule could be considered to be the distinguishing trait between the living and the non-living state. But then what about the following data and experiments?

The red blood cells of mammals lose their nucleus soon after they are formed and they also lose their mitochondria (which also contain DNA): such erythroplastids nevertheless continue to live for a few more weeks.

Even more surprising is the fact that in the eggs of both sea urchins and amphibians, cleavage and even partial blastula formation can occur *in the absence of a nucleus*, though no one would ever consider such cleaving eggs not to be alive! Eggs lacking functional chromosomes can be obtained as follows. A frog is fertilized with sperm that has been irradiated with X-rays to destroy the DNA. Such irradiated sperm can fertilize an egg. Next, the egg nucleus moves to the egg surface, where it can be mechanically removed. The chromosomes of the irradiated sperm nucleus degenerate but the sperm aster functions to contribute to the formation of the mitotic apparatus. Such eggs go through early cleavage at a somewhat retarded rate until a partial blastula is formed, though development ceases before gastrulation (Barth, 1964).

Similarly, when an ameba is enucleated, it is not instantly 'dead'. It continues to move and carry out metabolic reactions for at least several more minutes. The reason it takes some time for an enucleated cell to die is that such a cell still contains some information-carrying molecules, such as the different types of ribonucleic acid (RNA). For a limited period of time, this RNA continues indirectly coordinating the energy production of the cell in the mitochondria. However, when one kills the mitochondria of any cell type with an appropriate toxin, death of the whole cell ensues instantly, despite the presence of intact DNA and RNA. The reason is not that the DNA in the mitochondria is more important for life than that of the cell itself, but rather that with the elimination of the essential respiratory enzymes in the membranes of the mitochondria, the cell is no longer able to produce adenosine triphosphate (ATP). This form of chemical energy is required to build up an ionic-voltage gradient over the plasma membrane of the cell. It is the irreversible loss of this essential gradient that directly causes death, not the absence of DNA. This will be dealt with in greater detail in other chapters.

5.10 "Life is a machine."

In developing his philosophical system, René Descartes concluded that life itself is *automaton*-like. This led to the "life is a machine" metaphor, which remains a major conceptual force in biology even today. In his book *Life Itself*, Robert Rosen (1991) writes:

> "The question 'What is Life?' is not often asked in biology, precisely because the machine metaphor already answers it: "Life is a machine." Indeed, to suggest otherwise is regarded as unscientific and viewed with the greatest hostility as an attempt to take biology back to metaphysics."

Rosen further addresses the question as to why the machine metaphor is so attractive to many. He answers:

> "First it assures biologists that their subject is an analytical one, because it asserts that any machine is a set of parts. Second, it assures them that the same set of parts will solve all problems of fabrication and of physiology simultaneously. Third, it assures them that nothing happens in biology that is outside the ken of the physical universals (or rather of those fragments of physical universality necessary for the understanding of machines). As to the parts themselves biologists used to think that they were cells, but today they are molecules. And if biology is hard, it is simply because there are so many parts to be separated and characterized."

Rosen further states that the machine metaphor is not just a little bit wrong: it is entirely wrong and must be discarded. I agree with Rosen that it is an illusion to think that reductionism can lead to a plausible answer to the question "What is Life?", because a biological system as a whole is not simply the sum of its parts. It is more. And it is also of a different order. This will be explained later. Rosen uses a mathematical approach to define 'life'. His definition is:

> "Life is the manifestation of a certain kind of (relational) model. A particular system is living if it realizes this model."

In this book I will argue that, indeed, 'life' is not a machine, **but rather it is the activity of a machine**. At first sight this may appear to be a small, perhaps negligible nuance. In fact it reflects a different view of biology, which has far-reaching implications. The type of activity that life is all about will be explained in later chapters.

5.11   Other approaches: "Feeling alive"

So far I have only dealt with the major approaches that have been taken based on the 'exact' sciences such as biology, chemistry, physics and mathematics. The conclusion is that none of them is adequate for defining 'life'.

Hemmerlin (1996) - and probably many other thinkers in fields such as sociology, psychology and philosophy - will argue, not totally without justification, that the failure of the exact sciences to define life follows inevitably from their purely 'exterior approach', which both ignores the *interior feeling* of being alive and fails to explain it. Hemmerlin holds that even when we have problems with our self-regulation and self-organization, and even when we are no longer capable of self-reproduction, we are nevertheless still alive. The exact sciences are incapable of revealing why, even when totally paralyzed, we *feel* alive.

In Hemmerlin's view, death is obedience to mechanical causality, life is obedience to internal causality. Life is not really definable in and of itself. It is a property of being.

It is certainly true that in contemporary biology there is hardly any room for emotions and feelings: any textbook on general biology will confirm this proposition. In my opinion, this is one of the reasons why many students who are required to take a course in biology find that the subject does not contribute much to a better understanding of what we *feel* to be 'life'. One of the goals of this book is to help narrow the gap between the external and the internal views on the subject.

5.12 A list of classical approaches to the definition of life

In 1996 the University of Padova in Italy published a book (M. Rizzotti, Ed.) entitled: "Defining Life, the central problem in theoretical biology". Unlike many that publish in this field, each of the authors of the Padova book clearly formulates his or her own definition of life. This yields the following list of verbatim definitions.

André Brack: "Life is a chemical system capable to replicate itself by autocatalysis and to make errors which gradually increase the efficiency of autocatalysis."

Camilo J. Cela-Conde: "Living things are beings which process information in such a way that in the sequence <environmental stimulus → knowledge construction → motor response →, possible results (motor responses) in terms of inputs (environmental stimuli) cannot be mechanically predicted."

Lorenzo Colombo: "Life is temporally-based macromolecular order."

Sidney W. Fox: "Life consists of proteinaceous bodies formed of one or more cells containing membranes that permit it to communicate with its environment via transfer of information by electrical impulse of chemical substance, and is capable of morphological evolution by self-reorganization of precursors, and displays attributes of metabolism, growth and reproduction. This definition embraces both protolife and modern life, i.e., life."

Francesco S. Gaeta: "Life consists in the capacity to reproduce and evolve, possessed by open thermodynamic systems that have developed with their surroundings interactions consenting a continuous acquisition of order."

Tibor Ganti: "The living system – at the cellular level - are proliferating, program-controlled fluid chemical automatons the fluid organization of which are chemoton organization. And the life itself – at the cellular level – is nothing else but the operation of these systems."

Hyman Hartman: "If one starts with the definition of life as any set of entities which are able to replicate, mutate, and be subjected to natural selection then viruses are living. Biological viruses are compared and contrasted with computer viruses. The role of viruses in speciation and in the horizontal transfer of genetic information is considered to be important in evolutionary processes. Finally a brief consideration of the relevance of this definition for the origin of life is examined by a comparison of the RNA world with the Clay world."

Abir U. Igamberdiev: "Life is a self-organizing and self-generating activity of open non-equilibrium systems determined by their internal semiotic structure."

Pier L. Luisi, Antonio Lazcano, Francisco J. Varela: "A physical system can be said to be living if it is able to transform external energy/matter into an internal process of self-maintenance and self-generation. This common sense, macroscopic definition, finds its equivalent at the cellular level in the notion of autopoiesis. This can be generalized to describe the general pattern for minimal life, including artificial life. In real biological life the autopoietic network of reactions which leads to self-maintenance and self-generation is under the control of nucleic acid and the corresponding proteins."

Haboku Nakamura: "Living things are defined as systems simultaneously having three characteristics: (i) self-maintenance, (ii) self-reproduction, and (iii) evolution in interaction with the environment."

Pietro Omodeo: "The living being is an open system, cellular, self-reproducing, with self-regulated flows of matter, energy and information which run through it and control its growth and steady state. Because of its attributes such a system is capable of evolution, by adapting itself to changing environmental conditions."

Martino Rizzotti (in his Afterword: "What is Life not?): (I) Living things are not self-reproducing automata *sensu* von Neumann, and (ii) Living things are not dissipative structures *sensu* Prigogine."

The general approach taken by these twelve authors contrasts sharply with that taken in "What is Life? The next fifty years". Speculations on the future of Biology (Murphy and O'Neill, 1995), a symposium compendium co-authored by eminent authorities who were invited to commemorate the 50th anniversary of Schrödinger's 1943 lectures on "*What is life?*" at Trinity College in Dublin. I find it remarkable that none of the speakers at the Dublin conference made a serious effort to come up with a plausible definition. I have the impression they all found it a hopeless enterprise. Nobel laureate Manfred Eigen even writes (p. 9 in Murphy & O'Neill, 1995):

> "What is Life? Not only is this a difficult question; perhaps it is not even the right question. Things we denote as 'living' have too heterogeneous characteristics and capabilities for a common definition to give even an inkling of the variety contained within this term."

This is a discouraging statement for those like me who nevertheless are attempting to uncover the very nature of life and to formulate a plausible definition.

5.13 Still something missing? Where is the 'intellectual dimension' of 'life'?

So many characteristics of life are already enumerated in all the definitions and approaches we cite that it may seem improbable that anything important has been overlooked.

But what should we think of the following approach that I initiated already in chapter 3? "*In contrast to non-living matter, living matter can decode coded messages and do something with them, such as solving problems.*" This statement refers to the 'intellectual' properties of life that, except in the definition of Conde, are largely ignored in the above definitions.

In the following chapters I hope to make clear that, at least in my opinion, this ability to handle coded messages is the central issue in understanding the very nature of 'life'.

ESSENTIALS

TRADITIONAL DEFINITIONS OF LIFE

1. Biology is the study of 'life'. Any discipline should clearly define the object it studies.
2. According to Schejter and Agassi (1994), an adequate definition of 'life' should meet the following conditions. "Apart from its not being trite and uninformative (circular, to use a traditional term), it should be neither too wide nor too narrow; it should not exclude living things and it should not include dead ones. Furthermore, it should not make biology part-and-parcel of chemistry and physics."
3. Over a dozen possible definitions of 'life' have already been proposed but none seems to meet the above conditions.
4. The 'life is a machine' approach is latently present in nearly all-contemporary textbooks on general biology. The basic philosophy in this approach is that the whole equals the sum of its parts. Thus, by characterizing the properties of the different subunits of living matter (e.g. cell organelles, cells, tissues, organs, etc.) and adding them up, one should obtain the full picture of 'life'. This is obviously not the case: the whole is more than the sum of its parts. According to Rosen (1991), the "life-is-a-machine approach is totally wrong and should be discarded".
5. This means that a basic philosophy is needed that can account for the fact that whole is more than the sum of its parts.
6. But if life is not a machine, then what is its essential nature? In other words what is missing in all the definitions?
7. What seems to be missing is that only living matter has the ability to decipher coded messages and to generate an energy demanding response to them. One could also say that only living matter can solve problems, for the most part unconsciously, though in some systems consciously as well. This implies an element of 'intelligence'. Thus living matter differs in terms of its *activity* from non-living matter.
8. Hence, it remains to be verified whether or not this special problem solving activity is *the* crucial difference that may yield an acceptable definition of 'life'.

CHAPTER 6

ALIVE, NO LONGER ALIVE.
THE DEDUCTION OF A DEFINITION OF 'DEATH' AS A KEY TO THAT OF 'LIFE'

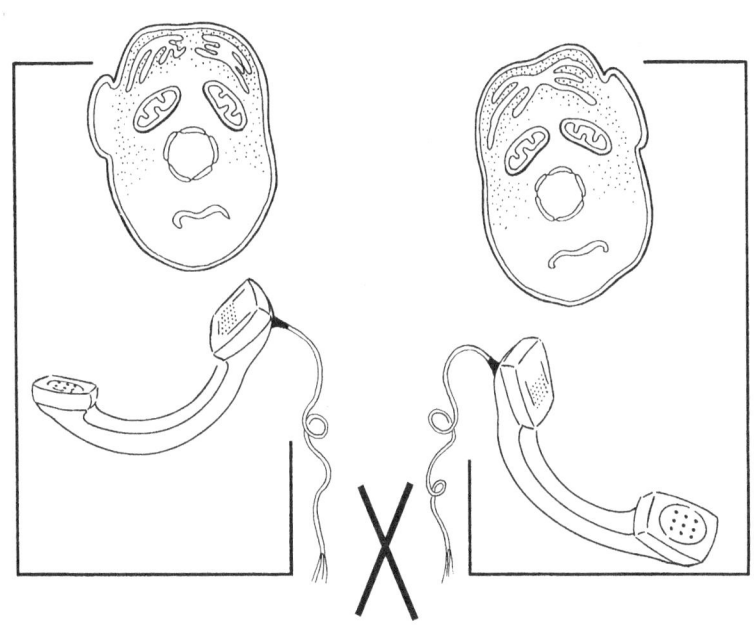

Death: the irreversible loss of communication at the highest level of compartmental organization

Contents

6.1 Accidental death (necrosis) and programmed cell death (apoptosis or cell suicide)
6.2 The deduction of a definition of 'death', the master key to a plausible definition of 'life'?
6.3 Ethical considerations

Essentials

6.1. Accidental death (necrosis) and programmed cell death (apoptosis or cell suicide)

The preceding chapter gives an overview of the broad range of different approaches that numerous researchers have taken in their efforts to define 'life'. Some people consider life to be a machine, others a state, still others an activity. The most often cited properties of prime importance to living systems are the abilities to sustain themselves, to create order and to reproduce. The imperfection of most approaches is apparent from the fact that for almost any property one can usually find some non-living system to which the property also applies. Fire can propagate itself and convert energy, a crystal is ordered, etc.

If life cannot be defined in a relatively simple way, then we must content ourselves with an approach that involves listing as many properties as possible. On the other hand, we could just as well assume a critical stance and continue demanding whether the right questions have as yet been asked or whether the right approach has as yet been taken.

'Life' can only be defined if it is a property of something. But which property and of what? Is life really a machine? Or could it be the activity of a machine? And if it is an activity, what kind of special activity can this be that is executed only by living matter and is carried out incessantly as long the system is alive?

Over 2000 years ago, some ancient Greek philosophers argued that a concept can only be adequately defined by contrasting it with its opposite. 'Warm' has no meaning without 'cold', and neither has 'light' without 'dark', 'love' without 'hate', etc. Hence, according to this basic rule, the state of 'being alive' should be contrasted to 'no longer alive' or 'not yet alive'. Yet, this approach has not been used in biology to gain insight into the nature of 'life'. Dictionaries and vocabularies of biological terms seldom give a definition for 'death'.

To my knowledge, the book by William Clark (1996) entitled *Sex and the Origins of Death* is the only recent publication about 'biological death'. Clark distinguishes between accidental death (necrosis) and programmed cell death. Synonyms for programmed cell death are 'apoptosis' and 'cell suicide'. When an external factor damages a cell in one way or another, the cell may die by necrosis. If given enough time, lysosomes will digest the cellular contents. Necrosis is not a 'decision' taken by the cell itself.

By way of contrast to necrosis, apoptosis is true suicide, which the cell chooses for itself. During the last decade, apoptosis has been a very popular research topic (The Sirens' Song, Melino, 2001). Finkel (2001) describes the phenomenon as follows: " Apoptosis unfolds like a well planned military operation. Within minutes, cells collapse their structural supports, digest and package their contents into membrane-bound parcels, and disappear without a trace into the bowels of scavenger cells."

Although apoptosis is detrimental for the cells involved, it is a positive phenomenon for the well being of the multicellular organism as a whole. During embryogenesis, it contributes by shaping the form of some parts of the body, such as the spaces between our fingers and between our toes. It also eliminates damaged cells. According to Clark

(1996), apoptosis of somatic cells is the price that must be paid for the advantages of sexual reproduction.

My personal opinion is that there must be some basic rule that governs the decision-making as to whether or not to commit suicide, and this rule applies to all levels of compartmental organization. I think that a compartment - be it a cell, an individual organism, or even a (sub)population – decides to commit suicide in accordance with its feelings of contentment/discontentment. Such feelings allow an organism to continuously evaluate the quality of its own life. Suicide becomes an option when the feeling of contentment has so badly turned into discontentment that the cell/organism/compartment does not want to continue to live in such unbearable conditions. How a cell can 'feel' that its life is no longer 'worth living' is not known. It is known that some stress factors, such as a lack of growth factors or exposure to ultraviolet light upon reaching a critical threshold, can induce apoptosis. The mitochondria then either rupture or leak, ensuring their own demise and releasing a cocktail of factors that trigger protein-splitting enzymes called caspases (Brenner and Kroemer, 2000; Fink, 2001). Whether or not the mitochondria are the actual primary decision-makers is still a matter of debate. I do not think that cells commit suicide in order to help other cells. Cells are not altruistic.

Clark's book (1996) does not contain a definition of death. Neither does it explain what changes occur at the moment that a given compartment passes from the 'still alive' to the 'no longer alive' situation.

How can we find out what death really is?

6.2 The deduction of a definition of 'death', the master key to a plausible definition of 'life'?

For the moment, let us put aside all the different approaches we listed in Chapter 5 – the one more complicated than the other – and let us address a few simple questions, answering them as directly and objectively as we can and making certain logical deductions.

First question: What is missing in the following sentences?
- "There is joy because there is birth."
- "There is sorrow because there is death."

These sentences make no sense as long as they do not indicate who or what has been born (a baby boy, a baby horse, etc.) or has died (grandmother, the cat, our goldfish, etc). To make sense, 'death' and 'life' must be related to a concrete instance of what in general terms I refer to as a '**compartment**'.
The definition of 'compartment' is formulated in Chapter 3.

> **First observable fact**: 'Life' and 'death' are invariably linked to some particular 'compartment'.

Second question: What types of living compartments exist in Nature?

From the observation that a bacterium, an ameba, a muscle fiber, a heart, a fish, a bee colony, a fetus in its mother's uterus, a population of deer, etc. can be either alive or dead, it follows that there are numerous types of compartments in nature. The biodiversity of compartments is outlined in Chapter 4. Fig. 4.1 illustrates it.

> **Second observable fact**: There are different levels of compartmentalization, and the majority of compartments contain within themselves one or more levels of subcompartmentalization.

The next question(s) is (are) very basic:

Third question: At what moment does any living compartment/system cease to be alive?

Or: What is the difference between being alive, no longer being alive and not yet being alive?
Or: What crucial parameter changes the very moment 'life' ends and 'death' takes over?

Such questions can be answered by searching for the *common denominator* in the following examples, which correspond to four different levels of compartmentalization. In my 16-level classification system, a bacterium is at level 1, a eukaryotic cell such as a unicellular alga is at level 2, a chicken is at level 7, and a population of deer is at level 13.

a. An example at the level of the multicellular organism: a chicken

> Suppose that I decapitate a chicken (Fig. 6.1). Is this chicken then dead? One will probably reply that the chicken is indeed dead, even though it continues to move for a while. Suppose now that this decapitation is done in a laboratory equipped for cell/organ culture and that I succeed in bringing all cells and tissues of the decapitated chicken into culture and that it all works out well: i.e. all the cells survive. I then ask the same question again: Can we still claim that the chicken is dead? Even in this situation we will conclude that the compartment we call 'chicken' no longer exists, despite the fact that parts of it survive. The 'duality' of death becomes apparent in this example.

b. An example at the eukaryotic cell level: an algal cell

> Eukaryotic cells contain several different membrane-bound organelles, such as the nucleus, mitochondria, chloroplasts, etc. Suppose that we collect some unicellular green algae from a pond. With a laboratory homogenizer we can homogenize the cells in such a way that their limiting membranes, the plasma membranes, rupture but their cell organelles remain intact. Next we can use a technique called density centrifugation to collect the different cell organelles. Mitochondria and chloroplasts that are collected in this way continue to function for some time. The 'duality of death' becomes apparent once again.

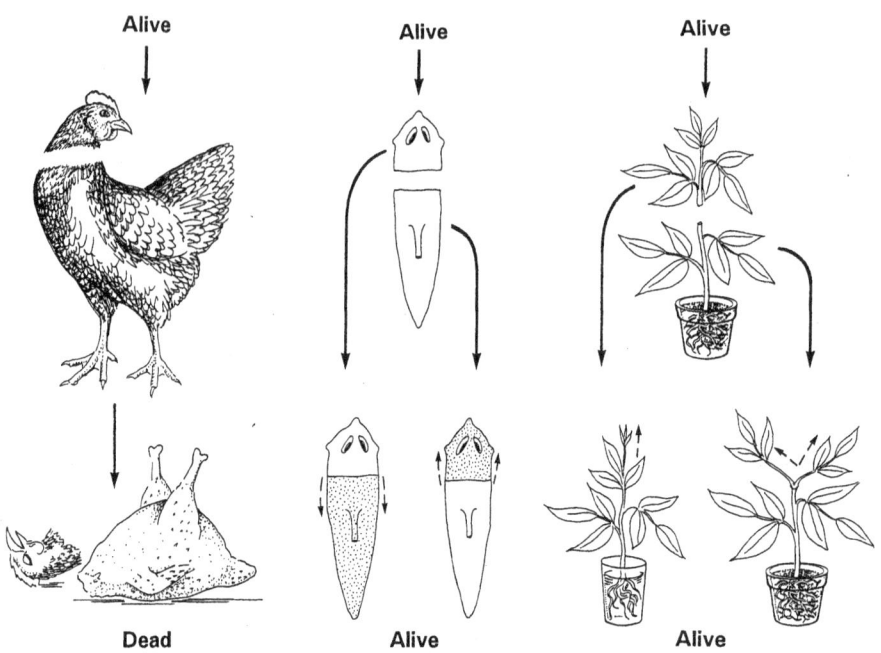

Figure 6.1 Death ensues the moment a given compartment *irreversibly* looses its ability to communicate at its highest level of compartmental organization.

c. An example at the level of the population: a small population of deer

    This example will clearly illustrate that the so-called essential traits of living matter are less essential than we usually assume.

    The same duality is also present at this high level of compartmental organization. Imagine that all the individual members of a small population of deer somehow get separated so far from one another that they completely and irreversibly loose contact. The individual members are still alive. They can metabolize, grow, cope with entropy, respond to a variety of stimuli and – within certain limits – adapt to short-term changes of conditions in the environment. Females that were pregnant at the moment of dispersal can even produce offspring. Despite all of this, we conclude that the compartment 'population' no longer exists.

d. An example at the level of the prokaryotic cell (or of membrane-limited cell organelles): a bacterium

    At the primitive end of organizational complexity, dual existence is no longer found. After homogenization nothing remains alive. A mitochondrion, a chloroplast or a prokaryote ceases to live after irreversible disruption of its limiting membrane (e.g. by ultrasonic homogenization). Under proper conditions, membrane fragments can still perform some metabolic functions. They are complex, but they are not considered to be alive. Disruption of the plasma membrane ends existing ionic-voltage gradient(s) across this membrane, and without such gradients, life is not possible, as will be explained in later chapters.

This irreversibility is important. Decapitation of the flatworm *Dugesia* does not result in death, while the same action kills a vertebrate. The explanation is that the flatworm's central nervous system has the ability to regenerate, an ability which is not present in most other animal phyla. The removal of the tip or some other part of a plant does not - at least immediately - result in death. This is because plants do not have a centralized communication system that is located in a well-defined area like the central nervous system in all the animal phyla except the coelenterates. Plants do have a diffuse communication system. A plant is not dead before its last cell is dead.

What is the *common denominator* in these four examples? What criterion did we use to conclude that 'life' has ended?

> **Third observable fact**: A compartment dies when it irreversibly looses its ability to communicate at its highest level of compartmental organization.

It is only the **highest** level of compartmentalization that matters: what happens with the lower levels, if present, is irrelevant. In a population, the highest level of organization is reached when individual members communicate. In the vertebrate, the central nervous system makes communication possible at the organismal level: this co-ordination system is irreversibly destroyed by decapitation. In the eukaryotic cell, the highest level of communication is at the level of the plasma membrane; and in the prokaryote or organelle of prokaryotic origin, communication occurs with the outside world across the limiting membrane. The loss must be **irreversible** to exclude situations of regeneration of the central nervous system, as in some flatworms (Fig. 6.1) or in the case of reversible coma in humans.

First deduction: *the definition of "Death"*:

'Death' is the irreversible loss by a given compartment of its ability to communicate at its highest level of compartmental organization.

This can be represented in terms of symbols as: $D(S, t) = -C_j$

in which D = death, S = compartment, t = time of death, C_j = communication at the highest level (j) of compartmental organization of the system under consideration.

Second deduction:

If 'death' is the irreversible loss of the ability of a given compartment to communicate at its highest level of compartmental organization, it follows that a given compartment starts to live from the moment it acquires the ability to communicate at its highest level of compartmental organization. **Thus, the very essence of what we call 'life' concerns communication. 'Life' is not an abstraction. 'Life' is an observable activity, and, like any activity, it requires energy.**

Fourth question: If communicating or not communicating makes the difference between being alive and not being alive, then what is the status of the other so-called typical properties of the living state listed previously in this chapter (order, adaptation, metabolism, movement, responsiveness, development, reproduction, genes, and evolution)? Must we conclude that these are not really essential?

The answer is 'No'. They are certainly essential. But to facilitate the understanding of the answer, let us first consider an analogy with music. To produce music, one needs an instrument (flute, violin, etc.) and a score, and one must also play. There are rules that govern the manufacture of the instrument (the hardware) and there are other rules for the production and use of the score (the software). The majority of the aforementioned properties are essential for the construction of the compartment as **an instrument for communication**, but all of this relates to the hardware only. These are necessary but not sufficient conditions for communication at the highest level of compartmental organization. Communication activity requires additional rules, as will be explained in later chapters, in which I will elaborate on the meaning of 'life as a double continuum'.

6.3 Ethical considerations

Medical doctors are often confronted with situations in which decisions have to be made either in favor of an individual patient or of the society in general. The prolongation of the life of terminally ill persons through the use of 'extreme measures' does not make sense. It helps neither the individual nor the population.

A coma patient is no longer living as a 'human compartment' from the moment that one can be sure that the coma has become irreversible: obtaining certainty about this is not always possible. If there is certainty, further medical treatment may be stopped from the ethical point of view.

The ethical code for organ harvesting for the purposes of transplantation, which nowadays is carried out in most countries, is correct from the biological point of view.

In the case of dementia, the ability to communicate at the highest level of compartmental organization certainly decreases. However, this can be no reason to deny such a person the right to live. As long as there is brain activity, even if it is not well coordinated, there is life.

In the case of autism, the functioning at level 11 (= social compartments) is minimal. However, such persons function well at level 7 and sometimes at 8 as well.

A hetero- or homosexual couple that divorces or separates falls back from level 10 of compartmental organization to level 8 (segmented animals with tools). Thus level 10 dies and that may cause sorrow and pain.

Although the biochemical mechanisms of apoptosis are well understood, the 'motivation' with respect to *why* cells engage in suicide is far from clear. One possibility is that cells are altruistic and sacrifice themselves for the benefit of others. I prefer the view that all living entities, cells inclusive, simply strive to live as comfortably as possible, and to 'enjoy life' as much as possible: organisms (or compartments in general) are 'comfort and pleasure-lovers', not 'masochists' (Chapter 15). When they commit suicide, they may be doing so because they experience their environment as so harsh that death becomes an acceptable option. They are driven not by 'altruism', but by 'despair'. Some readers may find my quotes too anthropomorphic. I do not have a more appropriate terminology. Motivation, feelings and emotions, although essential to life, have been banned so rigorously from classical biology that a non-anthropomorphic terminology is not yet available.

Western societies are having more and more problems coping with death as a natural biological phenomenon. From my definitions of 'death' and 'life', it follows that what we call 'death' only concerns the communication instrument, the body, and that, in addition to this, there is another continuum which can continue to exist after the death of the instrument. In funerals according to the Catholic rite, the priest says: "Nobody lives for himself, nobody dies for himself", a sentence, which expresses very well what is important. After the death of our body, we continue to be part of the planetary information continuum.

The definition of 'death' as the irreversible loss by a given compartment of the ability to communicate at its highest level of compartmental organization, is crucial for the further development of concepts that will be dealt with in subsequent chapters, because these concepts follow from logical deductions. If the reader does not agree with this definition, then it may not be worthwhile finishing the book. On the other hand, the inquisitive reader with the fortitude to continue is likely to find himself confronted with conclusions obtained by logical deduction that he would never otherwise have considered acceptable. I experienced this myself. Before I started writing this book, I was convinced that it was impossible to define 'life' and that neo-Darwinism sufficed to explain evolution. I had to change my mind once I came to the conclusion that 'death' *could* be defined as it is in this chapter.

ESSENTIALS

DEATH

1. The only certainty organisms/compartments have in life is that they will all sooner or later die.
2. A cell can either die from an accident (necrosis) or it can commit suicide. This cell suicide is also called apoptosis or programmed cell death. Apoptosis, a common process in embryonic development, involves a series of typical changes at the subcellular level that are not seen in the case of necrosis.
3. According to Albert Camus, "There is but one truly serious philosophical problem, and that is suicide. Judging whether life is or is not worth living." This applies to suicide at any level of compartmental organization, including the cellular level.
4. Apoptosis probably has nothing to do with altruism or sacrificing oneself for the benefit of others. Indeed, a lower level of organization is incapable of assessing the requirements for the well being of a higher level of organization. Hence it is likely that a cell or any other compartment decides to commit suicide when its overall assessment of the quality of its living conditions becomes so negative that deliberate termination of its own existence becomes an acceptable option.
5. Mitochondria seem to play a key role in initiating apoptosis.
6. We can start our search for a plausible definition of death by opposing two situations: 'still alive' versus 'no longer alive'. The question then arises: what exactly changes at the moment of transition from 'still alive' to 'no longer alive'?
7. The question whether or not a chicken is instantly dead upon decapitation illustrates the concept of 'duality of death'. This means that situations exist in which the highest level of compartmental organization seems to be dead while all lower levels of compartmental organization maintain all the classical properties of 'the living state'. A decapitated chicken is dead indeed, no matter whether all its constituting tissues and cells continue to live for a while.
8. We conclude that 'death' always applies to the *highest level* of compartmental organization (Chapter 4). Apparently in our interpretation of 'ceasing to be alive' we use as criteria the permanent loss of the ability to solve problems typical for this highest level and the ability to coordinate the activities of subordinate levels.
9. The example with flatworms that regenerate upon decapitation illustrates that a definition of death should include an element of irreversibility.
10. The following definition of death finally emerges from this line of reasoning: "The death of a given compartment j at moment t is the irreversible loss of the ability to communicate at its highest level of compartmental organization." A symbolic notation reads: $D = - Cj$.
11. A patient in irreversible coma is dead. It is acceptable to stop all machines keeping him/her alive. If there is doubt about the irreversibility, the patient should be given the advantage of doubt.
12. If the proposed definition of death is correct, then the following statement should be correct as well: "A given compartment starts to be alive from the moment that it acquires the ability to communicate/solve problems at its highest level of compartmental organization."
13. Thus, communication activity, in particular at the highest level of compartmental organization, presents itself as *the* essential element in a plausible definition of life.

CHAPTER 7

LIFE DEFINED IN TERMS OF COMMUNICATION AND PROBLEM SOLVING

$$L = \Sigma C$$

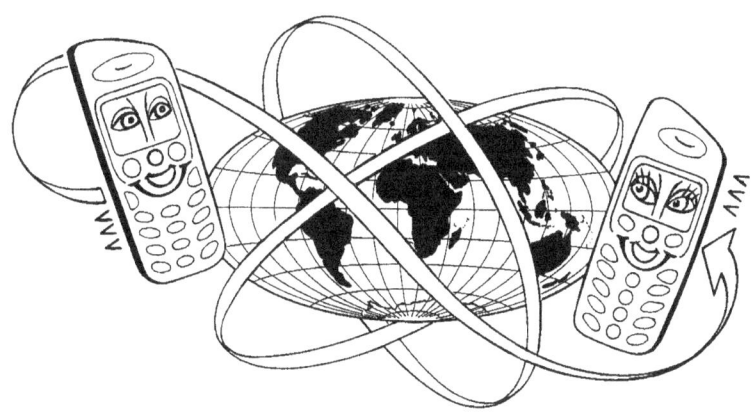

Life: communication at ever-higher levels of organization

Contents

7.1 Requirements for an acceptable definition of 'Life'
7.2 Underlying logic
7.3 Definition of an 'act of communication'. Example: control of muscle contraction
7.4 The multitude of possible messengers and responses in biological communication
7.5 Only 'active' acts of communication are important
7.6 The total sum of acts of communication
7.7 A symbolic notation of 'Life' (as an activity)
7.8 Definitions of 'Life' (as an activity)
7.9 Properties of 'Life'

Essentials

7.1 Requirements for an acceptable definition of 'Life'

In this chapter we come to the main point of this book: the formulation of a definition of 'Life'.
According to Schejter and Agassi (1994), an adequate definition of 'Life' should conform to the following restrictions.
"Apart from its not being trite and uninformative (circular, to use a traditional term), it should be neither too wide nor too narrow; it should not exclude living things and it should not include dead ones. Furthermore, it should not make biology part-and-parcel of chemistry and physics."

I realize, of course, that readers who are not familiar with the contemporary vocabulary of cell biology may have a hard time following some of the rather technical parts of this chapter. However, it would take too long to provide definitions of all the terms used.

7.2 Underlying logic

The conclusion of Chapter 6 was:
- "A given compartment stops being alive the moment it irreversibly stops performing acts of communication at its highest level of compartmental organization."

It then follows that:

- A given compartment starts to live the moment it acquires the ability to communicate at its highest level of compartmental organization.
Or:
- What we experience as 'Life' **is an activity:** it is the total sum of all *active* communication going on at the highest level of compartmental organization of the system under consideration. This does not mean that all the communication at the lower levels of compartmental organization, where present, is not important. To the contrary: we are unable to observe this communication at the lower levels as easily as at the highest level because the communication at the lower levels is concealed within the highest level. Communication at the highest level would not be possible if communication at the lower levels were to fail.

One could argue that in addition to the ability to communicate, there are several other properties by which living matter differs from non-living matter, as is outlined in Chapter 5. This is evidently true. I will not further elaborate on the validity of this counterargument in this chapter. Indeed, in Chapter 10, which deals with the question of whether viruses and computers are alive or not, the specificity for the living state of all these other properties will be dealt with.

7.3 Definition of an 'act of communication'. Example: control of muscle contraction

In Chapter 3, the following definition was formulated. An 'act of communication' is the pathway of events involving three main steps: (1) the release of a message; (2) the transmission of the message through a transmission channel; (3) the energy-consuming response that the perceived and decoded message elicits in the receiver. In my opinion, **this is the basic unit of function in biological systems.**

The following example from the physiology of muscle contraction illustrates how one act of communication can involve a number of chemical reactions. I hope that the accompanying figures sufficiently clarify the essence of the example.

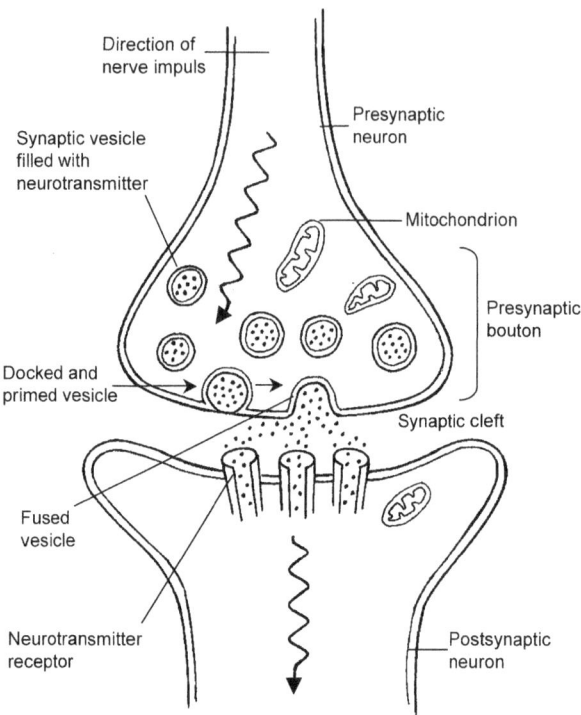

Figure 7.1. The structure of a chemical synapse. Neurotransmitter molecules are packed into vesicles that are transported along an axon of a neuron towards the presynaptic element or bouton. When an impulse arrives at the presynaptic element (= a sender), some vesicles make contact with the plasma membrane (= docking) and the membranes fuse. Upon being released the transmitter molecules diffuse through the synaptic cleft (= a transmission channel) towards the postsynaptic element (= a competent receiver) and bind there to a matching receptor. This binding causes fluxes of particular ions through the plasma membrane of the postsynaptic element, thereby causing electrical phenomena. Redrawn after Dobrunz and Galner, Nature 415, 278 (2002).

First some introductory information on the structure and functioning of a synapse (Fig. 7.1). A synapse is the point of communication either between one nerve cell and another or between a nerve cell and a non-neuronal target cell such as a muscle. The two main types of synapses are chemical synapses and electrical synapses (*Henderson's Dictionary of Biological Terms*, 1995). In a chemical synapse, (but not in an electrical synapse), the presynaptic element releases neurotransmitter molecules. These bind to receptors present in the postsynaptic element, thereby causing fluxes of particular ions through the plasma membrane. The electrical changes that accompany such fluxes will be explained in Chapter 9.

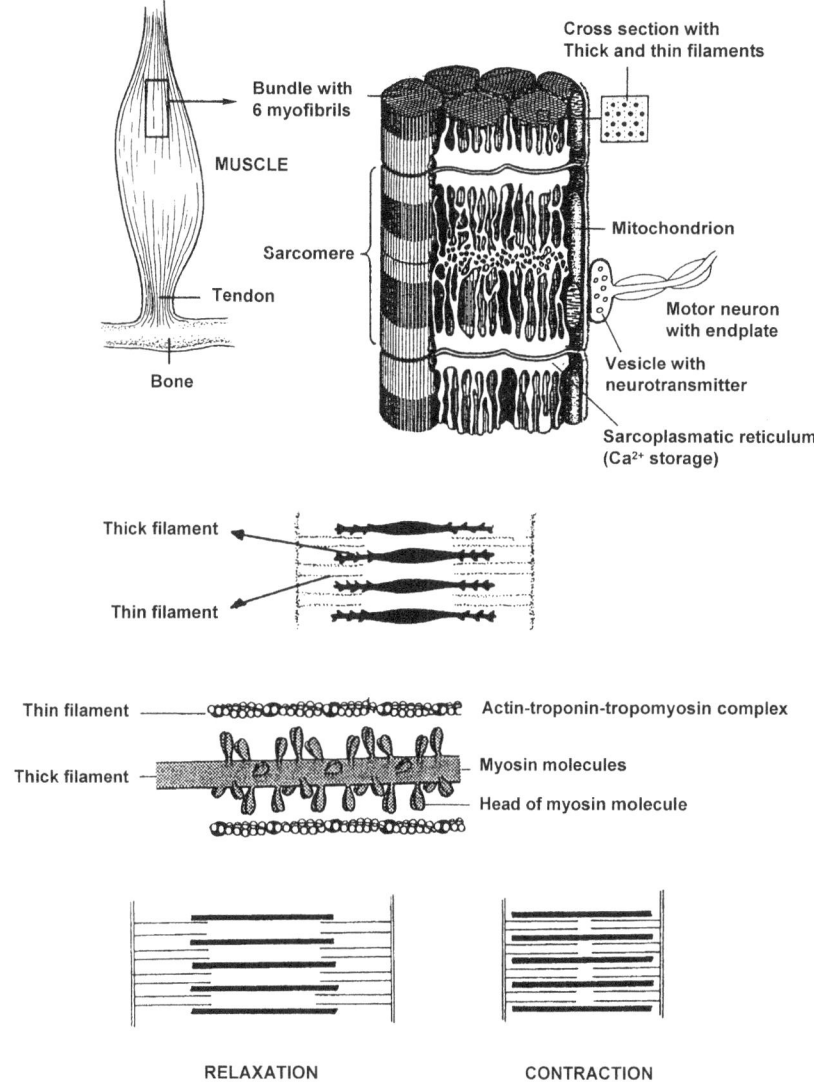

Figure 7.2 Schematic representation of the architecture and mechanism of contraction of a striated muscle. See text.

A motor neuron sends an axon to a skeletal muscle (Fig. 7.2). This axon ends in a motor end-plate, which contains acetylcholine, a substance that upon release incites the muscle to contract. One of the most commonly occurring excitatory neurotransmitters in motor end-plates is acetylcholine. Upon release from the motor end-plate (the *sender-encoder*), the acetylcholine molecules (the *messenger*) travel by diffusion through the cleft (the *transmission channel*) to the plasma membrane of the muscle fiber (the *receiver-decoder-amplifier-responder*), which contains receptor molecules for acetylcholine in its plasma membrane, as well as a signal transduction system.

Acetylcholine and other neurotransmitters are not 'energy-rich'' molecules like adenosine triphosphate (ATP), which can liberate part of the energy it contains upon losing one or two of its phosphate groups by enzymatic action. Thus a neurotransmitter does not deliver substantial amounts of energy. It acts in a different way. The binding of acetylcholine to its receptor facilitates the transition from the inactive to the active conformation of this receptor which is itself a Na^+/K^+ ion channel. This means that the ion channel suddenly opens, which enables the Na^+ ions – whose concentration is higher outside the cell than inside – to flow into the cytoplasm by means of diffusion. The opposite is true for K^+, whose concentration is higher inside the cell (i.e. in the cytoplasm) than outside. In other words, the binding of acetylcholine to its receptor disturbs the existing gradients of Na^+ and K^+ over the plasma membrane. This is associated with a local depolarization of the plasma membrane, the underlying chemical-electrical principles of which will be explained in Chapter 9. This depolarization starts spreading along the entire muscle fiber plasma membrane, the sarcolemma. This event, in combination with the flow of small amounts of Ca^{2+} into the cell through Ca^{2+} channels in the plasma membrane, provokes a massive release of Ca^{2+} from stores in the cytoplasm (the sarcoplasmic reticulum). Such a very steep rise in Ca^{2+} concentration is usually called a 'Ca^{2+} explosion'. This in turn induces contraction of the cytoskeleton through the interaction of actin and myosin molecules. The energy required for this contraction is delivered by ATP molecules, which have already been stockpiled in the cytoplasm before the onset of the contraction. After a short while, the acetylcholine is broken down by specific degrading enzymes. Very soon an ion pump called Na^++K^+-ATPase, which is present in the plasma membrane of the muscle cell, is activated. It starts pumping Na^+ out of the cytoplasm and brings K^+ in at the same time, so that the original condition is restored: the plasma membrane repolarizes. The contraction then stops, and relaxation follows.

The successive chemical reactions in an act of communication can be compared to the different links in a chain. The only major difference is that the reactions are dynamic events while the links in the chain are static objects.

7.4 The multitude of possible messengers, transmission channels and responses in biological communication

Figure 7.3 schematically represents some of the communication pathways at the cellular level that will be listed next. For a more complete overview of the many different possibilities, I would refer to the textbook by Barritt (1994): *Communication in Animal Cells*.

The sender and its coded messages

Physical carriers of information
- sounds (human speech, bird song, alarm cry, thunder, music, etc.);
- light signals (the image of food or of a sexual partner, light from bioluminescent organisms such as the Protist *Noctiluca scintillans*, fireflies, etc.);
- other physical stimuli: touching, stretching, etc.;
- extraorganismal electric fields (steady ionic currents) produced by organisms themselves (electric fishes, developing animal embryos, plant roots, fungi, etc.: see Chapter 9).

Chemical messengers

- olfactory and gustatory substances that can influence behavior: food odors and flavors, sex pheromones, kairomones, allomones, deterrents, etc.;
    - aggregation factors: e.g. cyclic AMP (adenosine 3', 5'-phosphate) in the slime mold *Dictyostelium*;
    - neurotransmitters and neuromodulators;
    - growth stimulators and inhibitors, morphogens;
    - factors causing programmed cell-suicide (apotopsis);
    - hormones in organisms with a circulatory system;
    - chemicals involved in defense mechanisms;
    - etc.

Some facts concerning the different types of transmission channels

Specialists in this field distinguish five different modes for the transmission of messages, each of which is suited to specific tasks. They are called transmission or communication channels. Several types exist.

a. Direct transmission

Some types of compartments can make direct contact with each other by means of a corridor through which information is carried from one compartment to the other. Gap junctions, when open, allow the transport of solutes with a molecular mass of up to somewhat over 2000 from one cell to the other. Two or more cells are sometimes interconnected by cytoplasmic bridges: such compartments are called 'syncytia'. Here the transported messages can be carried not only by simple inorganic ions but also by small and large organic molecules. In the case of fertilization, which takes place inside the body of the female, the messages can be sperm cells and signaling molecules present in the semen.

Some cell types with excitable membranes can make direct contact with each other and transfer a depolarization of the membrane in one cell directly to the other without the need for chemical signaling molecules. 'Electrical' or 'electrotonic' synapses are examples of this type.

b. Environmental transmission

Here, the message is more or less freely transported through the environment.

This was most probably the only transmission channel present when the first living cell came into being. It is still the major channel used by present-day prokaryotes and unicellular eukaryotes. At the single cell level it may have served to control certain simple behaviors such as chemotaxis (e.g. directed swimming towards a food source or away from a zone with unfavorable conditions) and the intake of some food types.

For intercompartmental communication, this is the means for transporting aggregation factors, sex attractants, a wide variety of olfactory and defense substances, sound waves (e.g. bird songs), visual signals and extraorganismal steady ionic currents. Such ionic currents are produced by a wide variety of compartments (Chapter 9).

The transmission of diffusible substances is usually slower in a watery environment than in the air. The more volatile a compound is, the faster its transmission in the air. Electric signals (e.g. those emitted by electric fishes) can only be transmitted in water. In some environments, gametes and viruses, which are two different types of compartmentalized messages, can also be transported by environmental transmission.

c. Volume and paracrine transmission in multicellular organisms

With this type of transmission, signals (e.g. neurotransmitters, growth factors, morphogens or inducers) spread more or less uniformly in all directions into the intercellular space(s), starting from their release site. The maximal working distance is usually not more than a few mm. This type is particularly suited for long-term actions such as the induction of a specific differentiation pattern in some regions of the body (e.g. formation of limbs under the influence of a morphogen such as retinoic acid or activin) or changes related to the plasticity of the brain (modification of wiring transmission in some brain areas). Hormones, which exert their effects on neighboring cells (paracrine effects), also belong in this category.

d. Wiring transmission

The best example of this type is the nervous system. Wires (axons, dendrites) connect neurons to other cells. The message, the impulse, which is ionic-electrical in nature, is transported along the wire. The key characteristic of the plasma membranes involved in this function is that they are excitable, which means that they have large numbers of ion channels that can open very quickly – an action which usually results in substantial changes in transmembrane potential. The plasma membrane of some other cell types is also excitable (e.g. those of muscle cells, some egg cells, etc).

Neurotransmitters are synthesized in the cell body of the neuron, packed into vesicles and then transported in this form along the microtubules, along the length of the axon, to their release site, which is the presynaptic element. At the molecular level, actin filaments, because of their properties as a polyelectrolyte, can serve as 'cables' that are capable of transporting electricity inside the cytoplasm (Lin and Cantiello, 1993, Chapter 9).

e. Hormonal transmission

Hormones are synthesized in a broad range of specialized cells and tissues (including gland cells in animals, the basal lamina surrounding some animal tissues, and the cell wall of plants in the elongation zone) and then released into a transporting medium (blood in animals, phloem fluid in plants) that conducts them to their target cells. In animals, the circulatory system is the principal communication channel. Paracrine transmission has already been mentioned. Hormones were a late 'invention' in evolution.

The receiver-decoder-amplifier-responder

This is usually the most complex part of a communication system. In addition to the appropriate form of stockpiled energy that must be available, there are three essential elements in such a subsystem (Fig. 3.1). First, there is a receptor to perceive that the message is needed. Second, the information acquired by the activated receptor is translated into a language that can be understood in the cytoplasm, where the use of specific forms of previously stockpiled energy can be controlled. The third step is amplification.

Receptors:

The term 'receptor' is generally reserved for a molecule that acts as a biological signal transducer. Receptors perceive the message at the boundary of the receiver. The following is a partial list of different receptors and their functions:
- receptors in the plasma membrane for a variety of signaling molecules: neurotransmitters, neuromodulators, hormones, olfactory and gustatory substances, attractive or deterrent food substances, foreign substances invading the body, etc.
- photoreceptors in light-sensitive cells;
- receptors for acoustic signals in the ear;
- receptors for tactile stimuli (e.g. touch), for heat, for stress, etc.;
- electroreceptors, which enable the perception of extraorganismal electric fields (e.g. in some fishes);
- non-membrane receptors (e.g. nuclear receptors for steroid and thyroid hormones).

The ernormous possibilities of the 'receptor system' are illustrated by the following example. The bacterium *E. coli* has only four chemoreceptors, yet responds to at least 100 different compounds with great sensitivity. The trick is inter-receptor communication (Gestwicki and Kiessling, 2002).

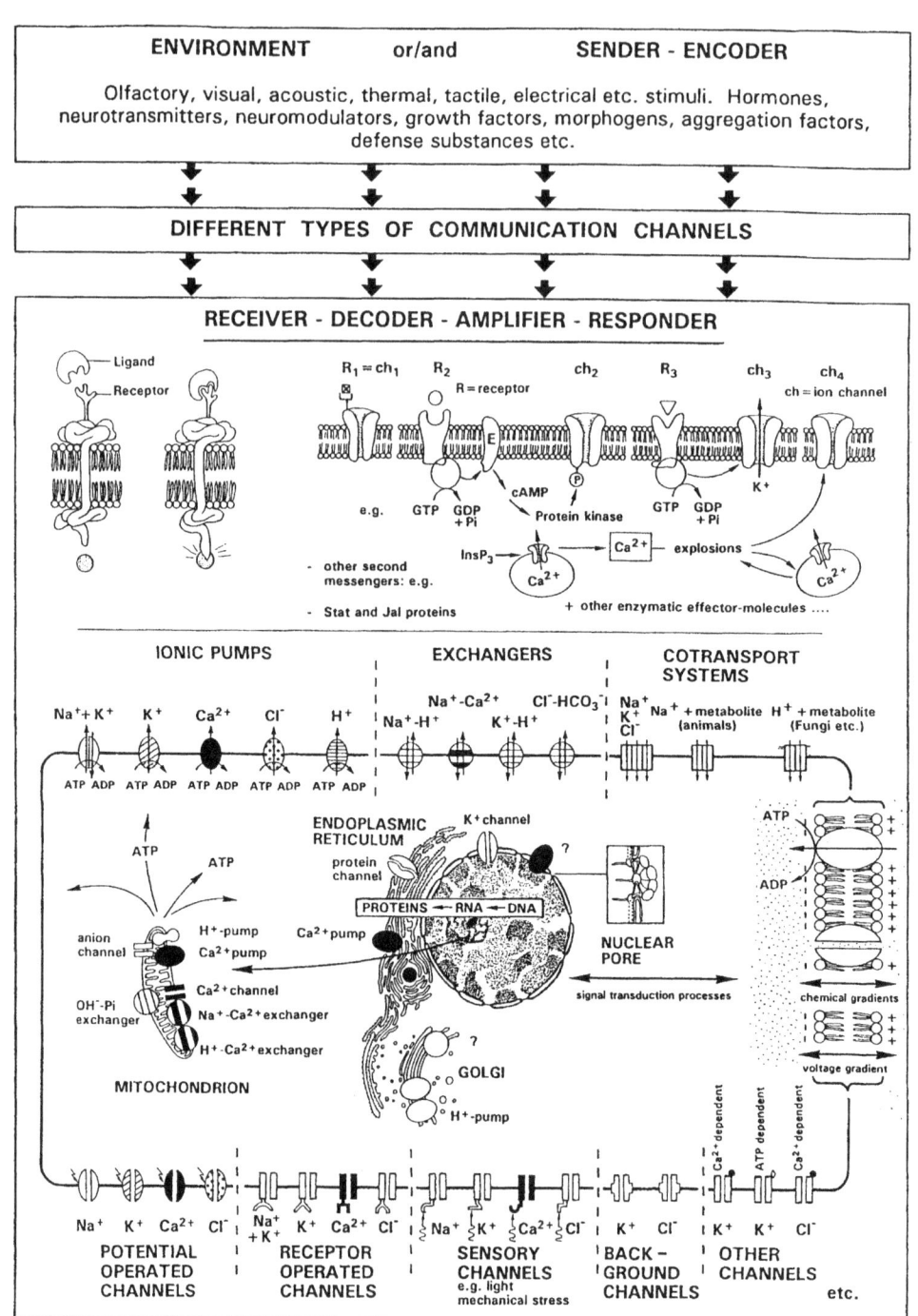

Figure 7.3 Schematic and partial representation of the complexity of the possible communication activity at the level of the eukaryotic cell.

Signal transduction systems perform a variety of functions:

- they couple plasma membrane receptors to specific ionic channels;
- they couple plasma membrane proteins to enzymes: e.g. G-proteins; enzymes for production of secondary messenger systems: cyclic nucleotides (cAMP, cGMP), Ca^{2+}, pH, inositol polyphosphates (e.g. IP_3) and diacylglycerols, protein kinases and phosphatases, intermediary metabolites (fructose 2,6-biphosphate, fatty acids in brown fat cells), etc.;
- they couple plasma membrane proteins directly or indirectly to structural elements of the cytoskeleton, viz.:
    - assembly and disassembly of cytoskeletal elements (the "function follows form" principle: De Loof, 1995);
    - endocytosis (phagocytosis, pinocytosis) and exocytosis;
    - contraction of the cytoskeleton for movement (behavioral effects);
    - transmission of electric signals along actin filaments (cable-like properties of actin: Lin and Cantiello, 1993, see Chapter 9);
    - transport of particles along microtubules;
- they signal the translocation of cytoplasmic receptors to the nucleus (e.g. receptors for steroid hormones);
- they serve as receptors in limiting membranes of organelles serving interorganelle cross-talking: nuclear envelope, mitochondria, endoplasmic reticulum (e.g. IP_3 receptor), chloroplasts, vacuoles, lysosomes, etc.;
- they serve as transcription factors in the nucleus to control gene expression.

Decoded messages bring about a number of effects

- they create, maintain and control some gradients;
- they constitute the electrical dimension of cells (membrane potential, transcellular ionic currents);
- they mobilize specific forms of energy for behavioral effects (contraction of the cytoskeleton and movement of flagella, cilia, undulipodia);
- they synthesize and metabolize macromolecules (nucleic acids, proteins, lipids, saccharides, etc.);
- they maintain osmoregulation and all other physiological functions;
- they perform the transport of body fluids, signaling molecules, etc.;
- they effect fertilization, cell division/growth.

If the receiver is directly linked to other 'compartments', the transduced signal can also affect other cells/compartments (e.g. through gap junctions, across cytoplasmic bridges, by electrical coupling). If the receiver produces chemicals or physical stimuli, these in turn can act as messages in feedback systems, and the receiver itself becomes a sender.

Multicellular organisms are collections of communicating compartments. This has been schematically represented in Fig. 3.1.

7.5 Only 'active' acts of communication are important

Why do we explicitly state that communication at the highest level must be 'active'? In Chapter 3 we defined communication as the transfer of information. It is conceivable that after the decoding of an incoming message the resulting information could be stored in the 'memory' of the system, but never be used to trigger any activity. A compartment that only stored incoming information but never amplified the decoded messages and produced an energy-consuming response would not be experienced as being alive, just as we do not say that a tape recorder is alive. The acts of communication that make a system alive are those that sooner or later result in the mobilization of some forms of stockpiled energy in order to do some kind of work (in the broadest sense). When I use the term 'acts of communication' later in the text, I have the 'active' forms in mind.

7.6 The total sum of acts of communication

In any cell, large numbers of 'acts of communication' are continuously occurring. However, these acts of communication are changing all the time. This means that 'life' is never constant. Therefore we must indicate at what moment (t) the acts of communication are being examined.

The more receptors a cell has in its plasma membrane, cytoplasm and nucleus, the greater the potential for acts of communication. Not only a cell, but any type of compartment (level 1 to 16) is continuously **adding up and integrating all ongoing acts of communication at any moment in time**. Furthermore, we must not forget that large organisms consist of billions of cells, and all these cells are continuously influencing one another in a multiplicity of different ways.

7.7 A symbolic notation of 'Life' (as an activity)

To avoid endless repetitions of terms and definitions, I will try to represent my ideas about the very nature of 'life' with some symbols, all the while realizing that such an approach has its limitations.

My colleagues who are more experienced in mathematics have told me that my approach is 'acceptable' (with a bit of goodwill). From my teaching experience I know that this approach contributes more to a better understanding of the concept than pages of printed text without symbols. The symbols are meant to function more as pictures ("worth a thousand words"!) than as mathematical statements.

Communication is the essence of life (L). In complex systems, multiple acts of communication (C) take place simultaneously. This can be symbolically represented as: ΣC. Thus, the symbolic notation for 'life', in its most simple form, becomes:

Life = total sum of communication acts
or
Life = communication activity

$$L = \Sigma C$$

If more elements are taken into account, i.e. if we consider that life is always a property of a given compartment at a given moment and that there can be several different levels of subcompartmentalization within a given compartment, then a more accurate expression of life can be derived. In the least complex compartments, namely in bacteria and cell organelles such as mitochondria and chloroplasts, there is only one level of compartmentalization. In this type, life is communication activity at the level of the plasma membrane. At this boundary, incoming messages from the environment are translated into a chemical language, which enables the cytoplasm to generate an appropriate response (e.g. making a flagellum rotate, activating protein synthesis, etc.).

A bacterium or cell organelle can also behave as a sender if it produces messages that can be perceived by other compartments (e.g. other bacteria, or other types of organisms).

A possible symbolic notation of 'life as an activity' at the lowest level of compartmentalization (level 1: the monomembrane or prokaryote type) at a given moment t is:

$$\text{Life (prokaryote, t)} = \Sigma \text{ Communication acts (prokaryote, t)}_{\text{at the limiting membrane level}} > 0$$

The higher the level of compartmentalization (Fig. 4.1), the more subcompartments are present and the more acts of communication which are required to make the whole function in a coordinated way.

Let us examine a few more examples at different levels of compartmentalization, as represented in Fig. 4.1. In a eukaryotic cell, which corresponds to level 2 of compartmental organization, there is communication with the outside world at the level of the plasma membrane. However, there is also communication between this membrane and the nucleus, between the nucleus and the mitochondria, between the endoplasmic reticulum and the Golgi system, between the Golgi system and the plasma membrane, etc. This communication with the subcompartments is essential for making communication possible at the highest level of compartmental organization, which, strictly speaking, is the only one that really matters.

$$\text{Life (eukaryote, t)} = \Sigma \text{ Communication acts (eukaryote, t)}_{\text{from cell organelle level}}^{\text{to eukaryotic cell level}} > 0$$

At the level of the multicellular animal, abbreviated here as organism (level 7):

$$\text{Life (organism, t)} = \Sigma \text{ Communication acts (organism, t)}_{\text{from cell organelle level}}^{\text{to organismal level}} > 0$$

At the level of the population (level 13):

$$\text{Life (population, t)} = \Sigma \text{ Communication acts (population, t)}_{\text{from cell organelle level}}^{\text{to population level}} > 0$$

The formulation, which is applicable to any level of compartmental organization (S) at moment t, becomes:

Life (as an activity) of a given compartment S at moment t occurs when the total sum (Σ) **of all acts of Communication (C)** at the highest level of organization of Compartment S exceeds zero. This sum at the highest level can only exceed zero if the total sum of acts of communication which take place at any lower level of compartmental organization (if present) also exceeds zero:

$$L(S, t) = \Sigma \text{ Communication acts } (S, t) > 0.$$
(to compartmental level j, from compartmental level 1)

The general symbolic notation of Life (as an activity) becomes:

$$L(S, t) = \sum_{1}^{j} C(S, t)$$

where **L** = Life activity, **C** = communication act, **S** = compartment, **j** = the highest level of compartmental organization, **1** to **j** = successive levels of compartmentalization, **t** = the moment at which the acts of communication are being considered. The conditions are that $\Sigma C^j (S, t) > 0$ and that, when adding up the acts of communication at successive levels, the same act is added only once.

Biological systems can only exist if they manage to integrate (add up) the wide variety of acts of communication which are going on all the time. These acts of communication can be both qualitatively and quantitatively different. A given act of communication can have a greater impact than other ones, etc. Some may even be redundant.

Is it mathematically proper to add up things that are both qualitatively and quantitatively different?

The answer is that one can add up anything as long as there is a common denominator. I will illustrate this with the classical example.

How much is 5 apples plus 3 pears?
5 apples + 3 pears = 8 *pieces of fruit*

Pieces of fruit is the common denominator – or 'collective/umbrella term – for apples and pears.

With one more parameter:

5 apples (102 grams each) + 3 pears (150 grams each) = 8 pieces of fruit (average weight 120 grams)

Again, an additional parameter:

4 red apples (102 g) + 1 green apple (102 g) + 3 yellow pears (150 g) = 8 pieces of fruit with an average weight of 120 g and with pigments in their outer layers reflecting wavelengths of light between, let us say, 470 and 700 nanometers.

One could introduce an additional parameter, for example an element of motion, by letting the pieces of fruit roll down an inclined plane.

Four red apples (102 g) rolling down an inclined plane at a speed of x_1 cm/sec + 1 green apple (102 g) rolling down an inclined plane at a speed of x_2 cm/sec + 3 yellow pears (150 g) rolling down an inclined plane at a speed of x_3 cm/sec = 8 pieces of fruit with an average weight of 120 g and with pigments in their outer layers reflecting wavelengths of light between 470 and 700 nanometers, rolling down an inclined plane at an average speed of x_4

If one continued scaling up the system like this, the common denominator would become very long and impractical. Instead one could a collective term and say: this is *the activity* of a fruit-sorting center.

It is necessary to systematically use (S, t) in both sides of the equation to eliminate mistakes. Indeed, 3 apples + 4 pears = 7 pieces of fruit. But 6 prunes and 1 cherry also make 7 pieces of fruit. However, this does not mean that 3 apples plus 4 pears equals 6 prunes plus 1 cherry.

This mistake cannot happen if one says:
(3 apples + 4 pears) lying in box 1 at 13.00 hr = 7 pieces of fruit lying in box 1 at 13.00 hr
(6 prunes + I cherry) lying in box 2 at 16.00 hr = 7 pieces of fruit lying in box 2 at 16.00 hr

If one were interested in the minimal requirements for life, one would only take into consideration the *essential* acts of communication. However, all of this is not really so very important (except perhaps for professional mathematicians) because it would make no sense at all to try to carry out the actual calculations. The description of all acts of communication in a 'simple' bacterium at a given moment in time - not to speak of a longer *interval* of time - would probably take much longer than a man's lifetime.

I must stress that the life activity of a given system is being considered at a given moment t (or during a time interval): time only matters (as a variable parameter) as long as the system S is alive (L (S, t) > 0). L = 0, when S is dead.

7.8 Definitions of 'Life' (as an activity)

In my opinion, the following definitions of 'Life' meet with varying degrees of accuracy the criteria mentioned in the beginning of this chapter.

The simplest definition is: **'Life' is communication activity.**

More elaborate definitions are:

The 'Life' of compartment S at moment t is an umbrella term denoting the total sum of all acts of communication being performed by this compartment, from its lowest to its highest level of compartmental organization, at any given moment t.

But the essence of communication is the deciphering of a vast array of coded messages and (sooner or later) providing energy-requiring responses to them. Thus, another definition could read as:

'Life' of compartment S at moment t is the total activity, at all levels of compartmental organization, consisting of the deciphering of coded messages and providing energy-demanding responses to them, all this occurring in a coordinated way.

In this variant, the term *coordinated* is most important because it is precisely the coordination, which makes the realization of the highest level of compartmentalization functional. This means, "*the whole is more than the sum of its constituent parts*", a crucial fact which is often overlooked in the philosophy of reductionist experimentation.

Because 'communication' largely overlaps with '(unconscious and conscious) problem solving', a general definition of 'Life' could also read:

'Life' is the total communication/problem solving activity of hierarchically organized communicating compartments.

I prefer the last definition.

7.9 Properties of 'Life'

If we wanted, at this point we could already list the majority of the properties of 'life' that we have deduced in the preceding chapters. Because there is more to come, however, we will draw up this list in the final chapter.

ESSENTIALS

LIFE

1. The requirements for a good definition of life have been outlined in Chapter 5.
2. A reminder: The basic unit of communication activity is the *communication act* (Chapter 3). This involves the production and release of a coded message by a sender, its transport through a communication channel, and the generation of an energy-demanding response by the receiver.
3. In general, the term 'activity' is used as an umbrella term for all kinds of acts in a given system.
4. Underlying logic: A deduction was made in Chapter 6 concerning the transition from 'still alive' to 'no longer alive'. It read:
"A given compartment stops being alive from the moment it irreversibly stops performing acts of communication/problem solving at its highest level of compartmental organization."
5 If this deduction is correct, then it follows that:
"A given compartment starts to live from the moment it acquires the ability to communicate/solve problems at its highest level of compartmental organization."
Or:
What we experience as 'Life as an activity' is the total sum of all acts of communication going on at the highest level of compartmental organization of the system under consideration. However, communication at the highest level is only possible if the communication at all lower levels is also functional.
6. From the viewpoint of problem solving, the definition then becomes: 'Life' is the total problem solving activity of hierarchically organized communicating compartments.
7. The symbolic notation of Life in its simplest form reads:
    Life = total sum of communication acts
    or:
    Life = communication activity
    or:
    $L = \Sigma C$
8. Because 'life' is invariably linked to a compartment, a more accurate notation reads:
    to compartmental level j
    $L(S, t) = \Sigma$ Communication acts $(S, t) > 0$.
    from compartmental level 1

9. The general symbolic notation of Life (as an activity) becomes:

$$L(S, t) = \sum_{1}^{j} C(S, t)$$

where L = Life activity, C = communication act, S = compartment, j = the highest level of compartmental organization, 1 to j = successive levels of compartmentalization, t = the moment at which the acts of communication are considered. The condi-

tions are that $\Sigma C^j (S, t) > 0$ and that, when adding up the acts of communication at successive levels, the same act is added only once.
10. From this approach, it follows that 'life' as an activity cannot possibly be constant. It changes incessantly. Furthermore, Life has both quantitative and qualitative aspects. It is different and unique for each communicating compartment. Other properties of 'life' will be listed in Chapter 16.
11. In a single cell, a large number of communication acts are taking place at any given moment. In an animal the size of a human being, with its billions of cells, the total sum of all acts of communication going on at any moment is incredibly high.
12. The basic question concerning the evolution of life thus becomes: How does 'life', (i.e. communication/problem solving activity) evolve in the course of time?

CHAPTER 8

MAKING CHOICES AND THE RESULTING VARIABILITY AND FREEDOM IN COMMUNICATION AND BEHAVIOR

Life's basic drive: "Solve problems to feel comfortable and contented!"

Life's drive: the choice is yours.
Evolutionary biologist: "My genes are selfish and struggle for life".
Molecular biologist: "My genes make proteins and mutate".
Integrative biologist: "My genes solve problems so that I can enjoy life".

Contents

8.1 An example of the interrelationship between the environment on the one hand and human communication and behavior on the other
8.2 The origin of complexity and variability in communication: illustration with simple whistle music
8.3 Translation of the whistle music principles into the chemical language of a simple biological cell system
8.4 More holes: the complexity of recorder and accordion music
8.5 Bifurcation points, the often-overlooked companions of mutations in generating communicational variability, offer possibilities but no certitudes. The (un)predictability of communication
8.6 Life's basic drive: "Solve problems if you want to feel comfortable and contented!"
8.7 Instinctive versus learned behavior. Not only humans have free will
8.8 Summary: freedom, responsibility and suffering

Essentials

8.1 An example of the interrelationship between the environment on the one hand and human communication and behavior on the other

Biological systems are incessantly changing. During embryonic development this change is mainly due to differential use of the genome by the different emerging cell types (Chapter 14). Sexual reproduction is an important source of variability because meiosis yields genetically different egg and sperm cells. The changes that populations undergo in the course of evolution are due to changes in the genes (mutations) and their frequency distribution. Physical factors in the environment can also cause changes in the phenotype (modifications). For example, the intensity of ultraviolet radiation coming from the sun has an influence on plant growth and on the tanning of our skin. Many biologists tend to stop the search for sources of variability here, thereby more or less assuming that these are sufficient to explain variability in all situations.

The following example may help to illustrate that there are situations which cannot be fully explained in terms of the above mentioned sources of variability. Imagine that identical twins are born in Leuven, the Dutch-speaking city where I live, and that they are separated early in life. One grows up in Leuven and the other in Tokyo. Despite the fact that they have identical genomes, the language they speak and the way they behave later on will be markedly different. Due to climatic differences, the tanning of their skin might be different as well.

If one asks why the twins behave so differently, the answer one can expect to get is that such variability has nothing to do with genes but that it is **epigenetic**, i.e. that it is *determined (imposed) by the respective local environments*. In biology, such changes are quite often considered to be of little importance because they are not inheritable. Furthermore, in the cited situation, biologists are likely to argue that this fits in the context of **cultural evolution** and that it is a clear example of **cultural development,** which is typical for the species *Homo sapiens*. In Chapter 15 this example will be used again in a slightly modified form, namely with a large enough number of identical twins to have a population to which the basic law of population genetics, the law of Hardy and Weinberg, applies.

In this example of identical twins, we are dealing with cultural *development* because the changes do not last longer than a lifetime. Evolution deals with changes in populations over longer spans of time. However, the mechanisms involved are very similar in the two cases.

Hitherto, hardly any attempts have been made to explain the physiological basis of such epigenetically controlled changes in behavior. In this chapter, we will show that this type of behavioral variability is likely to have its roots in the general principles underlying communication. We will try to illustrate this with analogies to 'music' produced with instruments of increasing complexity.

8.2 The origin of complexity and variability in communication: illustration with simple whistle music

Imagine that I am a teacher who is going to illustrate the principles used in a communication system by explaining step by step to an audience of students how sounds can be produced with a whistle (Fig. 8.1). I first put the whistle on the table and ask the students why it does not produce music. They will say that I have to blow into the whistle.

Next I take the whistle, close the outlet hole with a finger and blow (= loading the system with potential energy, stored in compressed air). The result - still no sound. The students will now say that I have to remove my finger from the outlet hole, and of course, if I do that, sound is produced. This illustrates that a functional communication system must have functional holes.

Now we have the basic requirements of a *sender* which delivers through a *transmission channel* (the air in this example) a sort of message that can be understood by the students, who are competent *receiver-responders* because their brain has the deciphering system (software) that is needed to 'understand the whistle music'.

After this demonstration, I can go one step further and ask a student whether he or she wants me to produce a long or a short piece of whistle music. Let us assume that the student says 'short" and that I do so. Next I can ask whether the student wants it loud or soft. Again, I do what the student says. We already have two variables: long-short and loud-soft. Now I continue by asking the student whether he or she wants me to produce a continuous signal or an intermittent one. If he or she says intermittent, I remove my finger from the outlet hole, put it back, remove it again, etc., as long as I manage to blow air into the inlet hole. Next I ask whether the intermittent playing should be regular or irregular. If the student says irregular, I have quite some possibilities for generating variability: I can produce long-short-short, or long-long-short, etc. Finally, I ask the student whether I should play loud and regular or soft and irregular. Then, after all these questions, I do exactly the opposite of what the student wanted. When the student protests, I reply that I have *forgotten* what he or she said.

What have I illustrated with this demonstration?

First, I have illustrated the basic architecture of a communication system, as well as the **high degree of variability** in music that can be produced by a very simple musical instrument. The variability comes from the 'strength' (steepness) of the air pressure gradient which is built up by blowing air into the inlet hole, from the time during which this pressure gradient is maintained, and from the opening and closing of the outlet hole. This last factor is of special importance within the context of this chapter.

Secondly I have illustrated how the environment, in this case a student in the audience, can influence the type of music produced by the whistle: **environmental control**.

Thirdly, we have seen that in music production, **choices have to be made all the time**: long-short, loud-soft, regular or intermittent, or combinations of these variables. This means that there must be mechanisms that enable such choices to be made, and that there are **imaginary points where the decisions are made**. Later in this chapter I will refer to such points as '**bifurcation points**'. Who made the choices in our example? Some were made by the student (= the sender or the environment), and others by the whistle player (= the receiver in this example). The central point, however, is that **the making of choices automatically implies some degree of freedom**. The freedom displayed here had nothing to do with changes in genes or in gene activity: it was based upon the principles of communication.

Fourthly, we see that structured music can only be produced if some data can be stored in **the cognitive memory** for some time. There are two situations in which the demonstration would not have been possible: first, if my brain memory had not stored information about long or short, loud versus soft, etc.; second, if I had not been able to retain the instructions given by the student in my memory until the time I had to execute them. At the current time, we do not know how our 'cognitive (brain) memory' works (Chapter 3) or whether any other cell type has the same sort of memory system. The fact that we do not know how it works, however, does not mean that we should overlook its importance in some disciplines of biology, such as in explaining the mechanisms of evolution (Chapter 15).

Fifthly, we see that feedback mechanisms are very common in communication. When I used the whistle and the students perceived the sounds, I acted as a sender and the students as receivers. When a student told me how I should blow and I responded, I became a receiver-responder at the same time and the student, who at first was only a receiver, became a sender.

8.3 Translation of the whistle music principles into the chemical language of a simple biological cell system

Cells are not filled with compressed air like the whistle, but rather with a watery salt solution. Upon expansion of the compressed air, an acoustic signal is produced in the whistle. One of the means cells use to generate a signal/message is to change the gradients of specific ionic species (e.g. Na^+ or/and K^+) over their plasma membrane. This is often accompanied by electrical phenomena, which can act as a message as explained in Chapter 9.

The equivalent of the 'whistle situation' in its most simple form might be a single cell with only one type of signal transduction system in its plasma membrane. We will use the ion pump/ion channel signal transducing system because it is a relatively easy system to graphically illustrate. As mentioned previously (Chapter 7), we should stress that this is only one of the possible signal transducing systems present in cells.

To continue with the situation of opening and closing holes in the boundary, we take as a first example a cell with only one ion pump and one corresponding ion channel in its plasma membrane, as represented in Fig. 8.1A. In this drawing we assume that both the ion pump and the ion channel are proteins with some sort of central hole. In reality, the spatial conformation of such proteins is much more complex. If both the pump and the channel permitted ions to pass at the same speed, no ionic gradient could be built up. Let us assume that both the ion pump and the ion channel change conformation under the influence of light of a given wavelength in such a way that the pump starts transporting ions against a concentration gradient and the channel closes. The result is that an ionic gradient is generated over the plasma membrane. We further assume that equilibrium is reached after some time. As a third step, we let a ligand (e.g. a neurotransmitter molecule) bind to an 'antenna' present at the environmental side of the ion channel, which results in the opening of the channel. If a competent receiver-decoder were present in the environment in the immediate vicinity of the channel opening, it would experience the outflow of ions as a message.

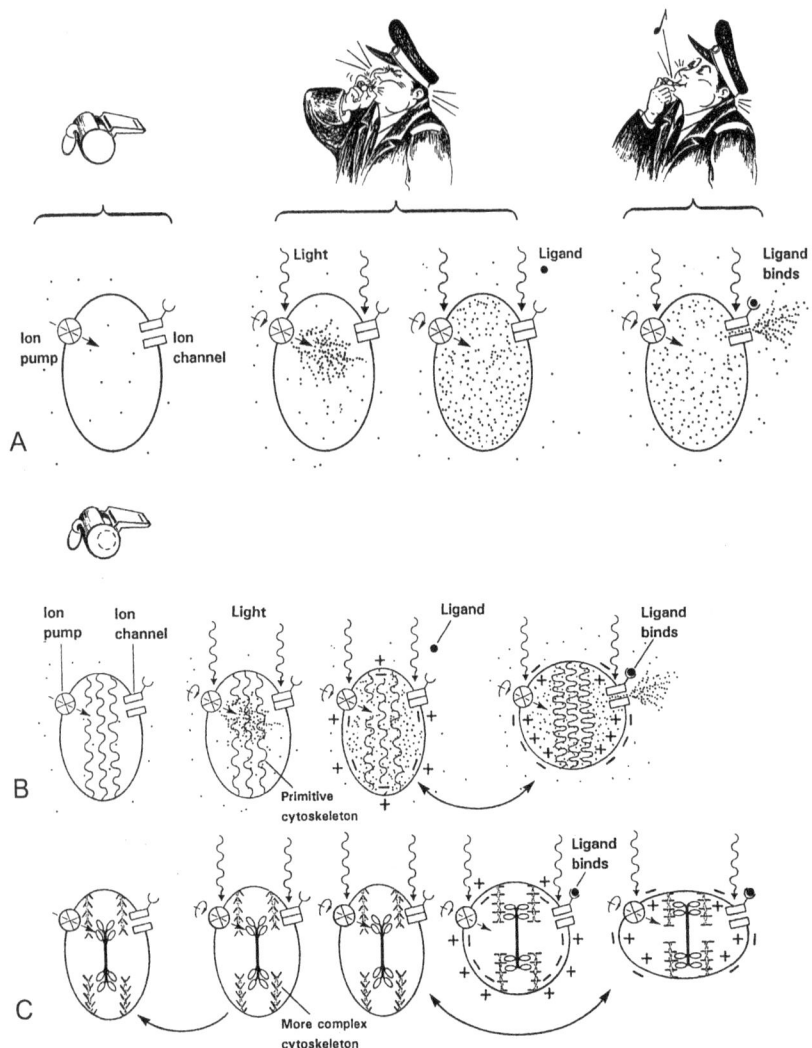

Figure 8.1 Making music (= coded message) with a simple whistle, and the translation of the principles involved in the whistle to a very simple biological system.
Top of the figure: The requirements for the production of 'music' (= coded message) with a simple whistle without a ball in its interior. When both the inlet and the outlet holes are open, no sounds will be produced. If the outlet hole is closed and air is blown into the whistle, thereby building up an air pressure *gradient* (higher inside, lower outside), no sounds will be produced either. The compressing of air means stockpiling the energy inside the recorder to be used later on. 'Music' is produced when the outlet hole is opened. The basic architecture of a sender-encoder can be deduced from this example.
A. A biological equivalent of the 'simple whistle situation' is a simple cell with only one ion pump and one ion channel in its plasma membrane and no cytoskeleton. For reasons of simplicity, we assume that the pump starts functioning and the channel closes when a physical stimulus, such as light in this example, reaches the cell. As a result, an ionic gradient is being built up until an equilibrium situation is reached. We further assume that the channel will open when a ligand binds to its receptor. At that moment a message, in the form of an ion flux, can be produced. This is represented here as an outflux of ions and a strong depolarization of the membrane.

B. and C. When a small ball is put into the interior of the whistle, a moving element is added and the complexity increases. The cellular counterpart of this system is a cell with a very simple cytoskeleton (B) or with a somewhat more complex cytoskeleton (C: sliding system of filaments as in muscle cells). When the proper ligand binds, membrane depolarization and contraction of the filaments result.

This simple system could mimic reality a little bit more by introducing an element that allows movement (Fig. 8.1B and C). So far, we have used the example of a whistle that was empty inside. A somewhat more complex situation is that of the 'normal' whistle of a soccer referee or of a policeman. Such a whistle contains in its interior a small ball that starts rotating when air is blown through the whistle, thereby having an influence on the sounds that are produced. In our cellular example we introduce some simple cytoskeleton as an element that allows movement (Fig. 8.1B and 8.1C). When the neurotransmitter binds to the receptor, the ion channel opens and the membrane depolarizes. This depolarization somehow acts as a triggering mechanism to make the cytoskeleton change its spatial conformation. In fact, this principle is used in a modified and more complicated form by muscle cells (Chapter 7, Fig. 7.2), which usually act as receivers-responders.

8.4 More holes: the complexity of recorder and accordion music

In reality, cells have more than one ion pump and one channel and other signal transducing systems in their plasma membrane. We will use the music produced by a recorder – which is still a relatively simple instrument – as a model to illustrate how fast the complexity of a communication system increases with the introduction of additional parameters that increase the variability.

Imagine a musician playing a recorder on which the location and number of the outlet holes can be changed. At first the recorder has only one outlet hole (Fig. 8.2), as in the example of the whistle. Let us assume that in our example the tone that is produced corresponds to the musical note 'do'. Next, a second outlet hole is made at another position and the tone becomes 're', and so on, until one has also 'mi', 'fa', 'sol', 'la', 'te' and the next 'do'. Thus, if only one outlet hole is present, it makes a difference where it is located. The resulting music will be rather monotonous, as in the example with the whistle.

If more outlet holes are introduced, it also matters where they are located. But now we are confronted with a most important aspect of music production, namely making decisions as to which tone to produce at any moment, for how long and with what intensity. Intuitively, one might think that the more outlet holes a recorder has, the more complex the music that will be produced. This is not necessarily the case, as can be illustrated by blowing air into a recorder with ten outlet holes which are all left open. Only one tone will be produced, as in the case with the recorder with only one hole (Fig. 8.2). The complexity of the music increases only from the moment that the musician starts alternating the opening and closing of the different outlet holes (Fig. 8.3).

By doing so, the recorder-player continuously decides which holes should be open or closed and for how long. He also has to decide how much air he will blow into the recorder at any moment. Thus, the musician can make use of bifurcation points to a higher degree than in the example of the whistle. In a recorder with two holes, such an imaginary bifurcation point is located between the two holes.

Additional bifurcation points are possible; namely the ones located in the holes themselves. Here the decision (fully-partially) open-closed can be made. A third location of a bifurcation point is the inlet hole where decisions about the influx of the air (flow rate, duration) have to be made. Thus, the doubling of the number of outlet holes from one to two does not simply double the possible complexity of the music that can be produced; rather, it multiplies it. The introduction of a third outlet hole again results in a much higher level of complexity. The more outlet holes, the greater the possibilities. There is an upper limit, however, which is imposed by the simple fact that the musician has only ten fingers.

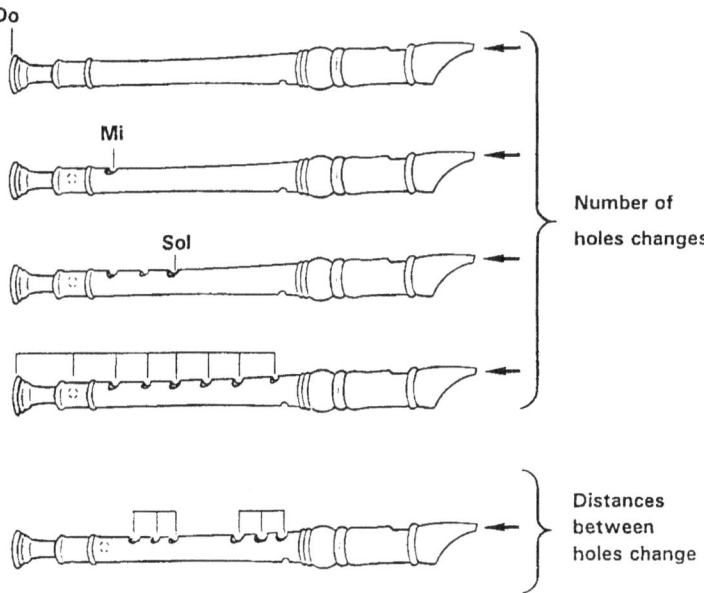

Figure 8.2 The complexity of the music that can be produced by a recorder. The instrument can be changed and made more complex by introducing more and more holes, and by changing the form and/or location of the holes.

What additional information about the complexity of music did we obtain from the analysis of how recorder music is produced?

i. The relative location of inlet and outlet holes matters.

ii. The addition of an extra outlet hole *can – though does not necessarily –* result in a drastic increase in the possible complexity of the music that is produced. It is not the addition of the hole itself that leads to increasing complexity, but the alternating opening and closing of the different holes by the musician. In other words, the increase in complexity only comes in when the different outlet holes are alternately and irregularly used, or in other words, **when the musician starts to make use of the imaginary points where decisions have to be made (bifurcation points)**.

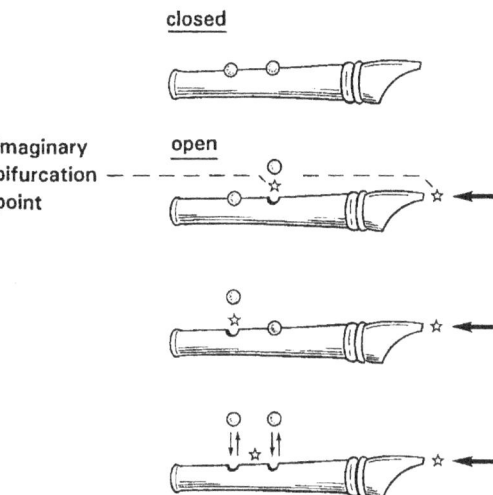

Figure 8.3. Imaginary points (indicated by stars) where decisions have to be made when playing the recorder. The player must decide how much air he or she will blow into the inlet hole. Furthermore, he or she must decide whether to let air escape from outlet hole number 1 or from hole number 2, or from both at the same time. Finally, the decision must be made as to how long the holes will be left open (short, long, intermittent, etc.).

iii. The addition of an extra hole to a simple instrument, as for example making a second outlet hole in a recorder that had only one, is likely to have a greater impact than the addition of an extra hole to an instrument that already has many holes. In the latter case, such an addition (e.g. from 10 to 11 outlet holes) may have little effect. From a certain number of holes on, the addition of extra holes can even become counterproductive because the musician does not have enough fingers to use them all. Thus, a given system may have an optimal number of points where choices can be made. Once the optimal number of holes is reached, the instrument can perhaps still be slightly improved by changing the size or relative location of the holes.

iv. Another conclusion is that *a musical instrument need not be very complicated to produce an infinite number of melodies with it*. All human beings on earth, if asked to produce some sound combinations (music) with the same recorder with only 10 outlet holes as in our example, would produce a different 'melody'. The number of melodies that can be played on a recorder is endless. In Fig. 8.4, I have attempted to illustrate how dramatically the number of possible tones increases with an increase in the number of outlet holes. At the cellular level, the following example illustrates this principle well. By using inter-receptor communication as the trick, the bacterium *E. coli* can respond to at least 100 different compounds with only 4 chemoreceptors (Gestwicki and Kiessling, 2002).

v. When a musical instrument has been optimized, there is no need to make any additional changes for the *music itself to further evolve in the course of time*. The evolution of the music itself is to a certain extent independent of the evolution of the instrument. The higher the possible number of 'bifurcation points', the less important are the changes in the instrument and the greater the possibilities for the evolution of the music itself. I will come back to this important issue when discussing the driving force of evolution (Chapter 15).

The recorder system can be compared to the system of a cell. The addition of extra holes in the recorder is like the insertion of extra types of ion transporting proteins (or other signal transducing systems) into the plasma membrane of the cell (Fig. 8.5 A-F). In this example, the cell has two different types of ion pumps and ion channels and two different operating systems. In the course of evolution, such a doubling could have arisen by a duplication of the genes coding for the original single pump and single channel, followed by mutations in the two genes. These mutations could have resulted in the ion transporting proteins acquiring different ionic selectivities, meaning that each protein would allow the passage of a different type of ion(s).

Even with such a simple system, the number of possibilities created by different ionic transmembrane fluxes and internal ionic environments becomes quite considerable. If we make it still more complex by adding one more pump and channel type, it becomes difficult to make and read such a figure, let alone deduce all the possible combinations (Fig. 8.5 G-I).

But three different types of signal transduction mechanisms is still far from the complexity of any eukaryotic cell type (Fig. 7.3). In terms of complexity, such a cell might be compared to an accordion containing in its interior other musical instruments corresponding to the different membrane-limited cell organelles that any eukaryotic cell contains. The possibilities are endless in such a complex instrument.

To achieve the complexity of a multicellular organism such as a human being, one must multiply the already very numerous possibilities of a single eukaryotic cell by the few billion cells that compose such an organism. In addition, one must take into account all the connections among these cells (e.g. in the brain, by hormones, growth factors, cytoplasmic bridges, etc.), and the fact that this vast complexity is so marvelously well coordinated, like a giant symphony orchestra.

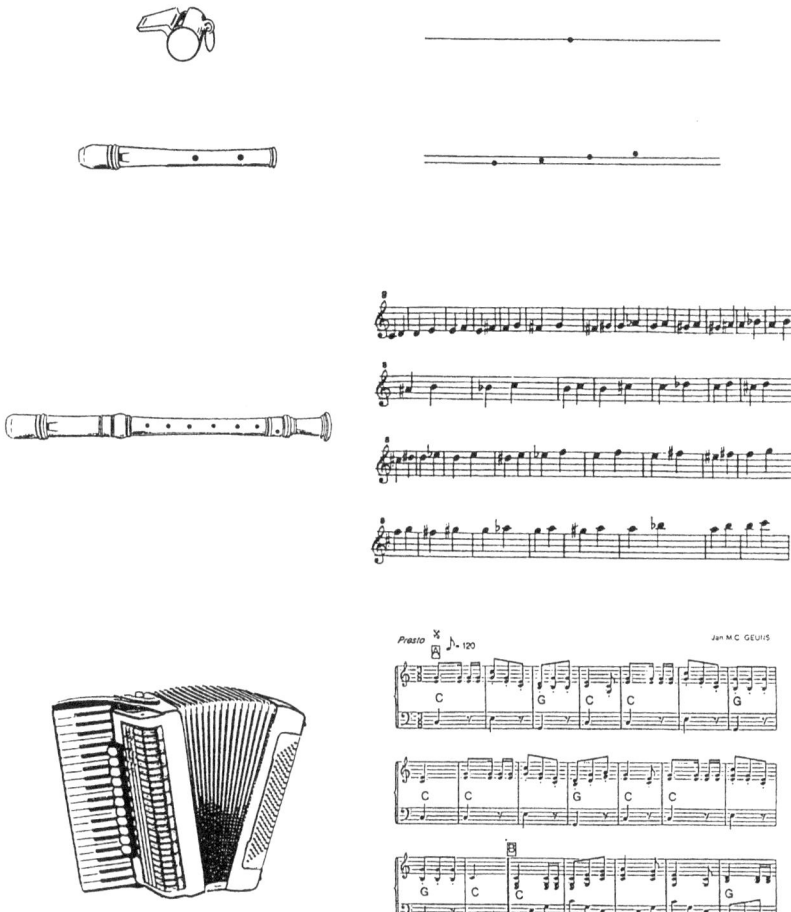

Figure 8.4. Only one tone is produced with an instrument having only one outlet hole. If two outlet holes are present, four tones can be produced. With a recorder with ten holes, a large number of tones can be produced. The complexity of accordion music is once again of another order of magnitude. It is evident that even with a simple instrument, one can make an endless number of melodies by combining again and again the different possible tones.

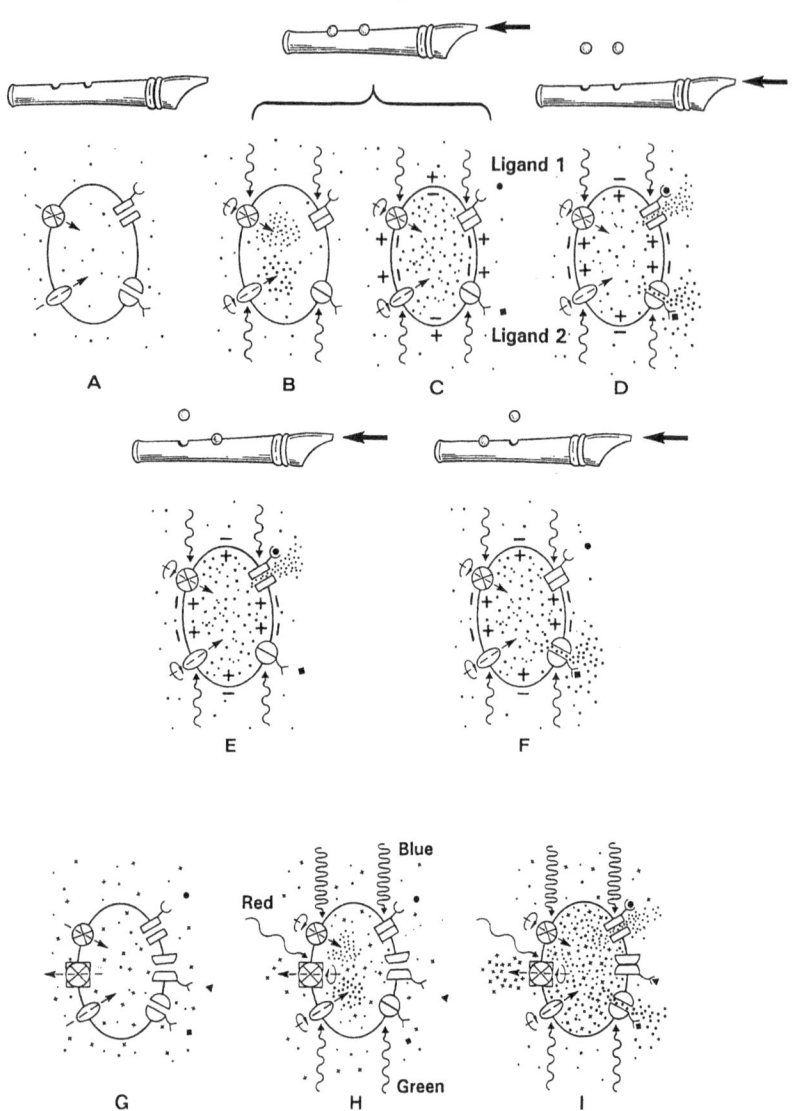

Figure 8.5. The 'recorder example' translated into a biological counterpart
A-F: The complexity of communication in a hypothetical cell with two ion pumps and two ion channels. We assume that all pumps and channels respond to light of the same wavelength, but that the opening of the channels requires a different ligand. For reasons of simplicity, we further assume that outfluxes of both ionic species result in depolarization of the membrane.
A: 'all open' situation.
B and C: in this theoretical example the pumps start working and the channels close when there is enough light. This corresponds to the situation in which a chemical and potential gradient is being 'built up' over the membrane as a means of stockpiling energy.
D, E, and F: binding of the ligands results in outflux of ions and membrane depolarization.
H-I: Hypothetical cell with three different types of ion pumps and channels. We assume that three different wavelengths are required to put all ion pumps to work and make all channels close.
G. Without light energy, the pumps are not functioning and the channels are open. No transmembrane gradients are being built up.

H. If the three different wavelengths of light are simultaneously present, then three different ionic gradients will be built up. In our example, two of the wavelengths result in higher ionic concentrations in the cytoplasm compared to the environment, and one results in lower concentrations.

I. A relatively large number of interactions between ion transporting molecules are possible, and this means that there are a large number of possible different cytoplasmic ionic environments.

8.5 Bifurcation points, the often-overlooked companions of mutations in generating communicational variability, offer possibilities but no certitudes. The (un)predictability of communication

From the examples of musical instruments of increasing complexity (whistle, recorder, accordion), it is clear that the complexity of the music produced depends both on changes that have been introduced in the instrument itself (e.g. more outlet holes), and on the differential use of inlet and outlet holes.

In biological systems, substantial changes in the cell, such as the addition of extra types of ion pumps or channels, are due to changes in the genome. Such changes can be grouped under the general term 'mutations' (point mutations, gene duplications, additions, inversions, deletions, etc.; see Chapter 2). However, as in music production, the variability in cellular communication also depends on bifurcation points.

The dictionary defines *bifurcation* as 'a division into two branches' (Fig. 8.6).

If *a* were a road, a traveler walking along this road would have to decide at the bifurcation point whether he should continue along *b* or *c*.

In the 'exact' sciences, the term 'bifurcation point' was first introduced - and is commonly used - in non-linear mathematical literature. *Bifurcation* in the rigorous mathematical sense refers to a situation in which beyond some critical parameter value a particular solution becomes unstable and the system spontaneously evolves into another more stable system.

Let us first analyze whether there are situations in biological systems where the term 'bifurcation point' in this original mathematical sense applies, and whether it might be correct to use the term in a broader sense as well, without loosing too much of its original meaning.

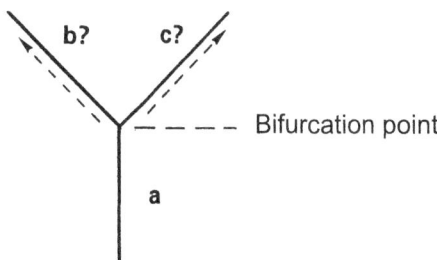

Figure 8.6. A bifurcation and a bifurcation point. Here, a choice must be made.

In Chapter 9 I will outline the electrical properties of living cells, and explain that there are both potential and chemical gradients over the plasma membrane that are essential for cellular communication. Furthermore, there are numerous 'holes' in this plasma membrane (ion channels and ion pumps) that can temporarily open and close.

Numerous chemicals and physical stimuli play a role in all kinds of communication acts. They cause changes in the permeability of the plasma membrane to particular ionic species. The ionic composition of the watery salt solution (here 'solution' has another meaning than in the mathematical definition on the environmental side of the plasma membrane is different from that on the cytoplasmic side. At rest, the salt solutions are stable on both sides of the membrane, despite the gradients that exist over the membrane. This stability is lost when a critical parameter value is reached. For example, in a postsynaptic element this happens when a certain quantity of a neurotransmitter molecule is released in the vicinity. This results in the pumping of an ion pump or in the opening of a responsive ion channel that is closed at rest, and the particular solutions on both sides of the membrane become unstable. The system will then tend to move toward a new stable state or condition. In cells, this second 'stable' state is usually of very short duration because it sets off yet other mechanisms that tend, insofar as possible, to restore the original state of 'rest'.
Thus, in the case of ion fluxes through membranes, the term 'bifurcation point' is close to its original meaning. It also applies to the case of air flowing through our throat, thereby producing sound waves. In the throat, which is a site with bifurcation points that are used in communication involving acoustic signals, the state of compression of the air coming from the lungs becomes unstable. The outflowing air soon reaches another state of equilibrium.

Not all acts of communication in biological systems involve ion or gas fluxes, however. There are other possibilities. In some acts of communication, phosphorylation reactions are involved. In others, changes in protein-protein or in protein-nucleic acid interactions occur. With respect to phosphorylation, the bifurcations involve the type of molecule that should be phosphorylated, etc. It is even much more complicated than all of this. As already mentioned, a cellular communication system with only three different pumps and channels, all with different mechanisms for activation, is already quite complex. Because of the high number of such ion transporting proteins in each cell and the different mechanisms for their activation, etc. (schematically represented in Fig. 9.4), the number of possible bifurcation points is very high. This, together with the other communication pathways present in each cell (e.g. the ones that make use of cyclic AMP, inositol polyphosphates, Ca^{2+}, etc.) and the great number of cells present in multicellular organisms, results in ***nearly infinite variability and low predictability***. No organism 'knows' what its acts of communication will be the next day, or even the next hour, minute or fraction of a second. It all depends on the interplay of internal and environmental circumstances that contribute to deciding which one(s) of the available bifurcation points will be involved and which one of the two forks in the road will be taken. *Bifurcation points offer possibilities, not certitudes*. This contrasts to mutations, which allow one to predict more or less the outcome, provided that the gene, which is affected by the mutation, is fully characterized and that all functions of the gene product are known.

It has been suggested to me that since the imaginary bifurcation points are in fact points where some sort of choice or decision has to be made, perhaps the term 'points of choice' or 'decision points' should be preferred over 'bifurcation points'. In my opinion, such terms suggest that it is always the system itself that deliberately makes the choice, which is very often not the case.

In the vast majority of instances, the choice of b or c [of + or -] is a totally unconscious event, comparable to the software programs running 'in the background' on a PC and the millions of bifurcation points that are processed every second in its microcircuits. The 0.0000000001% of the activity that is 'conscious' takes place 'on screen' and on the keyboard/mouse. The 99.9999999999% that is 'unconscious' and automatic takes place 'off screen', just like the functioning of the myriads of cells in our body.

To keep it as simple as possible, I will only use the term 'bifurcation point' to indicate all situations where, beyond some critical parameter values, instabilities arise that are essential for realizing communication.

When dealing with the evolution of biological systems, bifurcation points become especially important where elements of *'cultural evolution'*, as it has been called in classical biology, come into play. Such points are also important in understanding that the more complex a communication system is, the faster it can evolve (the phenomenon of accelerating evolution). An addition of one extra bifurcation point can sometimes result in a multiplication of the communicational complexity, as shown in Fig. 8.4. I will deal with these issues further in Chapter 15, which deals with the theory of evolution.

8.6 Life's basic drive: "Solve problems if you want to feel comfortable and contented"

We are unaware of the vast numbers of problems that the cells constituting our body are having to solve all the time – and all in a coordinated fashion. No doubt there are very often conflicting situations in terms of which problem should be solved first and how. The unconscious and conscious problem-solving capacity of our brain, the system that coordinates all other physiological systems, is enormous. Some problems can be solved in simple reflex arcs, such as retracting an arm upon touching a hot plate. Trying to restore one's equilibrium upon stumbling is already a very complex problem. Reading and understanding a text requires extremely complex problem-solving capacities. Communication, problem solving and decision-making go hand in hand. Problem solving makes up a large part of the over-all activity of any organism.

Why do we solve problems? Although problem solving requires an input of energy and is seldom pleasant, we nevertheless engage in it. Why? In order to be rewarded with something, most of the time unconsciously. In Chapter 3 I put forward the idea that not only human beings and every other living organism, but also every single living cell must have a system for continuously integrating its perception of temperature, pressure, availability of food, etc. into an overall feeling of 'contentment'. This feeling of contentment and the motivation to do something, in particular to solve a problem when contentment turns into discontentment, are often closely linked.

In my opinion, Life's basic drive (or impulse) is: "Solve problems if you want to enjoy comfort and feel contented!"

8.7 Instinctive versus learned behavior. Not only humans have free will

Behavior can be defined as the total sum of all movements an organism makes. At the cellular level, active movement involves contraction of the cytoskeleton. Because animal muscle cells have an 'exaggerated' cytoskeleton compared, for example, to that of an ameba, muscular contractions are more readily observable than cytoskeleton-based movements made by other cell types with less well-developed cytoskeletons. Movement by undulipodia (cilia) involves tubulin.

Neither a skeletal nor a smooth muscle cell can decide by itself whether or not it will contract. Such cells respond to nervous stimulation. Heart cells respond to an endogenous pacemaker. The cytoskeleton will automatically contract when the cytoplasmic Ca^{2+} concentration reaches a given value (around 10^{-4} M) because at such a high concentration some cytoskeletal proteins change conformation and interact with each other. Behavior is the response of a receiver-decoder-amplifier-responder to a given message coming either from the environment or from a sender-encoder.

The distinction between innate behavior and that which becomes possible as the result of learning processes is often difficult to make. The inconsistent terminology used in dealing with this issue has added to the confusion (Bateson, 1983; Papaj, 1993). Not only have different terms been used in inconsistent ways to denote predetermined behavior, but individual terms have acquired multiple and often conflicting meanings (Bateson, 1983). I will use the terms 'instinctive' and 'non-instinctive' behavior.

In Henderson's dictionary of biological terms (Lawrence, 1995), instinct is defined as "the behaviour that occurs as an inevitable stereotyped response to an appropriate stimulus, sometimes equivalent to species-specific behaviour". Instinctive behavior does not require 'learning' and 'cognitive memory'. If I burn my finger, I do not retract my arm because I was taught to do so and because I had stored that acquired information in my 'brain memory'. Instinctive behavior is predictable because there are no (or only very few) bifurcation points present in its whole communication-system cascade.

Non-instinctive behavior is much less (or sometimes not at all) predictable. If I travel to a city where I had never been before and start walking around, I have to decide continuously where to go, how fast, for how long, etc. If somebody starts talking to me, how the muscles involved in my speech production will contract and relax will depend on what the person tells me and how I think I should respond. Non-instinctive behavior only becomes possible when a choice can be made. This requires the presence of bifurcation points. The more bifurcation points there are, the higher the possible versatility. The predictability of the final outcome in behavior decreases as the number of possible incoming messages and the number of possible bifurcation points increases. We have illustrated how exponentially fast this decrease in predictability is with the example of simple cells with only two types (and then with three types) of ion pumps and channels, and with the infinite number of melodies that can be played on a recorder or an accordion. The more variables there are, the higher the contribution from the 'brain' memory that will be required.

Man often assumes that he is the only species that can freely decide what to do and that other animals only act by instinct. This is certainly not the case. For example, in searching for its food, a dog or a chicken can decide whether to go north or south, east or west. A parrot can decide whether or not it will recite what his boss taught him to 'say'. There are of course situations in which instinctive behavior will become dominant over free choice, but that is also the case with humans when they are in danger. The differences between the free will of man and that of other organisms are not absolute and qualitative, but rather quantitative. Because of the enormous number of possible bifurcation points in his brain, man has more opportunities to display 'free will' in more situations than organisms with less developed brains. The more cells the brain consists of and the higher the degree of interconnectivity among the nerve cells, the more points there are where choices and decisions can and must be made.

The human brain has such great versatility that any normal child is able to learn any language in the world and to behave according to any local cultural standards wherever on earth, provided that it grows up in an *appropriate environment*. In the example of the identical twins I used in the beginning of this chapter, it is the environment that tells the twins which bifurcation points in their brain should be addressed at any moment in order to behave and speak like a local inhabitant.

8.8 Summary: freedom, responsibility and suffering

In summary, it can be stated that 'life' and 'freedom' are inseparably interconnected. Freedom of action implies decision-making, which in turn implies responsibility for the outcome of the choices made.

The first central dogma, DNA $\rightarrow$ RNA $\rightarrow$ Proteins, is not a source of freedom of action. A gene does not decide by itself whether it will be transcribed into a messenger RNA. Neither is it the mRNA itself that decides to be translated into a protein. One of the reasons why many people do not readily accept the neo-Darwinian view of evolution, which relies almost exclusively on changes in genes, is that such a system does not allow for any decision making by a living organism. With respect to *Homo sapiens*, this is not at all an attractive worldview.

I have a very positive outlook regarding the basic drive of life, as seen from the purely biological point of view and thus without any ulterior religious motives. Life is not a vale of tears, it is a wellspring of joy. Joy, however, cannot exist without its opposites, sorrow and pain.
In times of individual or group suffering because of disease, war, climatic catastrophes or accidents, human beings who believe in an Almighty God may be inclined to ask the question why God allows all this suffering and why He does not intervene personally in our lives. My answer to this question is that God did not create a static universe, but a dynamic one that is based upon instabilities. This universe changes incessantly, sometimes in ways man does not appreciate. Without disequilibria, life would not be possible. Life without suffering would probably mean life without freedom.

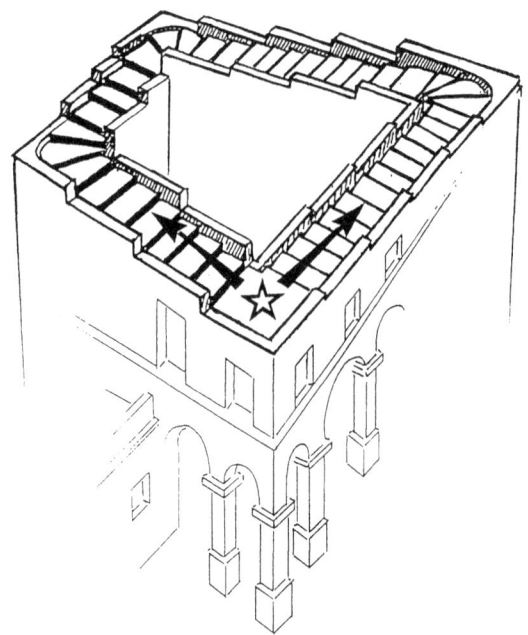

In life one has to make choices all the time.
After a lithograph by M.C. Escher ("Ascending and Descending").

In his evolution, *Homo sapiens* has not yet reached the status of a true social (eusocial) species, as some 'higher' social insects (ants, bees) have already done. Much of the suffering that humans inflict upon one another results from the fact that *Homo sapiens* is still in a pre-social phase of development. In human society, the maxim "One for all, all for one" has not yet been accepted as the general rule to be obeyed by all individuals. It is the task of *Homo sapiens* himself to eradicate the rule, "Me first and let the others help themselves", and realize a true social community. Let us hope that this will not take as many millions of years as it took some insect species to reach the eusocial stage.

ESSENTIALS

MAKING CHOICES. LIFE'S BASIC DRIVE

1. The (first) central dogma - DNA → RNA → Protein(s) - is very deterministic in nature. If organisms are reduced to the sum of their genes, very little freedom in behavior is likely to result.
2. All organisms are confronted with a variety of problems. Free-living organisms in particular must be able to quickly react to situations of danger, to sexual partners, to other members of the population, to changing climatic conditions, etc. Because the emerging problems very often cannot be predicted, decisions have to be made all the time, sometimes instantaneously, in order to solve the problems. Hence the question emerges: if decision making cannot be based on quick changes in gene activity, then on what other biological principle can it be based?
3. Decision-making happens at bifurcation points. These bifurcation points are present in communicating systems and are linked to the use of the 'compartmental software', as illustrated in the example of the identical twins that are raised in different cultures.
4. Bifurcation points offer possibilities, not certitudes.
5. The role and effects of bifurcation points can be easily illustrated with musical instruments of increasing complexity. With a simple police whistle, quite a number of different 'melodies' can already be made, despite the fact that there is only one outlet hole. In particular, the alternating opening and closing of the outlet hole(s) offers possibilities for variability. The environment plays a crucial role in governing which bifurcation points should be used.
6. The recorder example illustrates the fact that each additional hole (mutation) multiplies the number of possible melodies. It also illustrates the fact that simple instruments allow an endless variability in melodies, on condition that the possibilities offered by bifurcation points are utilized. In general biological terms, this means that there is no need for a large number of mutations in the genes coding for the hardware in order to obtain flexible communication systems. Thus, from the evolutionary point of view it is better to invest in additional bifurcation points than in additional mutations.
7. The plasma membrane of a cell contains a variety of functional holes, the best documented of which are the ion pumps and the ion channels. In concert, these functional holes build up a voltage (electrical) gradient over the plasma membrane. This is the urgradient of the living matter, which is still in use in any living cell.
8. Life's basic drive (or impulse) is: "Solve problems if you want to live comfortably and feel contented!"
9. Organisms/compartments use the integrative feeling of 'contentment' as a trigger to engage in problem solving. Any time this feeling turns into discontentment, they experience this as a warning signal - either unconsciously or consciously - that some problem has to be solved.
10. The more bifurcation points there are in the nervous system of an animal, the more room there is for learned behavior in addition to instinctive behavior.
11. Our freedom in behavior follows only partially from our genes. The possibilities offered by decision making should be taken into account in any theory that attempts to

explain evolution, and in particular the evolution of non-sessile species. 'Cultural evolution' is not possible without decision making.

CHAPTER 9

NO LIFE WITHOUT ELECTRICITY

THE PRINCIPLES OF BIOLOGICAL ELECTRICITY AS RELEVANT TO COMMUNICATION.
THE PRIMORDIAL GRADIENT OF THE LIVING STATE

To a large extent, 'Life' is an electrical phenomenon

Contents

9.1 Electric fishes are at the origin of the discovery of electricity
9.2 The nature of 'biological electricity'
9.3 Every living cell maintains a potential difference over its plasma membrane on the order of tens of thousands of volts per centimeter
9.2 All living cells actively generate and maintain differences in concentrations of specific ions over their plasma membrane. Ion pumps
9.5 There is also passive 'leaking' of ions through the plasma membrane. Ion channels
9.6 Cells manage to transduce a difference in concentration of some simple ions on the two sides of their plasma membrane into a potential gradient over this membrane. The principle of the concentration chamber
9.7 Calculation of the membrane potential. Depolarization and hyperpolarization.
9.8 Ion pumps and channels act in concert
9.9 Transcellular ion fluxes and extracellular electric fields
9.10 Self-electrophoresis: a means for generating intracellular potential gradients and gradients of charged macromolecules under specific conditions. The concept of 'the cell as a miniature electrophoresis chamber'
9.11 Actin filaments in the cytoskeleton: electricity-conducting wires in between the plasma membrane, the cytoplasm and the nucleus? DNA and electricity
9.12 Excitable membranes. Impulse conduction along axons. Electrical control of gene expression
9.13 Inorganic ions: one of the tools for generating order out of 'chaos'
9.14 General conclusion: to a large extent, life is an electrical phenomenon

Essentials

9.1 Electric fishes are at the origin of the discovery of electricity

This chapter is quite technical. Its reading and understanding will no doubt require quite some effort from many readers. Yet a book about 'life' does not make sense if it does not include the principles of biological electricity. In my opinion, the second most impressive 'invention' relating to 'life' ever made in the universe, (the first one being the genetic code), was the production of biological electricity starting with nothing other than water, a few simple inorganic salts, a membrane that was selectively permeable to certain ionic species and, finally, a source of energy that was needed to fuel certain ion transporting enzymes. The principle is very simple. But, as is often the case in nature, a simple but good principle has enabled the creation of such a great variability that the end result obscures the simplicity of the underlying basic principle. In my opinion, the principle of biological electricity is as important as the first central dogma DNA $\rightarrow$ RNA $\rightarrow$ Protein(s). Therefore, this chapter is worth the effort the reader may have to put into it.

Medical doctors use machines to measure certain natural electrical phenomena in our body. Electrocardiography is a technique that allows us to visualize electrical currents that are emitted by the heart every time it contracts. An encephalogram is a picture of the gross electrical activity of the brain. Electromyography is used to search for disorders in muscular contractility. If our heart and brain stop producing these electric currents, we are dead: no self-generated electricity, no life.

In this chapter I will explain how any living cell manages to produce this electricity and why it is so vital for communication and life. Only if one has some insight into the principles underlying 'biological electricity' is it possible to understand the statement that "life is to a large extent an electrical phenomenon". It then also becomes clear why living systems necessarily are 'dissipative systems' that consume energy to maintain an ordered and functional state that is far from thermodynamic equilibrium.

The study of electricity began a couple of centuries ago when some inquisitive people tried to understand the nature of the 'shock' that some fishes produce in order to defend themselves or to paralyze prey. The electric eel, *Electrophorus electricus*, is not a 'true eel', but rather a species belonging to the family of the Electrophoridae. Such fish are difficult to catch, transport and maintain in captivity. Hence, they are far from ideal practical models for unraveling the nature of the phenomenon. According to the literature, the voltage delivered by an electric eel differs between individuals and can range from between 450 and 600 volts, which is enough to occasionally kill an animal as big as a horse.

Rapid progress was made from the moment when the fathers of electricity -- Volta, Faraday and others -- discovered ways to produce electricity using non-living systems. The invention of the first battery by Volta was a big step forward. From then on, research efforts were concentrated more and more on 'physical' electricity rather than on the type produced by the electric organ of electric fishes, which later was found to be a modified muscular system. The study of 'biological' electricity got marginalized, but research continued. Ernest Solvay, one of the founders of the Belgian chemical company with the same name, was fascinated by 'biological' electricity. Already in 1894, he reported on what we now call 'action potentials' in a nerve. However, it took another half a century before the mechanisms by which all living cells produce electricity were discovered, and before the role of electrical phenomena in muscular contraction and in the conduction of impulses along nerve cells was elucidated.

9.2 The nature of *'biological electricity'*

Electricity is the movement of charges. The electricity from the socket that we use in our daily lives is the movement of charges carried by **electrons** through conductors and this occurs at very high speed (300 000 km/sec or the speed of light in an 'ideal' conductor). Most of the electrical phenomena in biological systems have nothing to do with the transport of electrons, but rather with simple **inorganic ions** such as K^+, Na^+, H^+, Ca^{2+}, Mg^{2+}, Cl^- and HCO_3^-. Electrons can move through solid conductors such as metal wires; ions cannot.

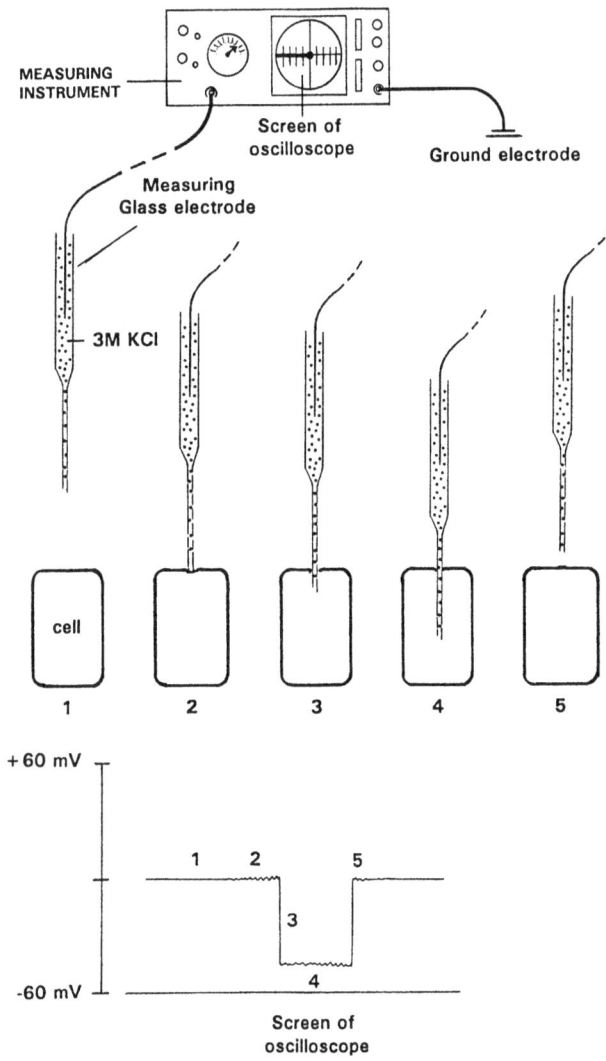

Figure 9.1 Schematic representation of the methodology which is followed to measure the membrane potential of a cell. For explanation, see text.

The movement of ions in their watery environment and through membranes is much slower than that of electrons. The basic equation of electricity, Ohm's law, is:

$$I = \frac{V}{R}$$

where I = the strength of the current expressed in amperes, V = the potential difference expressed in volts, and R = the resistance expressed in Ohms. This equation applies equally as well both to electron-borne and to ion-borne electricity: a charge is a charge, no matter what its vector is.

9.3 Every living cell maintains a potential difference over its plasma membrane on the order of tens of thousands of volts per centimeter

The physical basis underlying the potential difference over the plasma membrane, which is called the 'membrane potential', was uncovered in the course of studies intended to elucidate the mechanisms involved in impulse conduction in neurons. The experimental procedure for measuring a membrane potential is as follows (Fig. 9.1).
First one makes a glass electrode with a very fine tip. A narrow-bore glass capillary tube is positioned in a machine called an 'electrode puller'. The middle of the capillary tube is heated so the glass will melt, and in the mean time the machine pulls at one end of the capillary tube. One thus obtains two micropipettes with tips that are so fine that they can serve as electrodes to be pushed through the plasma membrane of cells with a micromanipulator. The micropipette is filled with a conducting solution (e.g. 3 M KCl). An Ag/AgCl wire is inserted into this solution in the micropipette and connected to an electrometer. The signal can be visualized on an oscilloscope screen or on a recorder. The changes in potential that occur when the electrode enters the cell can be followed.

The second electrode, which is needed to close the measuring circuit, is grounded. The measuring electrode is positioned in a micromanipulator. The cell whose membrane potential is to be measured is kept in physiological saline, where can be observed using a microscope. After the system is equilibrated, measurements can be taken. With the micromanipulator, one brings the measuring electrode nearer and nearer the surface of the cell. As long as the tip of the electrode does not touch the cell surface, a flat zero line is observed on the screen of the oscilloscope. At the moment of contact, this baseline becomes somewhat irregular.

When the tip of the electrode is pushed through the plasma membrane into the cytoplasm, the moving light spot on the screen of the oscilloscope makes an instant jump, usually to a value which in animal cells is a few tens of millivolts negative, and a new baseline is established. The magnitude of the potential difference depends on the cell type, the ionic composition of the bathing solution of the cell, the temperature, etc. The average membrane potential of animal cells is around 40-50 mV (cytoplasmic side negative). In plant cells, values of over 100 mV are common.

A potential difference, for example, of 50 mV over a membrane may look insignificant at first sight. One has to realize, however, that the plasma membrane is only 8 to 10 nanometers (1 nm = 10^{-9} meter) thick. Thus a membrane potential of 50 mV over 10 nm is equivalent to 50,000 Volts per centimeter. Very few engineers ever have to handle such large potential differences.

9.4 All living cells actively generate and maintain differences in concentrations of specific ions over their plasma membrane. Ion pumps

The giant squid axon is a model system that enabled a breakthrough to be made in the elucidation of the principle underlying the generation of a resting membrane potential as well as an action potential. Squids, like snails and mussels, belong to the phylum Mollusca. The central nervous system of mollusks, when compared to those of species belonging to most other animal phyla, consists of relatively few neurons, but these neurons are usually large cells. Squids have a pair of giant neurons. This model permitted the determination of concentrations of common inorganic ions in the cytoplasm of neurons. It turned out that there were very substantial differences between the cytoplasmic (= in axoplasm) concentrations and the hemolymph:

Ionic species	K^+ mM	Na^+ mM	Ca^{2+} mM	Mg^{2+} mM	Cl^- mM	isethionate mM
Hemolymph	20	440	10	54	560	
Axoplasm	400	50	0.4	10	40-150	250

There are steep gradients of all mentioned inorganic ionic species across the plasma membrane. The concentrations of K^+, Na^+, and Cl^- are much higher than those of the divalent cations. Especially important is the fact that the concentration of K^+ is roughly 20 times higher in the cytoplasm than in the hemolymph, while the concentrations of Na^+ and Cl^- are much lower in the cytoplasm than in the hemolymph. The total concentration of Ca^{2+} is low compared to that of the monovalent ions. However, in cells at rest, most of this Ca^{2+} is bound, which means that the concentration of free Ca^{2+} is even much lower, namely around 100 nM or 10^{-7} M.

Thus, the squid giant neuron manages to build up and maintain concentration gradients of inorganic ions across its plasma membrane. This is not at all an exceptional situation: all cells manage to do so. Since gradient formation requires energy, cells must have very good reasons to do so; otherwise it would be a tremendous waste of energy.

How do ions pass through the plasma membrane?

The plasma membrane is a fluid lipid bilayer, in itself impermeable to ions, in which hydrophobic molecules can be trapped. Many proteins have hydrophobic parts that are big enough to allow them to reside in the hydrophobic environment of the plasma membrane. In the context of this chapter, ion pumps and channels are especially important.

An ion pump is an enzyme that can transport specific ionic species from the side of the membrane where the concentration of that particular ionic species is lower to the other side where the concentration is already higher, as shown in the diagram in Fig. 9.2.

Such 'uphill' transport, or **active transport** as it is commonly called, evidently requires energy. ATP (Adenosine TriPhosphate) is a very common source of energy for ion pumps (H^+-ATPases, Ca^{2+}-ATPases, $Na^+ + K^+$- ATPases, etc., see Fig. 7.3, Fig. 9.2 and 9.4). Even if the energy supply were continuous, an ion pump could not pump indefinitely: at some point it would reach equilibrium, as will be explained later.

An ion pump can be *electrogenic* or *electroneutral*. It is electrogenic when there is a net transport of charge across the membrane when the pump makes one activity cycle. An H^+- pump that pumps one H^+ across the membrane is an electrogenic pump. A pump that, in one cycle, can pump one K^+ into the cell and one Na^+ out would be an electroneutral one. The common Na^+- K^+ ATPase in all cells of our body transports, in one cycle, 2 K^+ from the outside into the cytoplasm and at the same time 'ejects' 3 Na^+ from the cytoplasm into the environment. Thus there is a net transfer of charge in each cycle of such a pump: each cycle of this pumps makes the cytoplasmic side of the membrane more negative. This pump, which is often called the $3Na^+ + 2K^+$- ATPase, plays a crucial role in building up the high K^+ concentration in the cytoplasm while at the same time keeping the Na^+ concentration low. There is always some passive leakage of ions through the plasma membrane, not triggered by specific mechanisms. This means that ion pumps have to be at standby all the time and start working any time the cytoplasmic ionic concentrations trespass a certain value. A substantial proportion of the heat produced by our body is a side product from the action of ion pumps. Such pumps usually require ATP to do their job. This ATP is produced in mitochondria. The energy, which is stored in ATP, originates from the burning of glucose in the cell: 1 glucose molecule yields 38 ATPs.

9.5 There is also passive 'leaking' of ions through the plasma membrane. Ion channels

A second group of ion transporting proteins is present in all living cells, namely **ion channels**. An ion channel is also a protein residing in the plasma membrane and it also allows the transport of specific ions through itself. However, with ion channels the transport is from the side where the concentration of the ions is higher to the side of the membrane where it is lower (Fig. 9.2). This is **passive transport** (= downhill), which does not require energy.

If ion channels were fully open all the time, a cell would have a hard time building up an ionic gradient. The ions pumped in by an ion pump would immediately leak out through the open channels, provided their 'gate' was big enough. No communication and no life is possible under such circumstances. The usual situation is that the ion channels are pretty well closed but there are specific factors/mechanisms (chemicals that bind to specific receptors, physical stimuli, etc.) which can temporarily open the channels, thus allowing a flow of ions through the membrane. On the other hand, a lipid membrane completely devoid of ion pumps and channels would be impermeable to ions and would not allow communication. Thus, if the boundary of the first cell that appeared on earth had been as impermeable to ions as glass or metal, life as we know it today would not have come into existence. Consequently, a moderately 'leaky' membrane was a prerequisite for the living state to arise. In other words, *life started as the result of an imperfection in the impermeability of a lipidic plasma membrane to a given solute (probably an inorganic ion), combined with a gradient of the same solute over this membrane.*

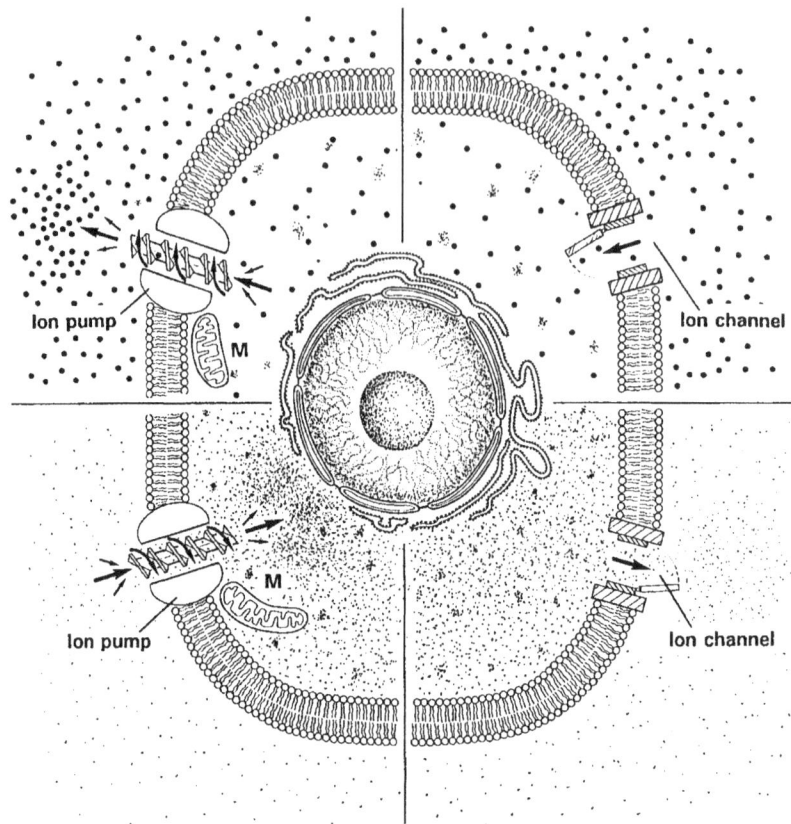

Figure 9.2 Diagram illustrating the principle of active transport by means of an ion pump and of passive transport (diffusion) through an ion channel. Only active transport requires energy, which is usually delivered by the mitochondria (M) in the form of adenosine triphosphate (ATP).

9.6 Cells manage to transduce a difference in concentration of some simple ions on the two sides of their plasma membrane into a potential gradient over this membrane. The principle of the concentration chamber

This key relationship between concentration gradient and potential gradient follows from the properties of what in physics is called a "*concentration chamber*" (Fig. 9.3). Such a chamber is made as follows. A container is divided into two compartments by a membrane that is differentially (or 'selectively') permeable to a specific type of ion (X^+ in our example) and not to ions of opposite charge (Y^- in our example). The two compartments are filled to the same height with water containing different concentrations of the salt X^+Y^- (for example KCl). Compartment 1 is filled with the more concentrated salt solution, and compartment 2 with the less concentrated solution, which in our example (Fig. 9.3) is pure distilled water (i.e. zero concentration of solute).

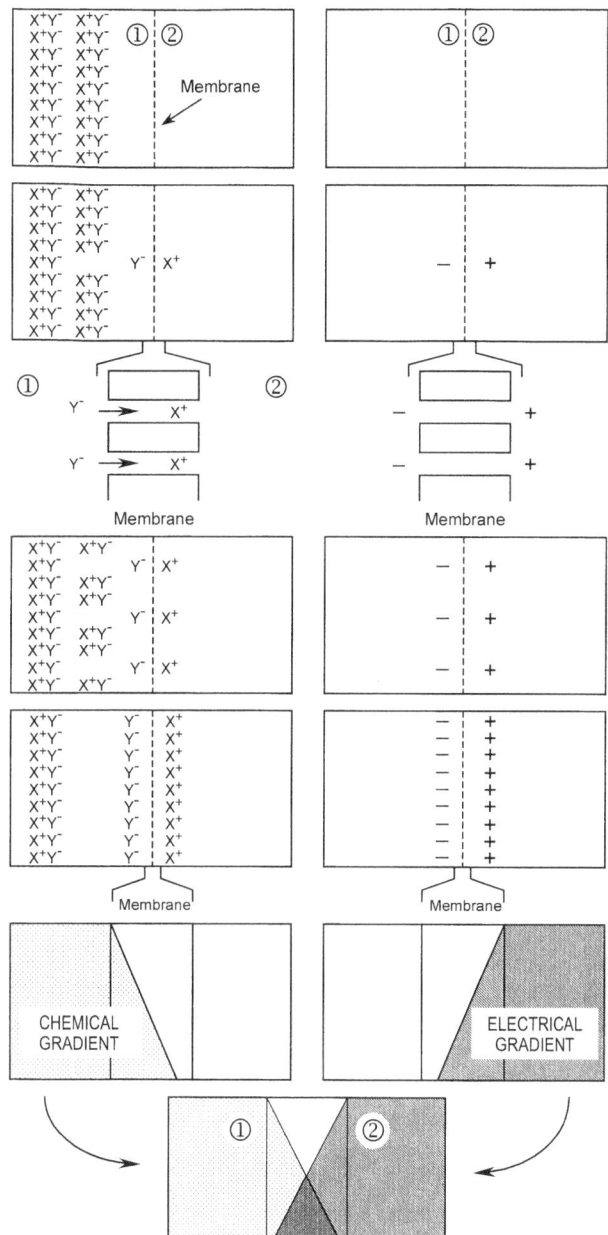

Figure 9.3 Schematic representation of a concentration cell or chamber, showing how a difference in the *concentration* of a given ionic species can yield a *potential difference*. The membrane in between compartments 1 and 2 is differentially permeable: certain ionic species can pass through the membrane while others cannot. In the end a state of equilibrium is reached in which the electrical gradient balances the chemical gradient. It is assumed that within a compartment there is an isopotential. The figure also illustrates that the potential difference only exists over the membrane.

One of the measuring electrodes of a sensitive voltmeter is introduced into compartment 1, the other into compartment 2. The voltmeter will show a voltage difference between the two compartments.

Because the salt concentration is higher in compartment 1 than in compartment 2 and because the membrane is permeable to X^+, a certain quantity of X^+ from compartment 1 will diffuse down the gradient to compartment 2. Thus, positively charged ions are taken away from 1 and used to charge the membrane capacitance, and a potential difference between 1 and 2 is set up. Compartment 2 therefore becomes positive in relation to compartment 1, although 1 still contains many more X^+ than 2. As more X^+ move from 1 to 2, the potential difference becomes greater or, in other words, the potential (electrical) gradient becomes steeper. But the higher the potential difference, the more difficult it becomes for X^+ to move against the electrical gradient. After some time, a state of equilibrium is reached in which the electrical gradient (which tends to move X^+ from compartment 2 to 1) exactly balances the concentration - or chemical - gradient (which tends to move X^+ from compartment 1 to 2).

Thus, the transmembrane potential gradient (membrane potential, measured in millivolts) is the automatic result of a difference in ionic concentrations on the two sides of a **selectively permeable** membrane. If the membrane were not selectively permeable, there would be no transmembrane potential. The term concentration chamber stems from the difference in concentration of the diffusible ion in the two compartments. If one replaces X^+ by K^+ and Y^- by Cl^-, the situation resembles that in a neuron (and other animal cell types) at rest. How living cells produce the ionic concentration differences between the two sides of a membrane will be explained later.

9.7 Calculation of the membrane potential. Depolarization and hyperpolarization

The pioneers in the field of electrophysiology found ways to calculate the membrane potential. To do so, one needs to know he concentration of the different ionic species present in the cytoplasm and in the environment (e.g. in the blood) and how permeable the membrane is for the different ionic species (the permeability coefficient). Some authors use the abbreviation E_m for the membrane potential, while others use another symbol. I will use E_m.

It is beyond the scope of this book to elaborate on the mathematics involved. Readers who are nevertheless interested in this fascinating aspect of physiology should consult textbooks on general physiology or of electrophysiology, such as *The Vital Force: A Study of Bioenergetics* by Harold (1986).

A membrane is said to depolarize when its membrane potential becomes less negative, e.g. from -70 mV to -30 mV. A membrane hyperpolarizes when the potential difference increases, such as from -70 mV to -75 mV.

The remaining membrane potential of animal cells usually varies between -40 and -70 mV. In plant cells it is often over -100 mV.

9.8 Ion pumps and channels act in concert

The activity of pumps and especially that of channels is well regulated. Considered over a sufficiently long interval of time, the pumps should be able to transport the ions faster than the ions can leak away through the channels. The major triggering mechanisms for opening channels are also represented in Fig. 9.4

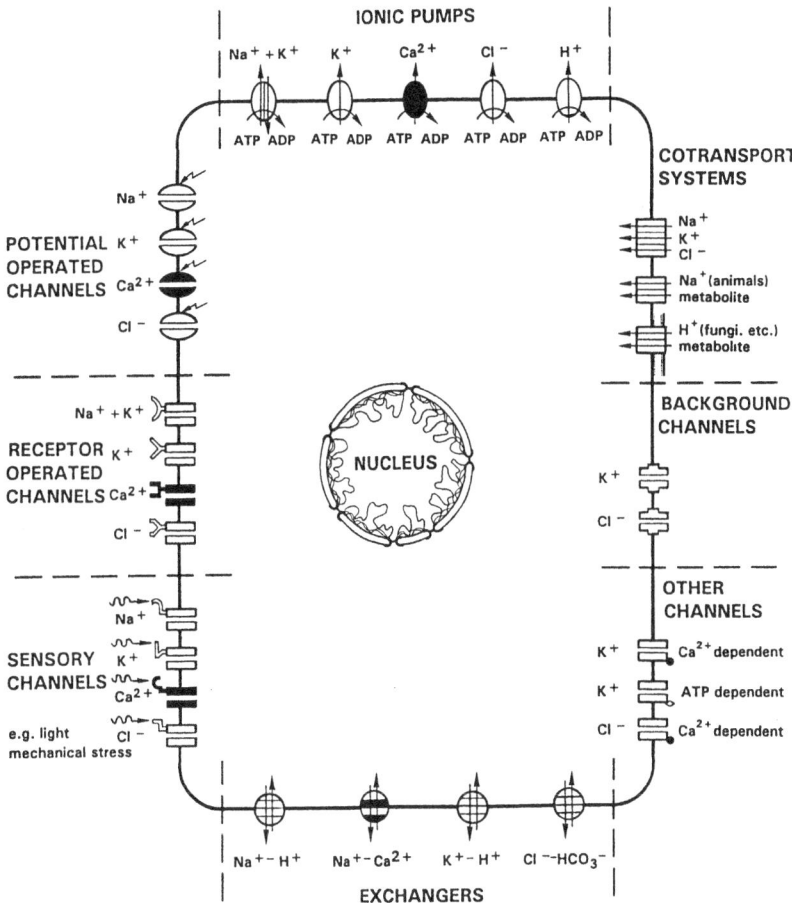

Figure 9.4 A large number of ion transporting proteins have already been described in a variety of organisms. This figure only lists the major groups. It is evident that not all of these protein types are present in every cell. Different cell types have different combinations of such ion transporting molecules in their plasma membrane. If one compares the complexity of this system with that of an instrument with many holes in its boundary, the picture of a cell as a very complex accordion with many inlet and outlet holes emerges.

Again, for the readers who are not familiar with electrophysiology, the following example from neurophysiology may help to clarify the system. In a neuron at rest, the concentration of K^+ in the cytoplasm is much higher than outside the cell. The opposite is true for Na^+. At rest, the Na^+ channels are closed, which means that the plasma membrane is relatively impermeable to Na^+. Upon stimulation, for example by the release of an excitatory neurotransmitter from a presynaptic element (or the motor endplate in a muscle, see Fig. 4.1), Na^+ channels in the postsynaptic element are opened. Because the concentration of Na^+ outside the cell is much greater than inside, there will be a sudden influx of Na^+ ions. The membrane potential thus becomes less negative, and the membrane is *depolarized*. This can be calculated from the so called Hodgkin-Katz-Goldman equation. When the cell loses its negative potential due to the Na^+ influx, the K^+ channels are also activated and the large K^+ concentration inside the cell will push K^+ out. This repolarizes the membrane, thus restoring the original situation: positive on the outside and negative inside. The $3Na^+ + 2K^+$-ATPases will then pump out the unwanted Na^+ and bring in K^+ again, thus restoring the concentrations. The influx or outflux of a single ionic species sets off a cascade of events involving other ionic species and leading to a new equilibrium.

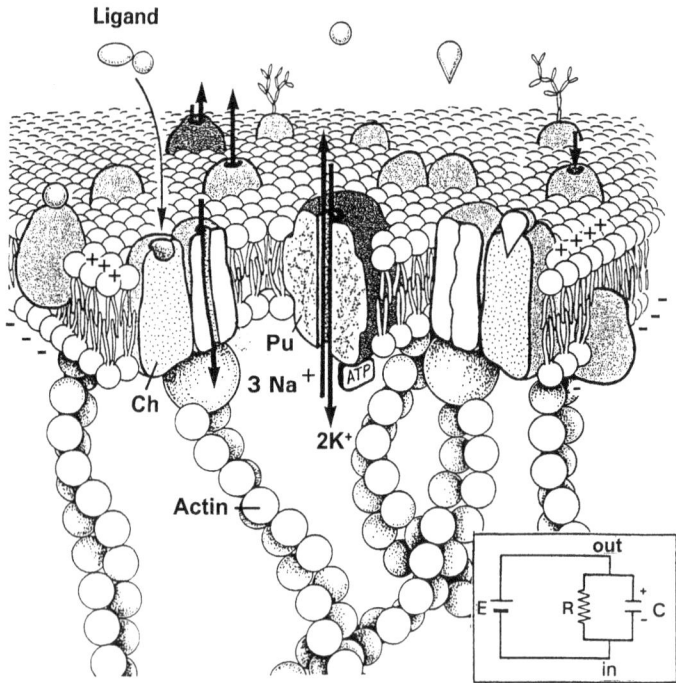

Figure 9.5 Some ion transporting proteins are anchored to the cytoskeleton by means of certain connecting proteins. The segregation of ion pumps (Pu) and channels (Ch) is especially important in epithelia. It enables the transcellular transport of water and certain solutes. The window in the bottom right of the figure indicates that the membrane behaves electrically like a network of a resistor (R) and a capacitor (C) shunted in parallel. The depicted ion pump is a Na+K-ATPase that in one cycle simultaneously transports 2 K^+ from the extracellular compartment (blood e.g.) to the cytoplasm and 3 Na^+ in the opposite direction. Thus in each cycle, there is a net transport of charge. Thus the 3 Na^++2 K^+-ATPase is an electrogenic ion pump. The cytoplasmic side of the plasma membrane is negatively charged (-) while the extracellular side carries positive charges (+).

The fact is often overlooked that there are also ion pumps and/or channels in membrane-limited cell organelles such as mitochondria, chloroplasts, lysosomes, the endoplasmic reticulum (which is an important storage site for Ca^{2+}), and the nucleus as well. Some of these pumps and channels are schematically represented in Fig. 7.3. Thus, the coordination of the activity of all these ion-transporting molecules present in eukaryotic cells is highly complex.

9.9 Transcellular ion fluxes and extracellular electric fields

To generate a plasma membrane potential, it is of no importance where the ion pumps and channels are situated in the plasma membrane: a random distribution will do. However, such a random distribution (spherical symmetry) is most unusual in nature. On the contrary, cells are almost invariably polarized, and this is necessary for carrying out their proper functions.

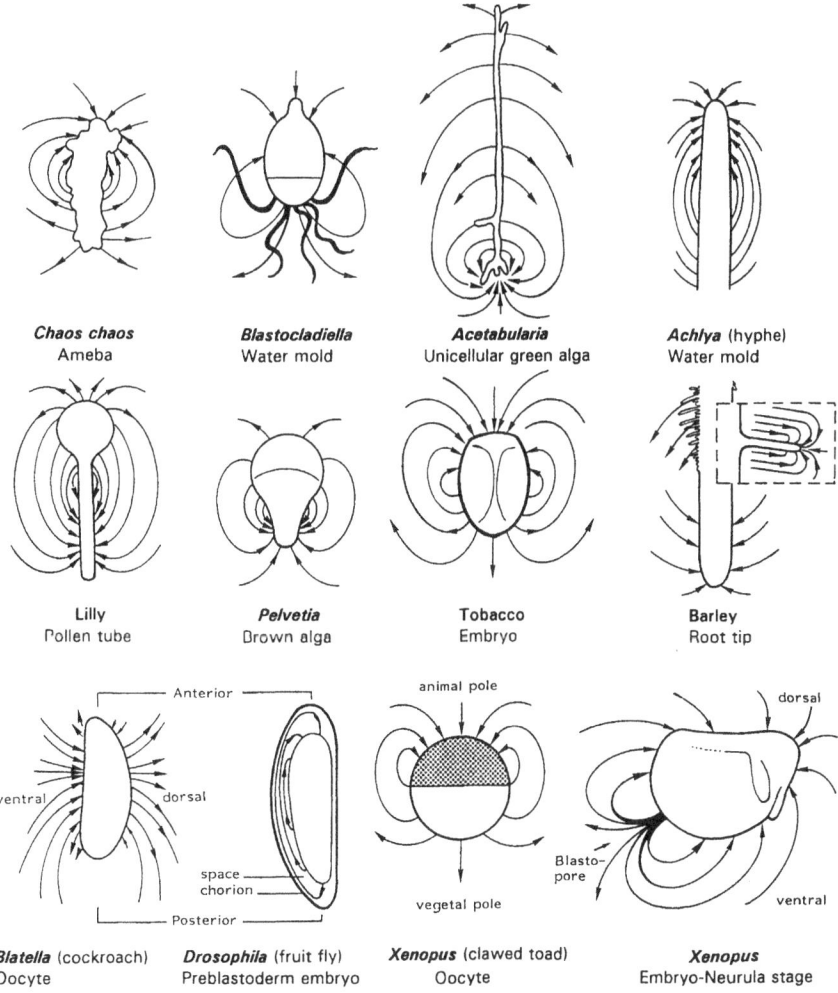

Figure 9.6 Schematic representation of the electric field around a variety of biological systems, as measured by the vibrating probe technique. The electrogenic ion fluxes are produced by the systems themselves. Redrawn after several authors: for references, see Jaffe and Nuccitelli (1977) and De Loof (1986).

One of the consequences of this polarity is that at least some of the ion transporting proteins – which without this polarity would be freely moving in the plane of the membrane like icebergs in an ocean – are anchored (e.g. by the cytoskeleton) in well defined areas (Fig. 9.5).

Typical examples are epithelial cells. They would not be able to carry out their typical function of transcellular transport of water and solutes if their ion pumps and their channels were not segregated at a sufficient distance from each other (e.g. one type in the apical and the other in the basolateral part of the cell). There are numerous examples in which it has been experimentally proven that an electrogenic ion flux enters the cell or organism at a particular site and leaves at another one (Fig. 9.6).

In some cases the ion pump and the ion channel activities are located far from one another. Examples of this include the situation in the basolateral and the apical parts of the plasma membrane of an epithelial cell, and the site in an ameba where pseudopods form, which is far removed from the contracting rear part (the uroid) of the ameba. If the cell drives an electrogenic ion flux (e.g. an electric current carried by ions – in short, an ionic current) through itself, the ionic current will enter at one side and leave at the other. By convention, the direction of the current is that in which the positive charges move. The efflux of net positive charge and the influx of net negative charge are both depicted as outgoing current. But in systems that drive an electric current through themselves, the current loop must be closed. This means that there must be an electrogenic flow of ions around the compartment that drives an ionic flux through itself. This is indeed the case: extracellular ionic currents can be measured by a very sensitive device called the vibrating probe, which was originally developed by Jaffe and Nuccitelli (1974) and has been constantly improved upon since then. In Fig. 9.6, a few examples of such extracellular electric fields are shown.

9.10 Self-electrophoresis: a means for generating intracellular potential gradients and gradients of charged macromolecules under specific conditions. The concept of 'the cell as a miniature electrophoresis chamber'

If cells drive an electrogenic ion flux through themselves, as many do, is it conceivable that they could generate intracellular potential and/or concentration gradients? Theoretically, intracellular potential gradients could be significant (a few millivolts) on condition that the resistivity of the cytoplasm is high enough. Even with the very sensitive equipment available today, it is not easy to make accurate measurements of intracellular potentials in such small cells. In classical electrophysiological research the precise part of the cytoplasm where the measurement is made is often not ascertained, as it is generally assumed that the cell's interior is at isopotential.

For most cells, and especially for somatic epithelial cells, their small size is the limiting factor for accurately establishing whether or not there is a continuous potential gradient in their plasma membrane. There are always the possible leakage artifacts due to the penetration of the membrane with microelectrodes. Another problem is the possible leakage of electrolytes from the interior of the microelectrode (e.g. 3M KCl) into the cell, with the consequent alteration of the intracellular electrolyte concentration and/or cell volume.

Oocytes, however, are giant cells and therefore better suited for such measurements. Self- electophoresis has been very elegantly visualized in the oocyte-nurse cell complex of a moth by Jaffe and Woodruff (1979) and Woodruff and Telfer (1980) (see also De Loof, 1986). In this moth, oocytes and nurse cells are connected by a relatively wide cytoplasmic bridge. A potential difference of several mV, which is huge in such a system, was measured over the bridge. By means of very elegant experiments in which the movement of micro-injected neutral and charged fluorescently labeled proteins was recorded, self-electrophoresis could be established. Self-electrophoresis could be a means for generating intracellular gradients of charged macromolecules under specific conditions. The concept of 'the cell as a miniature electrophoresis chamber', as formulated by De Loof (review 1986), involves the idea that perhaps all cells can drive an electrogenic ion flux through themselves at certain stages of their development. It would take us too far afield to elaborate on this interesting topic. In Chapter 12, which deals with the coming into existence of the Progenote, the non-random distribution of ion pumps and channels over the plasma membrane will be taken into account.

9.11 Actin filaments in the cytoskeleton: electricity-conducting wires in between the plasma membrane, the cytoplasm and the nucleus? DNA and electricity

As charged poly-electrolytes, actin filaments may contain a proportion of their surrounding counterions in the form of a dense or 'condensed' cloud about their surface, which may be highly insensitive to large changes in the ionic strength conditions of the surrounding saline conditions (Lin and Cantiello, 1993).

Provided that it is electrically shielded from the bulk solution, this tightly 'bound' ionic cloud will allow significant ionic movements along the polymer's length. The counterionic cloud about an actin filament could be, therefore, a highly conductive medium. As a consequence, electrically forced ions entering one end of the polymer will result in ions exiting the other; the ionic gradient lying along the polymers' length.

Using a variation of the 'patch-clamp' technique applied to isolated actin filaments in solution, Lin and Cantiello (1993) studied ionic (= electrical) currents along the surface of electrically stimulated actin filaments. Such currents were observed about the polymer's surface under both high (100 mM KCl) and low (1 mM KCl) ionic strength conditions. Counterionic waves were highly nonlinear in nature and remained long after the electrical stimulation of the actin filaments ceased. Upon pulse excitation, an input decomposed into a finite number of solitons and a low amplitude oscillatory tail.

The authors also observed that more net charge was transferred via actin filaments than via free solution. In solution, poly-electrolytes are known to electrically repel one another, thus minimizing the crosstalk between neighboring polymers. It seems possible that actin filaments could exist in a complicated dynamic network and still preserve electrical screening from one another. According to Lin and Cantiello (1993), one can envision an electric field fluctuation such as that elicited by a single channel opening to which an actin filament can be attached as a novel and efficient long-range intracellular signaling mechanism. Its significance will be dictated by the overall status of its organization. In the context of the mode of hormonal action, there are many hormones that elicit electrical phenomena at the level of the plasma membrane and affect the cytoskeleton itself (De Loof et al., 1996). If the electricity is generated at a site where contact is made with an actin filament, one may expect that such electrically stimulated actin filaments generate soliton-like wave patterns.

The actin network extends from the plasma membrane into the nucleus. The protein skeleton of chromosomes consists largely of actin (Sauman and Berry, 1994). Furthermore, one should bear in mind that at least in *Drosophila* (Agard and Sedat, 1983), the chromosomes are not distributed at random in the nucleus but rather they occupy very specific positions. Perhaps there are electrical connections between the actin filaments in the cytoplasm and the network in the nucleus. It may thus matter where in the nucleus a gene is located. It is unlikely that a cell would invest a lot of energy in keeping its nucleus and chromosomes in a very specific position if there were no good reasons to do so. I think that such a specific localization involves an electric signaling system situated between the plasma membrane and the genes, and I think that it makes use of actin filaments. As will be mentioned more explicitly further on in this chapter, it is well documented that the expression of a number of genes is electrically controlled (Vanden Broeck et al., 1992), but it remains to be elucidated how this works. Furthermore I think – though this is mere speculation – that the cytoskeleton and the electrical phenomena that take place within this complicated and flexible structure may perhaps play a very important role as a molecular carrier of non-genetic (cognitive) long-term memory. This has already been dealt with in Chapter 3.

The long-standing question as to whether DNA is an electrical conductor has been answered affirmatively by measurements of electrical transport through well-defined DNA molecules between metal nanoelectrodes 10 nm apart. Electrical conduction does occur through DNA beyond a certain voltage threshold, so here the DNA is acting as a large-bandgap semiconductor (Porath et al., 2000). Whether or not this reflects any physiological role, remains to be investigated.

9.12 Excitable membranes. Impulse conduction along axons. Electrical control of gene expression

An excitable membrane has the ability to propagate a locally induced depolarization over a certain distance in the form of an action potential. As previously explained, depolarization is due to changes in ionic gradients across the membrane. All neurons and muscle cells have excitable membranes. Another well-documented cell type with an excitable membrane is the oocyte. When the spermatozoon of a sea urchin penetrates the plasma membrane of the oocyte, the lysosomal enzymes from the acrosome of the spermatozoon make a short-lived little hole. There is an influx of Na^+ and Ca^{2+} and an efflux of K^+ through this hole. The result is a local depolarization that spreads over the entire surface of the egg, thereby causing a "Ca^{2+} explosion" in the cytoplasm. This is a rapid system for preventing any other sperm from penetrating (block to polyspermy). In plants, slow-action potentials have been recorded. It has been suggested that these might be involved in transmitting messages through the plant (e.g. to trigger the synthesis of specific defense proteins that plants start producing when they are being eaten by phytophagous insects).

Many signaling molecules (e.g. neurotransmitters and hormones) have been reported to alter the membrane potential of their target cells. Some nutrient molecules, such as certain sugars and amino acids, enter the cell by co-transport with an inorganic ion.

There is increasing interest in the study of direct electrical control of gene expression. A few years ago it became possible to study the synthesis of individual mRNAs under the influence of plasma membrane permeability to ions and of electrical activity. At the current time, a few examples of the 'electrical' control of gene expression are well documented. Examples include surface IgM in a pre-B lymphocyte cell line; preproenkephalin in rat adrenal medulla; c-fos in a phaeochromocytoma cell line; the acetylcholine receptor in chicken embryonic myotubes; defense proteins in plants (for references, see Vanden Broeck et al., 1992). More recently – using a knock-out mouse deficient in a particular protein that is essential for neurotransmitter secretion – it has been shown that embryonic neurons require the release of neurotransmitter and the concomitant electrical phenomena in order to survive (Verhage *et al.*, 2000).

The exact mechanism of the 'electrical' control of gene expression is as yet not fully understood. The suggestion has been made that ionic fluxes (especially of Ca^{2+}) affect protein phosphorylation systems and thus, indirectly, gene expression. A causal relationship between ions and puff formation in salivary gland chromosomes of the dipteran insects *Drosophila* and *Chironomus* was already demonstrated in the sixties and seventies. Ca^{2+} has been reported to affect chromatin conformation. The key ions in many signaling systems are Ca^{2+} and H^+, because especially these two ions have drastic effects on the conformation of many proteins. (For references, see De Loof, 1986; Vanden Broeck et al., 1992).

9.13 Inorganic ions: one of the tools for generating order out of 'chaos'

Solutions of simple inorganic salts such as KCl and NaCl have few, if any, elements of 'order'. Nevertheless all living cells manage to turn such solutions into an element of order and an important means of communication. They realize this by building up a concentration gradient of specific ions over a selectively permeable membrane and transduce the chemical gradient into a potential gradient. When ion channels open, the ions that flow through its gate carry charges, which means that there is an electric current. An electric current is already an element of order. Certain enzymatic reactions are dependent upon specific ionic concentrations. Eukaryotic cells seem to have the ability to generate different 'ionic environments' within their different membrane-limited organelles. I will focus again on this topic in Chapter 14, which deals with the principles of embryonic development.

Inorganic ions are a major source of **indeterminism** in living systems. This contrasts to proteins, the products of genes, which act in a much more predictable and **deterministic** way. This is important for understanding evolution based on communication (Chapter 15).

Communication requires gradients but gradients can only be maintained if energy is invested in the system. If no energy is invested, the ionic gradients run down because the ion pumps fail, and the system reaches thermodynamic equilibrium and dies. Any system that requires labile gradients for its existence is by definition an unstable one. *Living systems cannot be other than thermodynamically far-from-equilibrium because they absolutely require labile ionic gradients.* I can go even one step further and say that what we call 'communication' at the level of the plasma membrane of a cell is in fact to a large extent the *visualization of the incessant balancing/transition of the cell between order* (the cytoplasmic side of the plasma membrane) *and chaos* (the external environment of the membrane). Communication in which fluxes of inorganic ions are involved is an example of 'order out of chaos'.

Sometimes, one word can have several different senses. When biologists use the term "chaos", they usually have a situation in mind in which no orderly processes can be recognized. For them, chaos means disorder. When some physicists and mathematicians deal with 'chaos theory', they have another sense in mind in which there is room for elements of order.

9.14 General conclusion: to a large extent, life is an electrical phenomenon.

Electrical phenomena occur all the time in living systems. Some of them play an important role in communication (e.g. in impulse conduction in nerve cells and in other types of signal conduction). Therefore, we can say that 'life' is to a large extent an electrical phenomenon.

ESSENTIALS

BIOLOGICAL ELECTRICITY

1. Electric fishes lie at the origin of the discovery of electricity.
2. Electricity is the transport of charge. Electrons carry electricity from the outlet. In biological systems, electric charges are carried by simple inorganic ions, in particular by H^+, Na^+, K^+ and Cl^-.
3. Only three simple components are needed for making electricity: water, some inorganic ions, and a membrane that is made differentially permeable to particular ions by means of ion pumps and ion channels. Ion pumps are special proteins residing in the plasma membrane that are able to transport particular ions through the membrane from the side where the concentration of these ions is lower to the side where their concentration is higher, and thus against a concentration gradient. Such transport is called 'active transport'. It requires energy. Ion channels are also proteins residing in the plasma membrane. These proteins enable the passive transport of particular ions down their concentration gradient.
4. Every living cell maintains a potential difference on the order of up to several dozens of millivolts over its ± 100 nanometer thick plasma membrane. This corresponds to a potential gradient of ten thousands of volts over a thickness of one cm.
5. All living cells actively generate and maintain differences in concentrations of specific ions over their plasma membrane by means of ion pumps. There is also passive leaking of ions through ion channels in the plasma membrane. Ion pumps and channels act in concert.
6. Cells manage to transduce a difference in concentration of some simple ions between the two sides of their plasma membrane into a potential gradient over this membrane. How this is achieved is explained by the 'concentration chamber principle'.
7. A membrane is said to depolarize when the cytoplasmic side of the membrane becomes less negative in relation to the outside. It hyperpolarizes when the opposite happens. Nerve cells, with their excitable membranes, experience such changes all the time.
8. When ion pump and ion channel activity are spatially separated far enough in the plasma membrane, the possibility emerges for transcellular ion fluxes and extracellular electric fields. This situation is rather the rule than the exception in all cell types.
9. Self-electrophoresis is an elegant means for generating intracellular potential gradients and gradients of charged macromolecules under specific conditions. The 'cell as a miniature electrophoresis chamber' concept says that probably any cell at some stage of its development is able to drive a self-generated electric current through itself.
10. The role of 'intracellular electricity' is not yet fully understood. The expression of some genes seems to be electrically controlled.
11. Electricity plays a role in cognitive memory. Whether the remarkable ability of actin filaments (a key structural element of the cytoplasmic and chromosomal skeletons) to conduct electricity in a very special way plays a role in cognitive memory, remains to be further explored.

12. Inorganic ions represent a very important tool for generating order out of 'chaos' in biological systems.
13. General conclusion: Life is to a large extent an electrical phenomenon.

CHAPTER 10

VIRUSES, PRIONS, TOOLIZATION, COMPUTERS, MAN-MADE LIFE

Contents

10.1 Introduction
10.2 Viruses are not alive
10.3 Prions
10.4 Man-made or artificial Life
10.5 Toolization. Self-communication
10.6 Refining the symbolic notation of 'life as an activity'
10.7 Electronic life and tool-aided (mechanical) reproduction
10.8 The future?

Essentials

10.1 Introduction

If you are interested in provoking a lively discussion, turn the conversation to computers and argue that you believe a computer is alive. From my own experience I know that many people are remarkably quick to reject the idea, often in a hostile way. There is little chance that a consensus will be reached, either pro or con. If afterwards you draw up a list of all the arguments that were used, they will probably match the list given later in this chapter.

Among biologists, the debate as to whether or not a virus is alive has not been settled unanimously either. In this chapter I will try to provide clear answers. I will do this from the viewpoint of communication. Just a reminder: 'life' relates to the activity consisting of deciphering coded messages and providing an energy-demanding response to them. Thus, for an entity to be alive, it must meet these criteria.

10.2 Viruses are not alive

Viruses are very often regarded as structures on the edge of being just barely alive or not yet quite alive. Tobacco mosaic virus can be crystallized, which is not at all a typical feature of living matter. However, when such virus particles, which can be stored on the shelf for long periods of time, are smeared over lightly abraded tobacco leaves, the virus will increase spectacularly in number inside the infected plant.

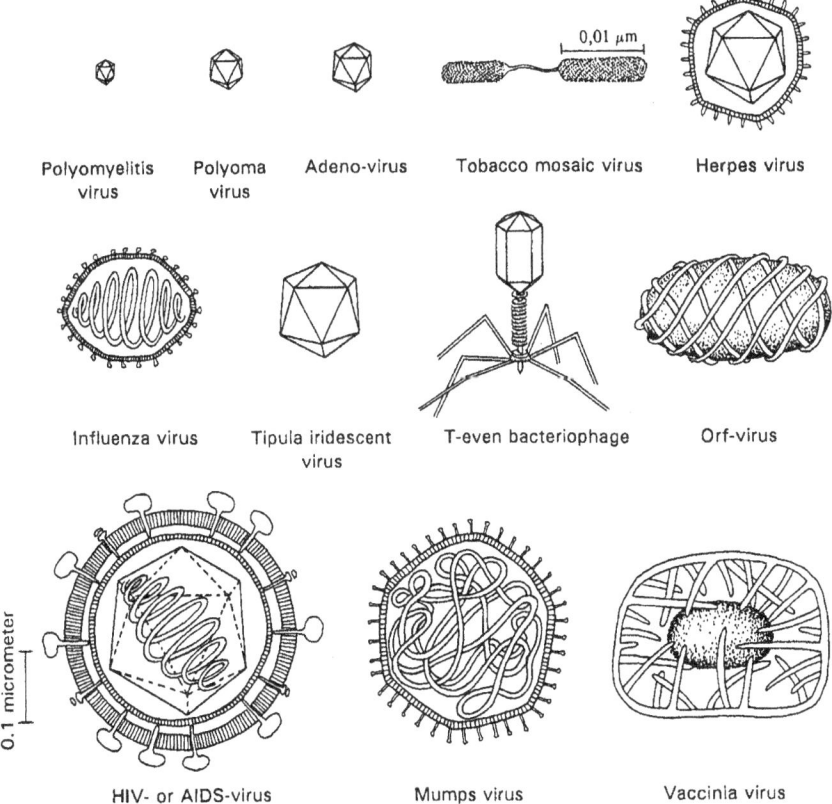

Figure 10.1 Schematic representation of the structure of a few viruses.

201

We now know that viruses can only multiply if they can make use of the protein synthesis machinery of the host cells. Reproduction is something typical for living matter but, contrary to what many people think, it is not essential for being classified as alive. Indeed, many organisms do not reproduce at all or stop reproducing at a given age though they obviously remain alive. In my opinion, at least the viruses that are not enveloped in a membrane cannot be classified as being alive because they have no means of building up a gradient that can be used for communication: no such gradient, no life (see Chapter 9). Years ago, some viruses were reported to be enveloped in a membrane. Recently, in some viruses this membrane has been shown to be an artifact due to inappropriate fixation procedures. It is unlikely, though very difficult to establish experimentally, that a viral membrane (if it is not an artifact) has ion pumps and channels, the tools that are necessary for building up an ionic gradient at this level of compartmentalization (Chapter 9). There is a hypothesis that asserts that viruses could be small clumps of nucleic acid/protein complexes, that escaped from the nucleus of a cell and started up a sort of parasitic existence. Thus, viruses are clumps of molecules, not organisms. In Fig. 10.1, the morphology of a few viruses is schematically represented.

10.3   Prions

The search for the nature of prions has been boosted enormously by the outbreak of the mad cow disease. For some time this infectious disease was thought to be caused by an unknown virus. The causative agent turned out to be a prion. Prions are proteins, thus chains of amino acids, that can undergo an abnormal folding and transmit this abnormality to other prion protein molecules in a sort of domino effect without using the complete normal DNA-RNA-Protein synthesis pathway. The crippled proteins clump together in insoluble aggregates that withstand all normal cellular breakdown mechanisms. Thus, once the abnormal folding starts, there is no way to stop it anymore. Protein molecules are clearly not alive.

Particular prions cause the deadly and potentially disastrous diseases known as the 'mad cow disease' or Bovine Spongiform Encephalopathy (BSE), 'scrapie' in sheep and the Creutzfeldt-Jacob disease in man. BSE destroys normal brain functioning in infected animals. In an infected animal, the prion can remain latent for months or even several years before the symptoms of the disease become apparent. The prion cannot be inactivated by normal sterilization procedures such as boiling, autoclaving or treatment with disinfectants like alcohol, bleach, etc. Only heating above 140 °C kills it. For this reason, the burning of the carcasses and offal, for example in cement factories, is the only safe way to dispose of contaminated material. A cow gets BSE not by contact with infected cows, but rather by eating food contaminated with BSE. Whether BSE can spread from cows to other mammals, including man, is still a matter of debate, but more and more data is pointing in this direction. To minimize the risks, European Union regulations require the immediate slaughter of the entire herd when a BSE-positive individual has been detected. For the time being (2002), none of the prion diseases can be cured, and no vaccines are available.

In recent years, more insight has been gained into the normal function of prion proteins. As long as their spatial organization is not crippled, they play an essential role in cells. Our brain makes use of normal prion molecules. In yeast, naturally occurring prion proteins can mask the stop signal in a coding gene, thus causing the gene to synthesize a different protein than what is specified by the genetic code. In a remarkable paper, True and Lindquist (2001) suggested that in yeast, prion proteins might play a role in facilitating evolutionary change in a way that had not been described before in molecular biology (see also Davenport, 2001). They also warn against overinterpretation of their results.

10.4 Man-made or artificial Life

If one were to ask a cross-section of 'educated lay people' which type of scientists might finally manage to produce an artificial form of life, the most frequent answers are likely to be: (1) molecular biologists; (2) computer specialists; (3) a team of both. It is unlikely that many people will think of membrane physiologists – researchers who study the permeability properties of biological membranes – as the ones who may do the job.

Molecular biologists

It is by no means the case that molecular biologists consider themselves to be little 'gods' striving to 'create life'. They are highly skilled 'ultramicroplumbers' (with all due respect), whose job involves cutting and soldering the DNA, etc., using no other tools than the ones they find already being used by living cells to carry out their normal functions and without which life would not be possible. By mastering the principles used by nature itself, molecular biologists can greatly influence the life of contemporary and future generations for good by producing all sorts of useful peptides and proteins, by correcting deleterious mutations, by improving the quality of plants used for food, etc.

Nevertheless, at the moment the last touches are being put on this book (2002), some people and groups continue to strongly oppose the use of molecular biological techniques to introduce foreign genes into the genomes of organisms or, in scientific language, to make 'transgenic' organisms. Their reasons vary, but they are all based on fear of the potential outcome: will molecular biologists create 'monsters' (whatever one might have in mind with this term), biological weapons (relevant in the aftermath of the terrorist attack on September 11, 2001), new organisms that will disturb local ecological equilibria, dangerous bacteria, etc.? Will these techniques harm the economies of developing countries?

Another reason – or rather motive – behind the activism and protest of some people and groups, unfortunately, is to profit financially from the creation of an atmosphere of fear for the unknown. Their appeal usually goes something like this: *"'Molecular biologists claim they have not yet harmed any species or the environment. But who knows what may be going on behind closed laboratory doors? Don't let the genie out of the bottle! Don't let them play with our grandchildren's future! Join us in opposing this global menace. No contribution is too small. All major credit cards accepted..."*

Another motive for protest can be a deep respect for God: man should not change what the Creator has made, nor should he create other forms of life than what God has intended. But what God created is changing all the time: children are not identical copies of their parents, siblings differ from each other unless they are identical twins, etc. Nature is incessantly preparing for the launching of new species, while other species are becoming extinct. Because molecular biologists can manipulate DNA, the universal carrier of genetic information, and because DNA is automatically linked to reproduction and the generation of new life, molecular biologists are suspected of wanting to 'play God'. But this is simply not the case.

Molecular biologists do not create 'new life'. Why can we assert this so firmly? *Because they do not make novel levels of compartmentalization*: only if this were achieved would there be good reason to speak of 'new' life. Making novel compartments is not a prime goal in molecular biology. What contemporary molecular biologists frequently do is amplify nucleic acid molecules by a method called the Polymerase Chain Reaction (PCR). In a few hours, a single copy of a nucleic acid molecule can be multiplied up to hundreds of thousands of times, which is orders of magnitude higher than any living organism can achieve in *in vivo* situations. The tools required to carry out this amplification are: (a) a thermostable DNA polymerase (= an enzyme derived from organisms that live in hot water springs, such as the bacterium *Thermophilus aquaticus*); (b) a DNA template with two oligonucleotide primers which are able to anneal this template, (c) free nucleotides and (d) the proper salts. One also needs a machine (a 'PCR machine') that can rapidly generate cycles of changing temperature conditions. Under the right conditions, this results in an exponential amplification of the original DNA template. The entire amplification process is performed in a test tube in roughly 20 - 100 microliters of reaction mixture.

If molecular biology is not the most promising approach to engineering life, are there any other more promising approaches? The answer is positive: man has already tried two approaches to creating new life: one is the making of artificial membrane vesicles; the other is 'computer-life'.

Membrane physiologists

Artificial membrane vesicles.

In my opinion, membrane physiologists working with artificial membrane vesicle preparations are the ones that have the tools to generate a very primitive form of life, as defined by biologists.

When we dealt with the question of when the first cell came into being (3.19), we concluded that life could not have come into existence earlier than when some compartmentalized aggregate acquired the ability to build up an ionic gradient over its limiting membrane and to communicate. This topic will be treated in more detail in Chapter 12. This communication required the presence of ion pumps and channels in the membrane. Today, with modern technology, this situation is already being mimicked to some extent. It is not that difficult to prepare artificial membrane vesicles: some appropriate lipids, a watery solution and a powerful homogenizer is about all that is required. Ion transporting membrane proteins can be obtained using the methods of molecular biology and biochemistry. Such protein molecules have hydrophobic parts, which means that they insert themselves into lipid membranes. In the right ionic environment and in the presence of adequate energy, the ion pumps may start to build up an ionic gradient over the membrane. If one introduced an ion channel with the right properties (selectivity, gating mechanism) into this membrane, (i.e. an ion channel through which the ion that forms the gradient could selectively leak away), then a primitive communication system would be realized. Such a system would use changes in ionic concentrations - and the resulting changes in plasma membrane potential – just as a 'true' cell does. The principles underlying the formation of an ionic voltage gradient over the plasma membrane have been explained in Chapter 9. If one prepared artificial membrane vesicles in a medium containing cytoskeletal proteins and DNA molecules, one would be close to the situation that may have prevailed in the first cell (see Chapter 12).

At the current time biologists are still far from making a functioning and self-replicating artificial cell. I doubt that it makes sense to try it at all; one would never be able to realize communication systems that are faster than the ones that occur in natural biological systems. The reason is this. Fast communication systems use electricity as an energy source. In biological systems *this electricity is carried by ions* (Chapter 9) and this type of transport is relatively slow. For example, the speed at which a nerve impulse moves along an axon is on the order of 100 meters per second, while the *electricity carried by electrons moves at speeds approaching the speed of light*, namely at 300 000 kilometers per second in ideal conductors. Therefore, it would be better to design communication systems and new 'life', which use electricity carried by electrons, rather than by inorganic ions. This is what computer specialists may have already achieved.

Computer specialists

By way of recapitulation: the symbolic notation of life (as an activity) that we arrived at in Chapter 7 was:

$$L(S, t) = \sum_{1}^{j} C(S, t)$$

Where L = Life activity, C = communication action, S = system or compartment, j = the highest level of compartmental organization, 1 to j = a given level of compartmentalization, t = the moment at which the actions of communication are being considered.
When deducing this formula we had 'biological life' in mind, not computers.

No one would consider a computer that has just been taken out of the box but not yet plugged in and turned on as being alive. It is just a machine that is not yet functioning. The distinction between being alive and not being alive is very clear here. As soon as the computer is connected to the electricity grid it is charged with energy produced by man. Thus the computer becomes integrated into the compartment comprising 'Man + his tools' (level 8 of compartmental organization, Fig.4.1). A first requirement for a communication system is thus fulfilled: the receiver must contain stockpiled energy in order to be able to do work. The second requirement is the installation of software in the memory of the computer. A screen, keyboard and mouse are also needed in order to make feedback possible.

Imagine now a computer with software installed in which the electrical energy comes from a solar cell. Such a setup has all the necessary elements of a communication system as outlined in Chapter 3: stockpiled energy, a built-in program to act as a receiver-decoder-amplifier-responder, and the possibility of using energy for carrying out specific types of work.

When a signal is sent via a keyboard, or by means of touching a touch-screen, the system will start working. In principle, such a system is not basically different from the one in which a neurotransmitter binds to its receptor (= the equivalent of typing the right code on the keyboard or of touching a touch-screen) and initiates signal transduction to mobilize previously stockpiled energy for the purpose of doing work. Intuitively everyone will reject the idea that such a computer is alive, although we sometimes use a vocabulary that implies it is living. For instance, we say the computer is 'dead', just as we say that the radio is 'dead', meaning that its communication system has failed.

Do we have such good arguments for claiming that a computer is not alive as we think we have?

Possible arguments are:

1. *"Computers cannot reproduce themselves, thus they are not alive."*

    This is the most common answer I get from my students when I ask them why they think that a computer is not alive. But if we claim that the ability to reproduce is an essential element of being alive, it then follows that a castrated bull or any fully differentiated neuron (which can no longer undergo cell division) is not alive. This is obviously not correct. Reproduction only comes into play when life is considered *over a period longer than one generation*: for time spans not exceeding the lifetime of a compartment, reproduction is *not* an essential parameter of the living state ('short-term' life versus 'long-term' life: De Loof, 1994).

2. *"A computer cannot come into being and remain in existence without the help of humans."*

    This type of argument relates to 'autopoiesis', which is assumed to be a typical characteristic of living entities (see Chapter5). It is true of course that a computer cannot come into existence out of nothing, but we could not have come into existence or stay alive as youngsters either without the help of our parents. A more valid argument is that our parents and we belong to the same (self-perpetuating) species, whereas the computer needs help from outside the 'computer-world'. But some parasites and symbionts cannot reproduce without a host either.

3. *"A computer has no DNA."*

    Red blood cells of mammals eject their nucleus when their cytoplasm gets filled with hemoglobin. Such 'cells' (erythroplastids) have no DNA but may nevertheless continue to live for a few weeks. The egg of an amphibian can be enucleated, thereby losing its DNA (though it retains the DNA present in its mitochondria), and irradiated sperm can fertilize it. Such eggs lack chromosomes but they still undergo a number of cleavages just as normal eggs do (Barth, 1964), though they will never develop beyond the blastula stage.

4. *"Computers use different principles of communication."*

   Although we do not know exactly how our brain works, it is likely that the basics of communication (sender-transmission channel-receiver, etc.) are rather similar in all communication systems. Computers are designed to take over some of the communication in our brain, and therefore it is logical that they should be designed in such a way as to be compatible with the principles of communication that our brain itself uses.

5. *"A computer cannot think, it is not autonomous and it has no emotions."*

   Can an ameba, a red blood cell, a plant or a fungus think? How autonomous are they, and do they have feelings? A computer equipped with a solar cell coupled to a battery for storing the solar energy can be autonomous for a much longer time than an ameba that has difficulties in finding enough food. Specialists in the field of artificial intelligence think that it will not take that long before 'emotional phenomena' will be generated in computers.

6. *"A computer cannot cope with entropy; it will break down after some time."*

   In this respect, a computer does not differ from living systems that age, get sick and finally die.

7. *"A computer does not carry out metabolic reactions."*

   Although this is true, it is not a good argument because the main purpose of metabolic reactions is to mobilize energy for performing work. In our example the computer uses light as its primary energy source while organisms usually use chemical energy. However, this chemical energy, provided in the form of food, is the same as solar light energy that plants transform into chemical energy during photosynthesis.

8. *"Computers are not based on carbon chemistry as true living beings are."*

   True again and a good argument. The basic chemistry of computers is inorganic, as opposed to the organic chemistry of 'biological' life. Computer chemistry is based on silicon and metals, with some organic polymers for insulation and for making the diskettes, etc. Since computers are very recent inventions, there is a possibility that in the future carbon chemistry may perhaps become more important in computer technology. It is not out of the question that some day even DNA will be incorporated for some purpose.

9. *"Computers use a different carrier of electricity."*

   True, and again a good argument. In biological systems, electricity, which is the movement of charges, is carried by inorganic ions, whereas in computers it is carried by electrons. Electrons can move at the speed of light, namely at about 300 000 km per second, while ion-carried electricity is many orders of magnitude slower. This is the major in ideal conductors, reason why computers are so efficient: they can work much faster than our brain, at least for some purposes.

10. *"Computers cannot adapt themselves to a changing environment."*

If one analyzes the success story of computer development, one cannot escape the conclusion that computers seem to adapt 'themselves' - though of course with the help of humans - much faster than organisms do. There can be no doubt that computers evolve, though the individual personal computer does not adapt: it becomes outdated. Research in the field of artificial intelligence shows that under certain circumstances man-made robots adapt themselves and even install a sort of hierarchy among one another without having been programmed to do so. It is not that difficult to make computer programs that allow changes and errors.

11. *"A computer is not alive because the sum of its parts equals the whole."*

This argument alludes to the enormous complexity of cellular structure and biochemistry and to the fact that it is impossible to bring homogenized cells back to life, while a computer can be disassembled into all its constituting parts and reassembled again and function again normally.

The comparison is not fully valid. First, the constituent parts of a computer are very large in size compared to the structures present in cells and can therefore be handled in an easier way than those of cells. If a computer were 'homogenized' as thoroughly as biochemists destroy cells with their laboratory homogenizers, the computer could never be reassembled. Secondly, before one starts to take a computer apart, one can shut off the electrical supply.

There is another counterargument. 'Life' is a property of a compartment and there are compartments that can be disassembled into their constituting units, thereby dying as compartments, but which can be reconstituted again and continue to live normally. It would be easy to 'kill' a small population of animals by catching all the animals and removing all the individuals far away from one another. Once all individuals have lost all forms of contact and communication with one another, and this in an irreversible way, then the population as such would be dead, as already explained in Chapter 6. But if we brought back all individuals after a day or so, then the population would be reconstituted and it would come to life again. It is possible to disassemble a freshwater sponge into its constituting cells by pressing the sponge through gauze or by immersing it in calcium-free water. Next, one can let the cells reaggregate and an intact sponge will result.

12. *"A computer is not irreversibly destroyed when it loses its stockpiled energy completely."*

If for some reason the electricity supply to a computer is cut, so that it loses all its stockpiled energy, it is not irreversibly destroyed. When the supply is restored, the computer becomes operational again. The 'information' that was not yet 'saved' in the memory is likely to be lost (depending on the degree of sophistication of the computer), but this does not mean that the computer is irreversibly destroyed.

If an organism is completely deprived of all its stockpiled chemical and electrical energy, it is dead and there is no means of reanimating it. However, there are situations where 'reanimation' is still possible, even after communicational activity has come to an end in a given compartment due to a decline in energy supply. Take, for example, a population of animals that stops functioning as a population when there is a severe food shortage and the individuals are lying nearly dead and scattered over a wide area. If the animals are once again fed so that they recover enough to restore contact with one another, then the death of the population has been reversed. Here we encounter once again the great difference between electricity that is generated and used by organisms and the electricity that flows out of an electrical outlet.

Conclusion: as long as a computer is supplied with electricity and is functioning, it is not easy to present unequivocal arguments against the view that at least in some sense(s) it is 'alive'.

What is the point of these arguments?

The point is that one should not too quickly and automatically reject the possibility that novel man-made forms of 'life' exist. Furthermore, we should all agree that one such form is already operational, namely 'computer life'. These statements have to be seen in the following context: *'life' is not a machine, but rather the communication activity of a machine*.

The truly relevant question is: "Can a computer perform acts of communication?" Can it decode incoming messages and come up with an energy-demanding response? Is feedback possible? The answers are no doubt affirmative. Indeed, that is what computers were made for. If one agrees that life is communication activity, then one has to accept that computers, because they perform acts of communication, are alive.

Nevertheless the feeling remains that something is wrong with this line of reasoning. This is due to the fact that most people, when confronted with the question as to whether or not a computer is alive, intuitively assume that only something that can exist autonomously, thus as an autopoietic system, can be alive. They assume that if a computer is alive then it should automatically be classified in the same category as a unicellular organism, namely Level 2 of compartmental organization (Fig.4.1). This is obviously not acceptable. In addition, man is tempted to compare the possibilities of a computer with the intellectual and emotional capabilities of human beings. That should not be done either. I have never heard anybody draw a comparison between a computer and a fungus or a plant.

By way of conclusion, the communication activities both of a biological system and of a computer are variants of the same theme. A computer is nothing else than a mechanical extension of a given level of compartmental organization, namely that of animals with their tools (Level 8, 'toolization'), that enables the brain to extend the range of acts of communication it can perform. From this point of view, the mechanical extension can be regarded as being alive.

10.5 Toolization. Self-communication

As outlined in Chapter 4 (Fig. 4. 1, level 8), *toolization* (= tool utilization) is a process by which a compartment enlarges itself *mechanically* by acquiring the ability to use tools.

Several animal species have already achieved this. One of Darwin's finches on the Galapagos Islands uses a cactus spine to collect insects. Beavers are quite skilled tool users. There is even a crow that shapes - or 'manufactures' - a tool until it has the right form and size.

Homo sapiens first used tools to collect food and to facilitate certain types of muscular work. Next he invented tools to perform certain brain functions. The entire human compartment has become populated with mankind on the one hand, and all its tools on the other. Computers are electronic-mechanical extensions of the brain that have been in existence for only a few decades. Nevertheless, they have become so sophisticated that they appear to be alive. A computer with a good speech recognition system is able to transform the spoken word into a written text, almost like a secretary would do. A computer voice sounds like a human voice. One could say that man has created a novel type of 'artificial' or 'man-made' life.

If one thinks about it, man uses his computers to a large extent for *self-communication*. Somebody who talks out loud, asking himself the question, " Did I formulate it well?" and who then answers himself either "Yes" or "No", is engaging in a form of self-communication. In the computer era, the sentence(s) is no longer spoken, but rather typed, and it appears on the screen. The author reads his own text and asks himself: "Did I formulate it well?" Only after the whole document is ready, will it possibly be transmitted to somebody else. From then on the activity becomes person-to-person communication.

The entire *Homo sapiens* compartment (Level 7 in Fig. 4.1) has reached a higher order of organization (toolization: Level 8 in Fig. 4.1) that contains 'organic chemistry-based' life forms, 'artificial' life forms (such as 'mechanical symbionts') and non-living elements (tools). The 'Life activity' produced by this supercompartment is therefore the result of the integration of actions of communication performed both by biological and by artificial (mechanical) components. Communication in our body uses organic chemistry and ion-carried electricity (= *ionic life*), while our electronic-mechanical 'companions' are metal- and silicon-based and their electricity is carried by electrons (= *electronic life*).

10.6 Refining the symbolic notation of 'life as an activity'

To make the distinction between 'truly biological' and 'artificial computer life or electronic life', one can make the symbolic notation of life listed earlier in this chapter more specific as follows:

$$L(S_{(\text{Carbon chemistry-based, ion-borne electric current})}, t) = \sum_{1}^{j} C(S, t)$$

In the case of communication by man-made communication machines such as computers, the notation for this other, artificial form of life would be:

$$L(S_{(\text{Silicon+metal chemistry-based, electron-borne electric current})}, t) = \sum_{1}^{j} C(S, t)$$

The general symbolic notation finally becomes:

$$\boxed{L(S_{(TC, TE)}, t) = \sum_{1}^{j} C(S_{(TC, TE)}, t)}$$

Where **L** = 'Life activity', **S** = a given system or compartment which uses a given **T**ype of **C**hemistry, **TC**, and a given **T**ype of **E**nergy, **TE**, to produce its communication actions **C**. The condition is that $\sum^j C > 0$ and that actions of communication are 'added up' only once.

In his evolution, *Homo sapiens* has reached the point where his coordination center, his physiological brain, has arrived at a high point of development – but this high 'point' turns out to be a 'plateau'. The brain's mutation-based evolution rate can no longer keep up either quantitatively (rapid increase in rates) or qualitatively (variability) with man's increasing and expanding information-processing capacities. Hence another (explosive) dynamic has taken over. It relates to the spectacular geometric rise in human development, for example, over the past few thousand years (technology: stone age → industrial age; Communication: oral→ written → paper → printing press → computer); over the past 500 years (European culture: 'Middle Ages', Renaissance, Reformation, world exploration, colonialization, 'Americanization', 'globalization', etc.); and over the past 50 years (vacuum tube, transistor, silicon chip, the 'computer revolution', the PC, the internet, etc.).

10.7 Electronic life and tool-aided (mechanical) reproduction

'Ionic life' is the only form of life so far that has achieved the capacity – or potency – to perpetuate itself by forming daughter compartments without the help of individuals from another species, (with the exception of certain parasites). The goal of any reproductive system is to perpetuate communication systems and information over more than one generation. The two 'classical' systems of reproduction are sexual and asexual reproduction.

By creating 'electronic life', man introduced a third type, namely *tool-aided* or *mechanical* reproduction.

In the past, species could only perpetuate their information and communication systems by means of ('classical') reproduction, and, in some species, by learning from other members of the population/species. Through the invention of certain man-made tools (books, audio tapes, floppy disks, films, etc.), a novel way has been introduced of transferring information to successive generations without needing to reproduce the compartment that produced the information. This is the second aspect of 'Life as a double continuum'. Less transmission errors are likely to be made in the process of copying the information onto a diskette than in the process of generating an offspring by sexual reproduction. Whether we like the idea or not, the fact is that some types of work can be carried out very efficiently by 'artificial' life forms and/or electronic-mechanical tools. This holds true, as well, for the transfer of some types of information from one generation to the next. There is an irony in the process of mega-evolution: the solution of problems at one level often lead inevitably to other problems at a higher level. The industrial revolution, capitalism and the silicon revolution have literally 'lightened the load' from man's back, but at the price of unemployment – or redundancy – in the deepest sense of the word. Just as the vast majority of eggs and sperm cells of all species are 'redundant', so also vast numbers of individuals (regionally) and even entire population groups (globally) have become 'redundant' in terms of the self-perpetuating ('reproductive') impulse of the current mega-system.

10.8   The future?

Prions are proteins, and viruses are complexes of nucleic acids and proteins. Neither the prions nor the viruses can be considered to be alive because they do not have all the elements that are required for functioning as a communication system.

It is interesting to pause and reflect on the reasons why so many people react with hostility towards the idea that a computer could be alive. Hostile reactions often emerge in threatening situations. The computer is perceived to be a threat to humans from the moment it acquires too many properties that are considered to be typically human. In the epoch of industrialization, the machines that took over muscular work were not at all welcomed by the workers. At the current time, such machines are no longer seen as a threat, but rather as a blessing. We could hardly live without them. Computers are now having the same effect in the realm of intellectual work. The fear of losing one's job is real for many people, though for others, the computer is creating new opportunities. Some people are predicting that only a few decades from now the World Wide Web will be the true decision-maker on Planet Earth. In this scenario, we are moving towards Cyber-Dictatorship and the human race is destined to slavery in the service of a Supercomputer.

We are only at the beginning of the development of 'artificial life'. In a few more decades, centuries or millennia, the distinction between 'computer life' and 'real life' may fade into a vague memory.

ESSENTIALS

OTHER LIFE FORMS

1. One commonly held view is that viruses originated as pieces of genetic information that somehow became detached from a regular chromosome in an organism and then escaped elimination by degrading enzymes.
2. The majority of viruses use the regular DNA language for their genetic code. Others, called retroviruses, use the RNA language.
3. Viruses cannot duplicate themselves. They depend on the replication machinery of the host cell they infect. Hence, viruses cannot be grown in culture media as is routinely done with bacteria.
4. Viruses are not alive, because by themselves they cannot decode messages and generate energy-requiring answers to them.
5. Prions are proteins with important regulatory functions in eukaryotic cells. In the brain of some mammals, some prions occasionally acquire a default spatial conformation and become infectious. Such crippled brain proteins are able to transmit their abnormal spatial configuration to other normally formed prion proteins, thereby causing a chain of reactions that can result in lethal brain diseases such as the devastating cattle disease BSE (bovine spongiform encephalitis). Protein molecules are obviously not alive.
6. Man invented tools to help himself solve all kinds of problems. Computers are the most advanced tools invented thus far. They are mechanical extensions of the human brain. They do not come into existence – and do not continue to exist - by themselves, but this is no reason to deny them the status of being alive. Computers can decode incoming messages and (depending on the properties of the preloaded software) generate energy-demanding responses. Such activity can be regarded as 'life activity'. It is a form of man-made artificial life that makes use of another type of chemistry and electricity than our brain.
7. The use of a personal computer is a form of self-communication.

CHAPTER 11

NO LIFE WITHOUT TIME

BUT TIME:
WHAT EXACTLY IS IT?

No life without time

Contents

11.1 Introduction
11.2 Time, space, space-time: an ultra-brief historical overview
11.3 The key questions
11.4 What do we mean when we say 'duration'? Parameters influencing 'duration'
11.5 A definition of time
11.6 Consequences of the definition
11.7 How to measure time? Time and motion
11.8 The speed of light
11.9 Bringing order into the multitude of different 'times'
11.10 The arrow of time
11.11 Conclusions: time as a special kind of 'elasticity coefficient'?

Essentials

11.1 Introduction

As I was writing this book, a colleague made the remark that a definition of 'life' can only acquire its full meaning if all elements of the definition are adequately explained. He argued that because my definition contains 'time', the nature of time should be explained. I must confess that I had not thought much about 'time' before. As most people do, I assumed that unraveling the nature of time was the task of astronomers, physicists and, perhaps, of philosophers as well. I also assumed that somebody should already have given a plausible definition of time to which I could refer. It turned out otherwise. The vast majority of textbooks on physics for undergraduates and even recent dictionaries on commonly used terms in physics (e.g. Chapple, 1999) keep silent about the nature of time, although t is used again and again in formulas. A search of the literature yielded several specialized books and research papers on the topic of 'time', but not the definition I had hoped to find. The non-interest of contemporary physics textbooks for the nature of time contrasts sharply with the over-all importance of 'time' in daily life.

In the course of the past decade I have had many discussions with colleagues, mainly in physics, about the nature of time and about my view of it. These discussions were seldom easy, and this was for a variety of reasons. Theoretical physicists and biologists seem to have a different perception of 'time'. Physicists need *time* to do their calculations. Biologists are more interested in the nature of the type of *time* that creatures of flesh and blood experience in daily life. This type of time was called 'vulgar time' by Newton (see below), who pointed to the fact that it does not seem to be constant. Physicists look at time on the much grander scale of the evolution of the universe. On such a scale, fluctuations of time, if they exist at all, can be neglected and time becomes an almost abstract concept. Newton referred to this type of time as 'absolute time' (see below). Such being the case, physicists are right when they say that there is hardly any need to understand the nature of time: whether we understand it or not makes no difference when we are using time in calculations.

In a book that is a compilation of a series of lectures given by Stephen Hawking and Roger Penrose on the nature of Space and Time (1996), Hawking says:
"He [i.e. Roger Penrose] feels that can't correspond to reality. But that doesn't bother me. I don't demand that a theory correspond to reality because I don't know what it is. Reality is not a quality you can test with litmus paper. All I'm concerned with is that the theory should predict the results of measurements. Quantum theory does this very successfully….".

Yet I have experienced that many people are puzzled by the unsolved question(s) of the nature of time. Did time come into existence with the big bang or was there already time before this event? Is time a given a priori? Is time an intrinsic property of matter? Of energy? Of space? Is time a property of some sort of system(s)? Is time inherent to energy conversions? Does each type of elementary particle have its own time? Is the nature of 'common' time different from that of 'absolute' time? Is time always linked to change or to motion? Although motion is undoubtedly a good means of quantifying time, does it provide insight into its nature? Can time change in the course of the development of the universe? Why does 'life' include a dimension of 'time'? Is an observer needed for time to exist or to make sense?

Numerous eminent thinkers have attempted to formulate definitions of space and/or time, such as Descartes, Mach, Bradley, Bergson, Russell, Minkowski, Einstein, Grünbaum, Prigogine, Penrose, Hawking, etc. Most of the key ideas are summarized in the books of Whitrow (1959), Smart (1964), Sklar (1974), Hawking (1996), and Hawking and Penrose (1996), where more detailed references can be found and from which most of the ideas in the following overview have been distilled.

11.2   Time, space, space-time: an ultra-brief historical overview

Concerning time:

One can imagine that the first humans were much less concerned about time than we are in our highly technological society. For them, time probably consisted of loose fragments, for example from sunrise till noon, the time needed to roast a piece of meat, to collect water, etc.

The ancient Greek philosophers were more intrigued by the concept of 'space' than by the concept of 'time'. One can imagine that, because it is so practical, the preferred method for measuring time has always been to observe the motion of something. Archyas of Tarentum, a contemporary of Plato, stated: "Time is the number of a certain motion and, in its general meaning, it is the interval of the natural order of the universe".

St. Augustine, attempting to define time, wrote in his *Confessions*:
"I once heard a learned man say that the motions of the sun, moon, and stars constituted time; and I did not agree....For what is time? Who can easily and briefly explain it? Who can even comprehend it in thought or put the answer into words? Yet is it not true that in conversation we refer to nothing more familiarly or knowingly than time? And surely we understand it when we speak of it; we understand it also when we hear another speak of it. What, then, is time? If no one asks me, I know what it is. If I wish to explain it to him who asks me, I do not know. Yet I say with confidence that I know that if nothing passed away, there would be no past time: and if nothing were still coming, there would be no future time; and if there were nothing at all, there would be no present time. But, then, how is it that there are the two times past and future, when even the past is now no longer and the future is now not yet? But if the present were always present, and did not pass into the past time, it obviously would not be time but eternity."

In his *Principia* (1687), Newton made the distinction between 'absolute' and 'vulgar' time as follows:
"Absolute, true and mathematical time, of itself, and from its own nature, *flows* equally with regard to anything external, and by another name is called duration: relative, apparent and common time is some sensible and external (whether accurate or unequable) measure of duration by means of *motion*, which is commonly used instead of true time; such as an hour, a day, a month or a year.
Absolute time, in astronomy, is distinguished from relative by the equation or correction of vulgar time. For the natural days are truly unequal, though they are commonly considered as equal, and used for a measure of time; astronomers correct this inequality for their more accurate deducing of the celestial motions. It may be that there is no such thing as an equable motion, whereby time may be accurately meas-

ured. All motions may be accelerated and retarded; but the true or equable progress of absolute time is liable to no change. The duration or perseverance of the existence of things remains the same, whether the motions are swift or slow, or none at all: and therefore it ought to be distinguished from what are only sensible measures thereof..."

Kant:
"Time is therefore, given a priori. In it alone is actuality of appearances possible at all. Appearances may, one and all, vanish; but time (as the universal condition of their possibility) cannot itself be removed."

Locke:
"Duration is fleeting extension. There is another sort of distance or length, the idea whereof we get not from permanent parts of space, but from the fleeting and perpetually perishing parts of succession: this we call 'duration', the single modes whereof are any different lengths of it whereof we have distinct ideas, as hours, days, years, time, and eternity... Time is duration set out by measures....This consideration of duration, as set out by certain periods, and marked by certain measures or epochs, is that, I think, which most properly we call 'time'."

Leibnitz: "Time is an order of successions."

Stephen Hawking, in the first chapter of his book, *A Brief History of Time* (1996 edition): "What do we know about the universe, and how do we know it? Where did the universe come from, and where is it going? Did the universe have a beginning, and if so, what happened before then? What is the nature of time? Will it ever come to an end? Recent breakthroughs in physics, made possible in part by fantastic new technologies, suggest answers to some of these long-standing questions. Someday the answers may seem as obvious to us as the earth orbiting the sun... Only time (whatever that may be) will tell."

In *The nature of space and time* (Hawking and Penrose, 1996), Hawking says that gravity 'curls' space-time so that it has a beginning and an end.

According to Prigogine (1980), time has always been. The big bang was only a phase-transition. Time is what glues the universe together.

Concerning space and space-time

In daily life, space and time are usually considered separately. One ordinarily thinks of objects as being 'in' space and events 'in' time. Early this century, however, Minkowski united 'space' and 'time' into the 'space-time' concept, which followed new developments in thinking from earlier centuries, for example by Newton and Leibnitz.

Newton made a distinction between absolute and relative space, just as he did with time.
"Absolute space, in its own nature, without relation to anything external, remains always similar and immovable. Relative space is some movable dimension or measure of the absolute spaces; which our senses determine by its position to bodies,...."

Leibnitz put it this way:
> "I hold space to be something merely relative, as time is; that is, I hold it to be an order of coexistences as time is an order of successions. For space denotes, in terms of possibility, an order of things which exist at the same time, considered as existing together, without inquiring into their manner of existing."

Minkowski:
> "Previously space was considered as being a set of points. But for spaces that are supposed to be more than mathematical abstractions, which are actually, somehow or other, to represent the world, what should we choose as points? ...We will take an idealized event as 'marking' a definite location in space-time. The points of space-time, then, will be all the locations of possible ideal events. Since the events are extensionless, so are their locations.... Henceforth space by itself, and time by itself, are doomed to fade away into mere shadows, and only a kind of union of the two will preserve an independent reality."

It is unimportant whether or not these events exist in the real world. Thus, the existence of space-time is postulated to be independent of the existence of any material objects.

Einstein:
> Time is relative. Newton's absolute clock does not exist. General Relativity is a theory postulating curved space-time. No need for a 'luminiferous ether' (see below).

Brane-worlds:
> According to Horava and Witten (1996), space-time has no less than eleven dimensions, namely ten spatial dimensions plus time. Six out of the ten space dimensions are beyond our experience.

Recent edition of Dictionary *of Physics* (Chapple, 1999):
> "Time is a base physical quantity in SI (= Système International d'Unités), symbol t. The SI unit is the second(s)."

'Time' as perceived in daily life

Notwithstanding the history of philosophical speculation and scientific theories (which often raise more questions than they answer), in our daily lives we intuitively sense that time:
1) flows continuously and equally;
2) forms the background for events (together with space);
3) is universal and the same everywhere;
4) is absolute and not influenced by physical phenomena (Dupré, 1997).

But, as Dupré stated, intuition may mislead.

Why is it so difficult to define time? Does the human mind lack the capacity to understand such a supposedly mysterious phenomenon? Are we overlooking an essential causal link? Are we misinterpreting certain facts? Or have we not even yet asked the right questions?

In an effort to gain more insight into the nature of time, in particular of 'vulgar' or 'common' time (Newton's terms), I propose approaching the problem by means of the same 'lateral thinking' that I used in earlier chapters to deduce the nature of 'life'. I first asked: is 'life' an abstract term or is it a property of some sort of system/compartment? Next I asked: of what type of compartment(s)? Next: what is the common denominator of the 'life property' in all the different types of compartments to which 'being alive' is applicable? Let us now replace 'life' by 'time' and 'being alive' by 'duration', and see if something meaningful can be distilled out of this new combination.

11.3   The key questions

"Is 'time' a fully abstract term or is it a property of 'something'"?

If the term denotes something completely abstract – a given a priori, as Kant thought, not linked to any system – then it does not make sense to try to formulate a plausible definition of time. Indeed, one can predict that it is a priori impossible to reach a consensus about its meaning because there is no conceptual framework to refer to, just as it does not make sense to discuss the gender of angels.

However, if time is a property of 'something', then one should find out what that 'something' could be. The following question may help us to find some answers:

"What can have 'duration': An object? A process? Something else?"

It is evident that a given *process*, which is a general term for a (natural series of) *change(s)* in a given system, always requires time or duration. However, it is less obvious that 'duration' could have meaning in relation to an object as such, in which no change whatsoever occurs. Our intuition links time/duration to change. But to what type of change? Some insight may be gained from the following example with a candle (Fig. 11.1).

11.4   What do we mean when we say 'duration'? Parameters influencing 'duration'

If we start by linking 'duration' to a process, the key question then becomes, "Why does one process have a longer or shorter duration than another?"

Imagine that we have two identical oblong paraffin candles with a mass of 100 grams each. If these candles could be stored at absolute zero temperature (-273°C), what would be their duration? The candles will last as long as the –273°C condition persists, thus infinitely. Intuitively, we say that only absolute time could apply here. One could also say that time/duration has no meaning under such conditions. If we keep the candles at room temperature, what is their duration then? Because we could hardly observe any change in a lifetime, we would say that the candles will last very long, perhaps forever. Eventually, the paraffin of the candles will very slowly evaporate and will be oxidized by the oxygen present in the air. Thus 'very long' is no longer 'infinitely' long.

Now one of the candles is lit. What is the significance of duration/time now? Let us assume the candle burns for a duration of 60 minutes. Time now has got a meaning and can be quantified if we have a standard (clock) to refer to.

Imagine that the second candle is cut into five pieces of 20 grams. What does duration/time 'mean' in this situation? Both at –273°C and at room temperature, there is no difference between the cut and the uncut candle. Next, the five pieces are simultaneously lit. They burn for let us say only 12 minutes, although the same amount of paraffin is involved as in the uncut candle.

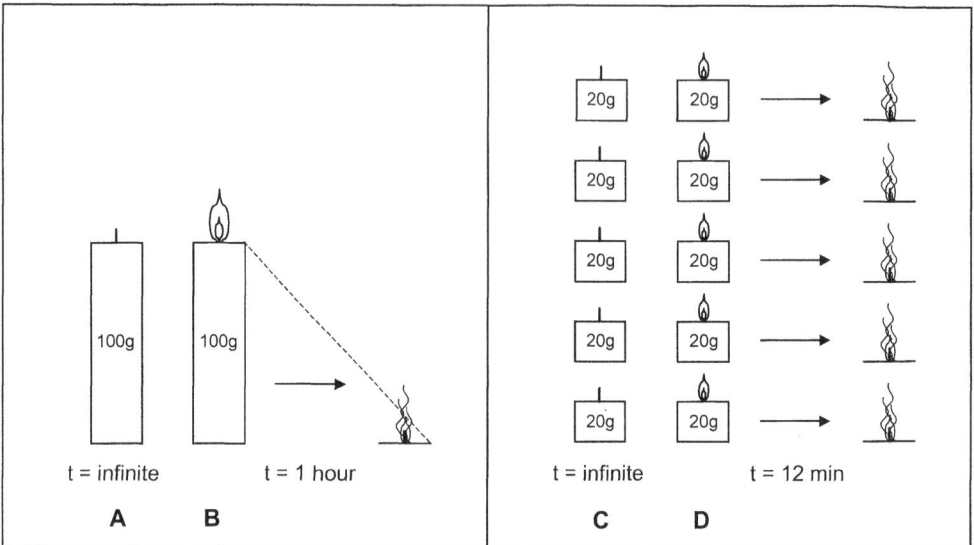

Figure 11.1 Illustration of the fact that in energy-converting systems not only the total amount of potential energy but also the form of the system matters with respect to 'time'. (A) The spontaneous oxidation of the paraffin of the candle by the O_2 present in the air is so slow that we do not observe any change. One has the impression that 'time' is infinite in this system, or that the term 'time' is meaningless here. (B) Our impression of time does not alter when the candle is cut into five pieces of 20 grams each. (C) Upon lighting the candle, we get the impression that 'time' begins to take on meaning: for example 1 hour, the 'time' it takes for all the chemical energy present in the paraffin to be converted. (D) Although the total amount of energy present in the five pieces, each weighing 20 grams, is not different from system (A), when all five are lit simultaneously, 'the burning 'time' that then elapses has a much lower value, namely approximately 12 minutes. Thus, changing the form of an energy-converting system influences 'time'.

This experiment tells us that in the system we described, duration/ time begins to have meaning only from the moment an observable energy conversion has been initiated within it. The form of the system is also important. The form is the only difference between the intact candle and the 5 pieces, not the amount of convertible energy present. Thus, two systems containing the same amount of convertible energy and in which the same chemical reactions take place can be linked to two different durations.

In my opinion, this approach can be applied to any type of energy converting system, for example the time it takes to heat an iron to a given temperature, to transport a car or a photon over a given trajectory, to produce electricity starting from a given amount of chemical energy stored in coal, to make the hands of a watch move a number of degrees, etc.

In the example of the candle at –273°C, duration lost its significance. This indicates that duration/time is not a property of an object as such, at least not at the macroscopic level.

Because my knowledge of the physics of elementary particles is very limited, I cannot address the question as to whether or not each type of particle has its own time dimension.

11.5 A definition of time

The definition which will be proposed makes use of the term 'inertia', which is well known from physics. An object at rest will not start moving by itself, and an object in motion will not spontaneously stop moving. Objects 'resist being disturbed'. The term 'inertia' denotes the ability of a system to resist being disturbed (*Henderson's Dictionary of Biological Terms*, 1995).

I think that what we call 'time' is invariably a property of an energy converting system. Time is the measure of the system's inertia (= resistance to being disturbed) with respect to the conversion of a given amount(s) of some form(s) of energy contained in a given system into another energy form(s). Thus, time is a measure of energy conversion inertia, i.e. of the 'laziness' by which a given system converts its energy. Its relative basic unit is the second.

The reader may now argue that 'energy' should be defined. The following definition is taken from *The Feynman Lectures on Physics* (1966).
   "Energy has a large number of different forms: gravitational energy, kinetic energy, heat energy, elastic energy, electrical energy, chemical energy, radiant energy, nuclear energy, mass energy. Conservation of energy: It is stated that there is a certain quantity, which we call energy, that does not change in the manifold changes which nature undergoes. That is a most abstract idea because it is a mathematical principle. It says that there is a numerical quantity which does not change when something happens."
Both stars and biological systems are experts in energy conversions.

11. 6 Consequences of the definition

Time has no meaning (it simply does not exist) for systems in which no energy conversion takes place. From this it follows that Newton's 'absolute time' does not exist and that there are as many different forms of duration - or time - as there are energy converting systems in the universe: i.e. infinitely many. In a universe without 'pockets' of energy conversion, time would not exist. It also follows that time is not necessarily constant in systems in which the way energy is converted can change. For example, a match will burn fast (a low number for inertia or duration) at the moment it is lit and then it will gradually burn slower.
Why do we nevertheless have the impression that there is an absolute time, which forms the fourth dimension of the universe and the background of all events we can observe? Because we are dealing here with the basic energy conversions on the scale of the universe. Here, changes in conversion rate on the scale of 'common time' are so extremely small that they can hardly be observed.

If we only consider time/duration on the scale of the sun, which is only a very tiny part of all energy-converting systems in the universe, one can understand why our intuition makes us assume that absolute time exists. The basic chemical reaction that takes place in the sun is the fusion of hydrogen into helium.

$$H_2 + H_2 \rightarrow {}^4He$$

As outlined in Chapter 1, a number of other elements up to atomic number 16 (oxygen) can be produced in stars the size of the sun through further fusion reactions. Our sun contains enough hydrogen to burn for about 10 billion years. Humans live for such a short time that during their lifetime they cannot observe any change in the burning of the sun. Hence they have the impression that the sun and the stars are there forever. This is the basis for Newton's 'absolute time'. In my terminology, absolute time is a measure of the inertia in the total sum of all energy conversions in the universe as a whole. The fact that about 12.5 billion years after the big bang (Sneden, 2001) the universe still consists for 98 percent of hydrogen illustrates well the inertia in energy conversion on this scale.

'Common ordinary' time is relevant to biological systems. Here, the basic chemical reaction is the conversion of glucose. This molecule, which is formed during photosynthesis, stores solar light energy in the form of chemical energy. During respiration in aerobic organisms, glucose (=$C_6H_{12}O_6$), in the presence of O_2, is converted to CO_2 and H_2O, a process in which part of the liberated chemical energy is stored in ATP (adenosine triphosphate) and part is lost as heat.

$$C_6H_{12}O_6 + 6\ O_2 \rightarrow 6\ CO_2 + 6\ H_2O + energy$$

Just as there are no two stars with exactly the same starting volume of hydrogen, so there are also no two biological systems with exactly the same conditions for their glucose conversion.

11.7 How to measure time? Time and motion

The most common changes we see around us have to do with motion. From ancient times, humans have used the apparent motion of the sun relative to the earth (or vice versa) to measure time. If nothing moves around us, we have the impression that time is standing still. The more and faster the movements we observe, the faster time seems to proceed. No wonder that motion is by far the preferential way for humans to measure time and that and that theoretical thinking on time has for the most part focused on motion.

My personal experience is that a lot of confusion about the nature of time, mainly in discussions with non-biologists, stems from the fact that we easily overlook the fact that the observer himself is an energy converting system. In addition, the principle of relativity comes into play.

A moving object does not necessarily convert energy. It is possible that a moving object is only reflecting light that was produced from a distant energy converting system. Does this imply that the definition of time I proposed is incorrect? Not necessarily. When one uses motion to measure time, one introduces an observer of flesh and blood. When this observer uses the motion of an object to measure time, he uses the conversion of a given amount of energy in his sense organs to quantify duration, not the motion of the object as such. This may require some further explanation because we are now dealing with a principle of physiology.

One should be well aware of the phenomena occurring in our eyes and brain when we are asked to answer the question, "Do we see motion or do we see light?". Although everyone is convinced that he or she can see motion, this is not the case. We do not 'see' motion; we see only light. It is our brain which makes us believe that we see motion. The explanation of this fact lies in the physiology of vision. The light-sensitive elements in our retina are the rods and cones. Following activation by incoming photons, electrical phenomena occur that are propagated into the visual cortex of the brain. An individual cone or rod can only transmit the signal 'light present or light absent'. It cannot detect motion. Thus, if our retina were to contain only one rod, then we would not 'see' motion, we would only discriminate between: light off – light on. But our retina contains millions of rods and cones. When the relative positions of the object and its projected image in relation to the observer and his retina start changing, the incoming photons start to activate other sets of rods and cones. These rods and cones then send their messages to the visual cortex of the brain, which interprets this displacement of the projected image along the retina as movement either of the object or of the observer.

A passenger in an airplane provides a simple illustration of the relativity principle: as long as the airplane's speed remains constant, he experiences no motion. For him, there is no relationship between the movement of the airplane and duration. Yet an observer on the ground can use the movement of the plane to measure time. The difference in perception of duration between the two situations is that the motion does not cause any energy conversion in the sense organs of the passenger, while it does for the observer on the ground.

In conclusion, although motion is a good means for an observer to quantify time, it is not the best model for uncovering the nature of time. Furthermore, the act of seeing motion involves a double energy conversion: one to generate the light that is either produced in or reflected by the moving object, and another that is electrical and chemical in nature and occurs in the brain of the observer.

11.8 The speed of light

A nearly ideal standard for comparing different times is the movement of light particles (photons) because the speed of photons in a vacuum is always 300 000 km/sec, irrespective of the wavelength of the light or the relative speed of the observer in relation to the light source. This means that an observer will measure the same speed of incoming light when sitting in a plane moving towards the light source as when moving away from it. This unexpected property of light was first detected by Michelson and Morley.

The behavior of light raises a few interesting questions. First, why is the speed of light limited to maximally 300 000 km/sec in a vacuum? Nobody knows. One explanation could be that it is an intrinsic property of any electromagnetic wave that they have a built-in maximal speed limit. Another possible explanation is that even in a vacuum, moving photons are hindered by something that is unobservable by our sense organs and contemporary measuring equipment. In the 19th and early 20th century, this 'strange phase' was called 'ether', a term referring to the hypothetical carrier of electromagnetic waves. Later on in the 20th century, the 'ether concept' was abandoned. Light can be slowed down by some media. Compared to its speed in a vacuum, the speed of light in air is 299 900 km/sec, in water it is 226 000 km/sec, and in glass it is reduced to 197 000 km/sec.

In addition to the limitation of its maximal speed, light also displays another unusual property. It does not obey the 'addition rule'. This rule says that speeds of a given object have to be added up. For example, for an observer standing in a train station, a passenger walking at a speed of 3 km/hr towards the front of a train that is itself moving at a speed of 100 km/hr, is himself moving at a total speed of 100 + 3 = 103 km/hr. But the speed of light (particles) in a vacuum is always 300 000 km/hr, irrespective of whether the light source that is emitting the light is standing still or is moving either away from or towards the observer. Nobody knows why.

In my lectures, I use the following illustration when trying to explain the idea of hypothetical hindrance in a vacuum by an as yet unknown mechanism. It should be emphasized that at present there is no experimental evidence whatsoever for this hindrance. Imagine that a train passes through a train station at a speed of 50 km/hr (= 14 m/sec) and that a passenger throws a handful of confetti through the window towards the platform. What does an observer on the platform see? A trail of papers whirling down on the platform at a speed which is much lower than that of the train, let us say 0.5 m/sec. Imagine now that a second train passes through the same station at a speed of 100 km/hr and that again a passenger throws a handful of confetti towards the platform. The observer now sees a longer tail of confetti whirling down on the platform, though at exactly the same speed as from the first train, namely 0.5 m/sec. Strictly speaking, the addition rule should have applied but apparently it didn't. Why not? Because the addition rule operates only for a short while after the confetti is thrown. Immediately, a second mechanism comes into play: the small bits of paper are slowed down in their movement by friction with the air. Thus the observer sees the papers whirling down at the same speed, independent of the speed of the trains. If the observer did not know what a gas phase (the air) is, he would not be able to explain the discrepancy between the actual speed of the confetti he observes and what the laws of physics predict its speed should be in a vacuum.

How to translate this to the situation with light? Light likewise consists of particles: photons are the equivalent of the bits of confetti. One could imagine that the photons would 'like' to displace themselves at an infinite speed if they got the chance to do so, just as a salmon might perhaps wish to swim around the world in one minute. The fish is unable to do so because even supposing it could generate sufficient energy to swim around the world in a minute (in a hypothetical vacuum), in the real world the resistance of the water would be too great. Perhaps, one could imagine that the resistance of something as yet unknown (formerly referred to as 'ether') exerts in one way or another such a high resistance that the speed of light can never exceed 300 000 km/sec. Again, the idea of a 'luminiferous ether' had been abandoned since Einstein (Bartusiak, 1955). As a result of the tremendous progress in elementary particle physics, the idea that the interstellar 'vacuum' contains 'something' is gaining more and more credibility again. However, the mystery remains.

Some physicists now claim that the speed of light is not necessarily constant. They theorize that under early big bang conditions the speed was higher. From spectral analysis of the light at the boundaries of the universe, there should be some evidence in favor of such a view. They hypothesize that this change in speed could help to explain some remaining mysteries in the development of the universe. However, if the dogma of the constant speed of light is being called into question, perhaps one day somebody will raise the question as to whether 'time' itself could have changed in the course of the evolution of the universe.

Perhaps on the day the problem of the missing mass in the universe is solved (see Chapter 1: dark energy), the cause of the strange behavior of light will be cleared up as well.

11. 9   Bringing order into the multitude of different 'times'

How can we bring order into the multitude of different forms of time, one for each energy-converting system? By introducing a standard upon which everybody agrees. Here, the practical advantages of measuring time by means of motion come into play, although a moving object is often not a system in which energy conversion takes place.

What we use in daily life as a basic unit for comparing the different forms of inertia in energy converting systems is the second, with its related units (minute and hour), the day (from which the second is derived), and the year. One second can denote the inertia involved in converting a given amount of energy that is stockpiled in the machinery of a watch (e.g. in the form of electricity in a battery or in the form of mechanical energy in a spring) to make the second hand proceed six degrees on a rotation scale of 360°, with a speed that all watchmakers have agreed upon. Initially, watchmakers aligned their clocks to the rotational speed of the earth. Now there is the more precise atomic clock. In this system, one second corresponds to the energy conversion brought about by 9,192,631,770 vibrations of a cesium atom in appropriate measuring equipment. The cesium clock, in turn, is being replaced by the mercury clock, which is ten times more accurate (1 second wrong per 2 billion years). One second can also be expressed as the inertia involved in the production of a given amount of chemical/electrical energy in our eyes by a bundle of photons 299 900 km long (corresponding to the speed of light in the air). There are endlessly many more possibilities for attaining an inertia factor equal to 1 second.

11. 10   The arrow of time

If time is a property of a process, then time has an arrow in every system in which a given form of energy can only be unidirectionally converted, in that particular system, into another form(s). The first law of thermodynamics says that the different forms of energy are convertible without energy being lost or gained. This law does not say that this conversion can take place in just any system or at just any speed. Energy stored in coal can be converted into electricity and heat, but the combination of heat and electricity does not yield coal. Entropy comes into play. The example of an ideal ever-swinging pendulum, which is often used as an example in which time has no arrow, is misleading. First of all, a pendulum does not start moving by itself. To start it moving, one has to bring it out of equilibrium, and this requires an input of energy. Secondly, in real life, any pendulum experiences friction and it will eventually come to a standstill. Thus a moving pendulum is an energy converting system and time has an arrow in it. Particle decay studies also reveal that time has an arrow (Kestenbaum, 1998)

11.11 Conclusions: time as special kind of 'elasticity coefficient'?

Although my approach makes it possible to provide a definition for 'common' time, not all the mystery is solved. The mysterious thing about time is why the transformation of one form of energy into another cannot be achieved with an infinitely low inertia or, in other words, in an infinitely short span of time. Yet, thanks to this inertia, the first energy conversion, which happened at the big bang, did not instantly yield a universal form of final energy. Instead, a cascade of energy conversions was started, each with its own inertia. One could say that we live in a universe which "prefers being lazy over being tired" and in which "not too fast!" is the message. Thanks to this fact, the universe and living organisms can continue to exist for some time to come. Time can be regarded as the equivalent, though now in energy converting systems, of the resistance against being stretched displayed by materials when they are subjected to mechanical stretching forces. In other words, time is a special kind of 'elasticity coefficient'.

Whether time will ever come to an end is unclear. If time can die, it will be at the moment when the last energy conversion in the whole universe irreversibly comes to an end. I do not see the conditions which could make this a plausible scenario in a universe in which the process of expansion is still accelerating.

ESSENTIALS

TIME

1. Over the centuries, numerous attempts have been undertaken to formulate a plausible definition of 'time', but all in vain. Not long ago, the famous Oxford astrophysicist Stephen Hawking spoke in terms of "Time, whatever that may be...".
2. Physicists use time as a parameter in their formulas and calculations. They assume that time is constant. In such a context, the nature of time is unimportant. In daily life, however, time is experienced as something real but not tangible and not necessarily constant.
3. It is not possible to come up with a definition of 'time' that everybody can agree on unless 'time' is a property of some sort of 'system'. How can we find an umbrella term to denote all the systems in which 'time' has a meaning?
4. Motion is a good means for quantifying time or duration, but it is not a good basis for understanding its nature.
5. The example of the candles of different lengths, whether burning or not, illustrates that in daily life 'time' is experienced as a property of an energy-converting system. Both the amount of energy present and the form of the energy-converting system influence the 'time' or 'duration'.
6. Time as experienced in daily life could be defined as "the inertia by which in a given energy-converting system, one form(s) of energy is converted into another." 'Time' is for energy converting systems what 'elasticity' is for systems that are subjected to stretching. In this view, there are as many different forms of time as there are energy converting systems, and thus infinitely many. This necessitates the adoption of standard units upon which everyone agrees. These units are the second, the minute, the hour, etc. In a battery-driven watch a second is a measure of the inertia by which a given amount of electrical energy stored in the battery is converted into another form(s) to make the seconds digit proceed 6 degrees of a full circle.
7. The basic chemical reaction underlying 'common, daily time' is the 'burning' of glucose into CO_2 + H_2O + Adenosine Triphosphate (ATP). The basic chemical reaction of interest to astronomers is the 'burning' of hydrogen in stars (= clouds of gas). The inertia in the conversion (by fusion reactions) of a large cloud of hydrogen and helium gas into other elements is so huge that this approximates 'absolute, constant time'.
8. If this view is correct, then time came into being when the very first energy conversion took place at the moment of the big bang. In the universe, time will exist until the very last energy converting reaction takes place, it that ever happens.
9. Why the speed of light, which is a good parameter for standardizing time, is limited to 300,000 km/sec is not yet understood. If interstellar space is not empty but rather filled with some substance of an as yet unknown nature (a concept that seems to be gaining acceptance again after having been rejected before), the hindering properties of the unknown material may perhaps help to explain the upper limit (cf. the train confetti example in this chapter).
10. Time has an 'arrow' because energy-converting reactions are usually uni-directional.
11. Life was able to come into existence because energy-converting reactions do not proceed with an infinitely low inertia. One could say that life is possible because

CHAPTER 12

THE BIG HELLO: THE VERY MOMENT THAT LIFE BEGAN

FROM PASSIVE TO ACTIVE UNIVERSE

THE GRADIENT PROVOKED - TRIPLE S PRINCIPLE

Then God said: "Let the membrane compartments now come to life."

Contents

12.1 Life came into being the moment the first communication system started to function
12.2 What was first: the DNA-, the RNA-, or the peptide world?
12.3 The *pregradient* era: the problem of generating "order" in the primordial conditions
12.4 The *cytoskeleton*: as important as nucleic acids for the emergence of the living state?
12.5 The importance of the achievement of the *transmembrane ionic/voltage gradient* and of transcellular ion fluxes
12.6 Requirements for the perpetuation of the first communication system
12.7 The problem of osmosis. The Gradient-Provoked Swelling and Self-Selection Principle (the GP-Triple S Principle) and how it might apply to social and economic compartments
12.8 Conclusions

Essentials

12.1 Life came into being the moment the first communication system started to function

By way of analogy with the 'big bang' or 'big splash' that marked the origin of the physical universe, the term 'big hello' might be appropriate to denote the coming into existence of communication activity, and thus of life.

If the 'death' of a compartment means the irreversible loss of the ability to communicate at its highest level of compartmentalization (conclusion of Chapter 6), it follows that a given compartment starts to live at the moment it acquires the ability to communicate at its highest level of compartmentalization. This deduction paves the way to answering the question: "When did life start on earth?" According to my line of thinking, the answer is simple: "At the moment that some primordial aggregate acquired the ability to communicate at its highest level of compartmental organization, thus when it started to decode incoming messages and to provide an energy-demanding response to them."

It may be worthwhile to have a second look at Fig. 3.1 and 3.2, which depict a communication system and its essential components. In biological systems, communication (life) requires **compartments with selectively 'leaky' boundaries, gradients, prestored energy, messages, a transmission channel,** etc. This means that before the ability to communicate could be acquired by some primordial aggregate, several stringent conditions had to be satisfied. This may have required a long period of *prebiotic evolution and selection*.

12.2 What was first: the DNA-, the RNA-, or the peptide world?

Certainty about the exact nature of prebiotic conditions on earth some 3.7 billion years ago will not easily be gained. Specialists in the field, such as Stanley Miller, argue that the first gases present in the primordial atmosphere could only engage in reducing reactions, not in oxidizing reactions. Furthermore, a variety of organic molecules such as formic acid, some amino acids, etc. might have been formed from CO_2, CH_4, NH_3 and H_2O (Miller and Orgel, 1973; Miller, 1987), on condition that these gases were indeed present (Horgan, 1991).

The living state, as we know it today, requires large macromolecules such as DNA, RNA and, in particular, proteins. Hence, it is not surprising that the first theories on the origin of life on Earth postulated a protracted origin by the self-assembly of high molecular weight structures, such as RNA, proteins and vesicles, in a cold prebiotic broth of preaccumulated modules (Fry, 2000; Wächtershauser, 2000).

At present, we live in a 'DNA world' in which proteins are synthesized downstream of DNA. But, in order to synthesize DNA, enzymes are needed. For the time being, nearly all enzymes are proteins. Hence the question emerges as to how DNA could be formed under conditions in which proteins were lacking. Or, in other words, can some enzymatic activities be executed by molecules that are not proteins?

T. Cech suggested a solution for this dilemma by proposing that perhaps the DNA world was preceded by an 'RNA world' (Cech, 1986; Cech and Bass, 1986; Gilbert, 1986). Perhaps the catalytic activity required to translate RNA into DNA was present in the proto-RNA itself, whether directly or indirectly. Indirectly would mean that RNA first catalyzed the synthesis of certain proteins out of pre-existing amino acids. Next, these proteins would have catalyzed the synthesis of DNA out of RNA.

RNA can indeed catalyze certain important reactions:
1. The ribosome, the structure that serves as the locus for protein synthesis, is a ribozyme. The catalytic activity required for joining amino acids through amide (peptide) linkages resides in the RNA moiety of the ribosome, not in its protein moiety (Cech, 2000).
2. Furthermore, Johnston et al. (2001) managed to develop *in vitro* an RNA that can copy itself, thereby 'repopulating the RNA world' (Strobel, 2001).
3. Finally, the spliceosome, which is the structure that removes intervening sequences (introns) from eukaryotic RNA precursors, is also an RNA enzyme (Valadkhan and Manley, 2001).

The way contemporary retroviruses solve their problem of replication may shed some additional light on the transition from an RNA world to a DNA world. Retroviruses such as the AIDS virus do not have DNA as carriers of their genetic information, but rather RNA. In order to be replicated by the normal machinery of the cell they infect, their RNA must be converted into DNA. However, eukaryotic cells lack an enzyme for catalyzing the conversion of RNA into DNA. Retroviruses solve this problem this way. In their own RNA genetic code they have a gene that codes for an enzyme called 'reverse transcriptase'. The infected host cell translates this RNA gene into the corresponding protein that next catalyses the transformation of the whole viral RNA genome into the corresponding DNA genome. Once this has been realized, the virus can start using the (first) central dogma: DNA $\rightarrow$ RNA $\rightarrow$ Proteins.

Thus, from the chemical point of view, the hypothesis of Cech is, in theory, acceptable. However, Miller and other investigators have never found RNA in laboratory experiments mimicking prebiotic conditions.

De Duve (1991) suggests that proteins could have been formed by thioester chemistry without the need for catalytic activity of RNA. Another possibility is that short chains of amino acids that could replicate themselves or other peptides (auto- and cross-catalysis) were first somehow formed. Such peptides seem to exist (Yao et al., 1998)

More recently, theories pointing at the origin of the first living cells in the immediate vicinity of hydrothermal vents on the bottom of the sea have gained ground. In these theories, life originated at extreme conditions of high temperature (above 100 °C), pressure, and, perhaps, pH (acid or alkaline) as well. According to Cody et al. (2000) and Wächtershauser (2000), a form of life that is based on rather low molecular weight constituents (thus not on large macromolecules) could have originated in such a world, such as in an iron-sulfur world.

12.3 The *pregradient* era: the problem of generating 'order' in the primordial conditions

Opinions are quite divergent concerning the way in which all organic (macro)molecules essential for the origin of life were synthesized. For an overview of the different, and sometimes conflicting, hypotheses and their limitations, see Horgan (1991), De Duve (1991, 1995), and Wächtershauser (2000).

By whatever reactions, all organic (macro) molecules necessary for finally yielding the living state were formed, life could not arise unless these molecules were packed in an orderly way. A mixture of all the constituents of a living cell, even if they are present in the right proportions, will not yield the 'living state'. This is due to the fact that gradients of solutes will not form spontaneously.

The physicist Hoyle states the problem very nicely: he says that it is more probable for a Boeing 747 to be suddenly assembled from all its parts by a tornado than for a functioning cell to arise from a mixture of its constituents. This is not a valid comparison, however, because it completely neglects the possibilities for self-organization which are clearly present in some types of organic molecules that are essential constituents of living matter.

The main chemical mechanisms of self-organization that might have been used in 'primordial soup conditions' to generate 'order' in the form of aggregates of organic (macro) molecules were most probably protein-protein, protein-nucleic acid, and hydrophobic interactions.

I suppose that somewhere and sometime on the young earth a variety of molecules interacted with one another. Probably some complexes were more stable than others and, as a consequence, were better conserved than others. Selection of the more stable ones was a result, not a cause. Only the 'ordered states' that were able to maintain their stability or managed to be precursors for novel ordered states (e.g. by accumulation of order) persisted over longer periods of time and were positively selected based on their stability ('fitness').

As I will explain next, I think that the cytoskeleton, in whatever form it may have had, may have played a crucial role in generating the type of order required to establish a functional communication system. It may have acted by linking important (e.g. enzymatic) elements and processes. Major contemporary cytoskeletal proteins like actin and tubulin can polymerize and depolymerize very fast under the influence of the ionic environment.

12.4 The *cytoskeleton*: as important as nucleic acids for the emergence of the living state?

In discussions between specialists in the field of the origin of life, the origin and early roles of the cytoskeleton and of the plasma membrane do not seem to be 'hot topics', (both being rather late acquisitions) (see e.g. Darnell and Doolittle, 1986). In most papers the cytoskeleton and the problem of the acquisition of 'cellular polarity' are not even mentioned.

At a given moment, proteins must have appeared in the primordial conditions, whatever these conditions might have been – hot tidal pools, deep sea hydrothermal vents, the surface of pyrite crystals (Wächtershauser, 1988a, b), clay environments, etc. Whether this happened before or after the appearance of RNA is not crucial in the context of this chapter.

There must have been mechanisms that enabled a variety of macromolecules to clump together, resulting in the formation of aggregates. I think that a sort of framework, like the actin-tubulin cytoskeleton that all contemporary eukaryotic cells use, might have been instrumental in aggregate formation. However, the absence in bacteria of actin, a universal cytoskeletal constituent of all eukaryotes, has puzzled researchers for a long time. Recently, this problem was solved by the discovery of an actin-like molecule named MreB in the bacterium *Thermotoga maritima* by van den Ent et al. (2001). The properties of MreB resemble very well those of actin. A bacterial homologue (namely FtsZ) of another major cytoskeletal protein, namely tubulin, had already been discovered before. Hence, the hypothesis that the Progenote already had a functional cytoskeleton is no longer unwarranted.

I suggest that the formation of the cytoskeleton, with its intrinsic ability to polarize because of its filamentous constituents, had to be realized *before* a cell, or perhaps even an aggregate, could be formed. Cytoskeletal proteins with contractile properties could generate some movement and could function as an anchor for a variety of macromolecules. Perhaps the special electricity-conducting properties of actin (Chapter 9) were also essential for life to emerge.

At a given moment, some aggregates got encapsulated by a lipid membrane: the first level compartmentalization was established.

12.5 The importance of the achievement of the *transmembrane ionic/voltage gradient* and of transcellular ion fluxes

A cell is dead from the moment that it irreversibly loses its electrochemical potential gradient. This happens when the plasma membrane over which the gradient exists is disrupted (Chapters 6 and 9). It is then logical that the realization of this gradient might have been *the* differentiating event between living and non-living. In my opinion the following course of events is reasonable (Fig. 12.1).

After encapsulation by a lipid membrane, a variety of lipophilic molecules became trapped in this membrane. Once trapped in a fluid membrane, proteins with a hydrophobic moiety could, at least in principle, float freely in the membrane as long as they did not become anchored to the ends of the cytoskeleton sticking into the membrane. Besides their hydrophobic moiety, some of these proteins probably also had hydrophilic parts that were in contact with the external environment or with the cytoplasm or both, as in the case of many contemporary transmembrane proteins. Apparently some of these transmembrane proteins could use energy to actively transport some ionic species (e.g. K^+, H^+, Ca^{2+}), a mechanism which would result in the development of an ionic gradient. Such proteins could have been the precursors of ion pumps. Other transmembrane proteins may have allowed the passive leakage of ions: these could have been the precursors of ion channels, which, depending upon the type of channel, open or close under the influence of a variety of factors. The combination of ion pumping activity with differential permeability of the membrane for some types of ions is likely to have resulted in the generation of a voltage gradient: ionic/electrical compartmentalization was achieved.

One condition for producing this gradient was that the ion pumps had to transport a specific ionic species faster than such ions could leave the compartment passively through the ion channels. Whether the formation of an ionic gradient was the last and decisive step for reaching the status of *short-term* living matter is difficult to say. I do not exclude the possibility that this may indeed have been the most primitive form of life.

However, I hesitate to be more affirmative for the following reason. To date, because of lack of sufficient data, it is as yet impossible to give a conclusive answer to the following two questions: "*Can a cell engage in de novo protein synthesis* (transcription, *etc.*) *without first starting to drive a flux of ions through itself?*" and, "*Can a cell store information in its non-genetic (cognitive) memory without first driving an electric current through itself?*" These questions might be less irrelevant than they might look at first sight.

Cells which are spherically symmetrical with respect to their plasma membrane-cytoskeletal complex are very rare in Nature, if they exist at all (De Loof, 1995): all cell types are somehow polarized, at least when they are using their protein synthesis machinery. Polarity is a crucial property of the living state in the present state of affairs, but a clear positive correlation at this current time is not a proof for a causal relationship at the time when life just was emerging.

This may be just a matter of coincidence, but in animals with what is called a 'mid-blastula transition', substantial *de novo* protein synthesis starts only when the animal embryo becomes organized as a blastula. It is exactly in this stage that the embryo starts functioning as a simple epithelium which is capable of transporting solutes through its cells (De Loof, 1992).

My opinion is that the answers to both questions are likely to be negative (De Loof, 1986; De Loof et al., 1993a; Vanden Broeck et al., 1992). It may not have been difficult for the cell to acquire the ability to drive such a flux through itself. I can imagine that some of these ion transporting transmembrane proteins would have been able to interact with the ends of cytoskeletal filaments in the plasma membrane, as is still the case today (Luna and Hitt, 1992). If the cytoskeleton was polarized, segregation of ion pumps and channels might have ensued. The result would have been that such cells could have also started driving a flux of ions (electrogenic or electroneutral) through themselves, with the possibility of generating *intracellular gradients* under proper conditions. I think that this transcellular flux may initially have been just an epiphenomenon resulting from the pre-existing polarity in the cytoskeleton, but that in due time it may have become indispensable to the cell for performing certain functions. A nice example of the possibilities of a transcellular ion flux is found in the polytrophic insect ovarian follicle, which behaves like a true miniature electrophoresis chamber, as already mentioned in Chapter 9 (Woodruff and Telfer, 1980).

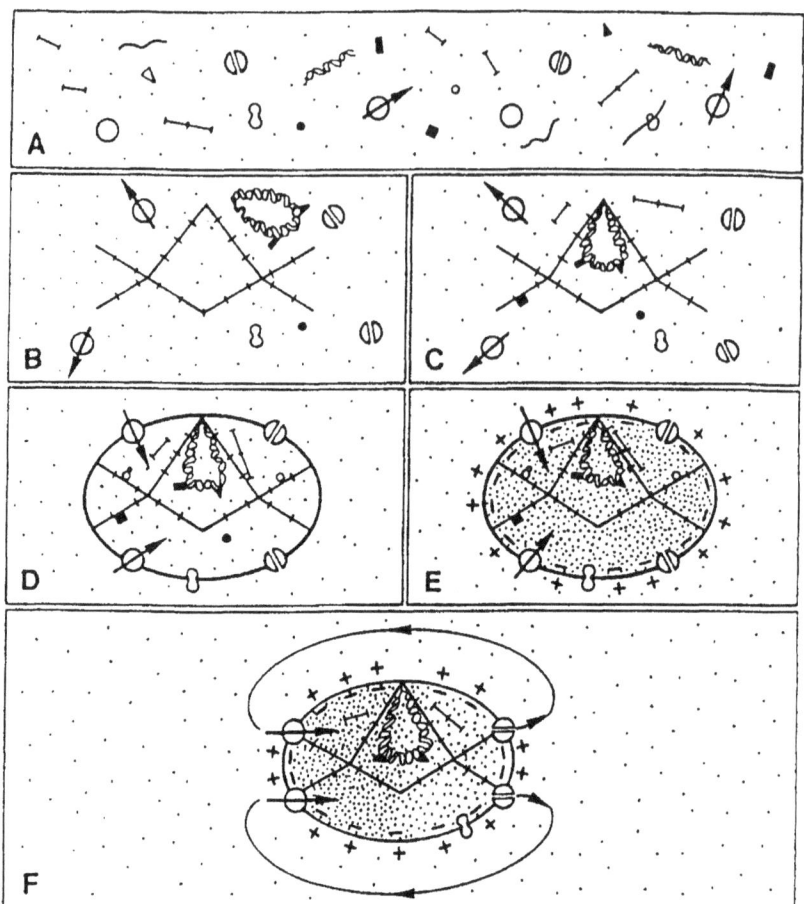

Figure 12.1 Schematic representation of a possible succession of crucial stages leading to the formation of the first living cell, as described in this chapter. From De Loof (1993a). At a given moment a variety of organic molecules that were essential for yielding the living state (in due time), must have been present in the 'primordial conditions' (A), whatever these conditions might have been (hot tidal pools, deep sea hydrothermal vents, etc.). Filament forming proteins may have been formed early (B) and may have been able to anchor a variety of other macromolecules (C), thereby forming 'polarized' aggregates. Sooner or later, compartmentalization took place by encapsulation in a lipid membrane in which hydrophobic proteins became trapped (D). Some of these proteins are supposed to have been primitive ion-transporting pumps (circles with arrow) and channels (circles with hole). In concert, they established an ionic/electrical gradient over the membrane (E). When some ion transporting proteins became (selectively) anchored to the ends of the cytoskeleton in the plasma membrane, the possibility for a cell to drive a flux of ions through itself was established (F), thereby making it possible to generate intracellular gradients under the right conditions.

The electricity conducting properties of actin filaments were also mentioned in Chapter 9. Suppose that the archetype may have contained filamentous proteins that were able to conduct electricity like actin does. As a consequence, the primitive cell would have contained a network of electricity conducting wires. But electricity will only flow along such 'wires' if there is an input of electricity at one of the ends of such a wire. If electrogenic ion transporting molecules were anchored to the ends of cytoskeletal filaments, the necessary ingredients for the cytoskeleton to be serving in conducting electricity would have also been present. Whether the bacterial homologue of actin, namely MreB, has similar electricity-conducting properties as actin remains to be investigated.

What could have been the functions of such a system? We now know that a variety of genes are electrically controlled (Chapter 9). How this works at the molecular level is still unknown. We also know that there are dozens of different proteins that can bind to present-day actin, that actin is a major component of the chromosome skeleton (Sauman and Berry, 1994), and that ribosomes and some messenger RNA's are associated with the actin cytoskeleton. Recently, the idea has been put forward that the cytoskeleton may play a role as carrier for the non-genetic memory,

Perhaps electrical current (in the form of soliton-like waves) travelling along the filaments may have caused changes in the interactions between the cytoskeleton and some macromolecules attached to it, thereby causing changes in some cell regulatory processes. Perhaps the driving of an electric current by a cell through itself is essential for storing non-genetic information somewhere in the cell. It cannot be excluded that initially transcellular ion fluxes were an epiphenomenon without real function and that essential functions were acquired later.

I share the opinion of those who think that the acquisition of the living state on earth was probably not a single 'flashpoint event' (De Duve, 1991). I imagine that it was a multistep, rather deterministic process, which may have taken millions of years and an endless number of trials and errors before a decisive spark made the difference. It is likely that this spark was due to the squeezing in of the last link in a chain that is called a 'signal transduction pathway'. Several such pathways are simultaneously operational in cells.

In summary, the following stringent conditions had to be satisfied before some primordial aggregate could have acquired the ability to communicate:
1. A scaffold to which a variety of organic macromolecules (proteins, nucleic acids) could be anchored in an ordered way, and which could carry out, in concert, a number of prebiotic metabolic reactions. Functionally, this primitive scaffold corresponds to the present day cytoskeleton.
2. An ionic environment suitable to the 'well being' of the aggregate.
3. A means of turning the aggregate into a compartmentalized unit. This may have been realized by encapsulation in a lipid membrane. This membrane had to be selectively permeable to a particular (or perhaps several particular) ionic species.
4. A means of acquiring and using the right form of energy (e.g. light, heat, chemical energy from adenosine triphosphate (ATP), etc.), so that a gradient could be generated over the limiting membrane which could be used as a means of communication. The simplest type of such a gradient is likely to have been an ionic/electrical one (H^+, Ca^{2+}, K^+, Na^+, Cl^-, or combinations of these). This type is still the basic gradient in all contemporary cells, without exception.
5. A system(s) was also needed for sensing changes in steepness of the gradient(s) over the limiting membrane and for transforming this information into a language comprehensible to the cellular metabolism.
6. Mechanisms which would have enabled the organism to cope with gradients, e.g. to counteract the *osmosis effects* resulting from the buildup of gradients of

solutes. This is very important and almost always overlooked when dealing with evolution and self-selection. We will come back to this issue of self-selection later in this and in other chapters.
7. A system that would have allowed some simple but active locomotion, especially for moving away from sites with unfavorable conditions or towards sites with food molecules.
8. A sort of 'cognitive memory' to be used in communication.

The acquisition of the ability to communicate is therefore likely to have occurred only after a long 'pregradient era' of evolution of organic molecules, and after many 'trial attempts' to assemble the essential molecules in an orderly fashion.

12.6 Requirements for the perpetuation of the first communication system

For the ability to communicate to become transmissible and sustainable *over generations* (heredity), there were additional requirements such as:
1. A molecular carrier for genetic information to serve as macromolecular memory in the long term (nucleic acids: deoxyribonucleic acid (DNA) and/or ribonucleic acid (RNA), depending on what was first);
2. A system, preferably imperfect, of replication of this information so that occasional errors could be introduced in the nucleotide sequence (mutations, see below);
3. A system to move the copy away from the parent molecule;
4. A system that can be activated at the right moment to pinch off a daughter compartment from the parent;
5. Such a daughter compartment must receive from the parent all the necessary traits to generate at least a durable ionic/electrical gradient (the basic gradient of the living state) over its plasma membrane, to grow, to continue to pinch off daughter compartments, to cope with osmotic problems, to replicate its DNA, etc.;
6. A system to duplicate the cytoskeleton in a conservative way.

12.7 The problem of osmosis. The Gradient-Provoked *S*welling and *S*elf-*S*election Principle (the GP-Triple S Principle) and how it might apply to social and economic compartments

The papers that have been published on the origin of the first cell rarely deal with the problems inherent to osmosis that results from maintaining different concentrations of solutes inside the cell compared to the outside world. The means of locomotion of the first cell are not known either.

According to the fossil record, the first living entities on earth were probably simple bacterium-like cells. Because gradients are not conserved in fossils, we have to make plausible guesses about the chemical nature of the gradients such cells might have had.
All contemporary organisms are thought to be descendants from one single cell, which was called 'Progenote' by Carl Woese. If in all contemporary cells the nature of the trans-plasma membrane gradient were the same, the most likely explanation for this fact would be that this type of gradient was used by the first cell and that all descendants of that cell inherited it and have continued to use it up to the present time. Present day bacteria use H^+ gradients to a much larger extent than animals and plants. This could suggest that the first ionic gradient used in communication may have been an H^+ gradi-

ent, which means a pH gradient. Some contemporary extremophile bacteria tolerate very low pHs. For example, *Ferroplasma acidarmanus* even thrives at pH 0 (Pennisi, 2000). In fact, it does not really matter what type of ions made up the first ionic-voltage gradient because the underlying principles remain roughly the same for any ionic species.

In this section I will focus on two often overlooked, but very important aspects of gradient formation, namely the osmotic effects and the effects of changes in ionic concentrations (e.g. in pH) on the conformation and reactivity of some macromolecules.

Imagine a compartment divided into two parts by a membrane that is permeable to water but not to sugar. Such a membrane is said to be differentially permeable. If compartment A is filled with distilled water and B with a sugar solution, water will passively move across the membrane from compartment A to B. In principle this will go on until the water concentration in both compartments is identical. In reality this will not happen in the system depicted in Fig. 12.2 because of the building up of hydrostatic pressure that counteracts the effect of osmosis. Indeed, after some time an equilibrium will become established in which the hydrostatic pressure and the osmotic pressure balance each other.

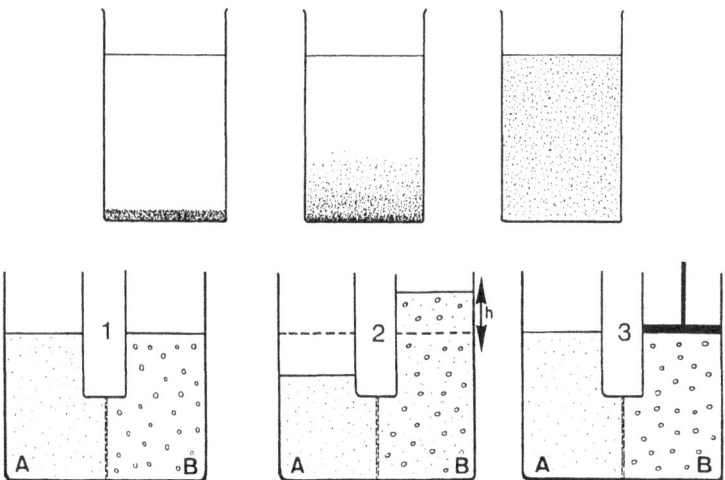

Figure 12.2 Schematic representation of the difference between diffusion and osmosis.

Diffusion (upper drawing). Diffusion is the movement of solute molecules from an area of higher to an area of lower concentration.
Osmosis (lower drawing)
1. Starting conditions: compartment A contains pure water, compartment B contains water with some solute (e.g. sucrose).
2. Water molecules diffuse from A to B through a differentially permeable (also called semi-permeable) membrane (= permeable for water but not for sucrose in this example). The height of the water column in B increases while that in A decreases. After some time the increased pressure exerted by the additional height of the water column h becomes so high that the water concentration in B no longer increases.

3. The osmotic pressure is the pressure that must be exerted in B (e.g. by the weight of a piston) to prevent the movement of water into B.

To clarify the point, let us assume that the first ionic gradient in the Progenote was exclusively based on hydrogen ions (H^+) and that these ions were pumped from the outside world into the cell (Fig. 12.3). We further assume that the cytoplasm had no buffering capacity and that the H^+ did not bind to macromolecules. In such circumstances, the interior of the cell would have became acidic, just as is the case today inside lysosomes. These are cell organelles that contain lytic enzymes and in which the pH is about 4.6 as compared to 6.8-7.2 in the cytoplasm due to the activity of H^+- pumps present in the limiting membrane. This acidification would have had a double effect. First, the increased H^+ concentration would have elicited osmotic effects: water would have started flowing into the cell, with swelling as a result. The more H^+ to enter the cell, the greater the increase in volume.

In theory, cells can counteract the effects of osmosis by
- residing in an isotonic environment;
- using a cytoskeleton to withstand the increase in internal pressure (turgor);
- encapsulating themselves in a rigid cell wall.

The second effect of the increased H^+ concentration would have been at the level of the organic molecules present in the cell, and especially on those with enzymatic activity. Many proteins are sensitive to pH changes. Proteins and other molecules that could not have coped with the acidic conditions would have been inactivated or even eliminated. Cells that were too badly damaged by the ionic gradient they had built up would have died: self-selection at work.

All this means is that a compartment that builds up an ionic gradient, whatever its nature, automatically swells (if the concentration is higher inside) and subjects itself to self-selection. This principle could be called *Gradient-Provoked Swelling or shrinking and Self-Selection* or the **GP-Triple S Principle**, for those who are not members of the SAB (Scientists Against Abbreviations) league.

In the example used, the cytoplasm became acidic but the principle would work as well with an alkaline environment provoked for example by an ion pump that exchanges K^+ for H^+ or with any type of gradient. If a gradient exists in the reverse direction, shrinking would result.

I think that the GP-Triple S Principle may apply to any level of compartmentalization as depicted in Chapter 4, social compartments included.

Imagine that a University attracts a new professor and provides him with some money to start working. The professor will start talking to potential collaborators and students, trying to make them enthusiastic about his research ideas. If for one reason or another the communication between the professor and the people he is going to hire does not 'click', for instance because the professor is too rude, the laboratory is not likely to get started and grow. Colleagues who have their labs on the same floor may claim the space for themselves and the new lab may even shrink. Self-selection and self-elimination based on communication was at work. If on the other hand the professor has excellent communication skills, he is likely to attract bright people and form a research team that is likely to grow for quite some time. This laboratory will soon be too small, and a call for extra space will be launched. Here, the process of 'osmotic swelling' is at work: the high quality, good working atmosphere, etc. incites others to join in the effort and thus the research team 'swells'.

The GP-Triple S Principle is very clearly at work in our economic system. Imagine that in a given street there are two almost identical shops, both selling home appliances. Their location, parking facilities, the brands they sell and the prices are nearly identical. Thus, there should be no reason why one shop would sell more than the other. But one of the shop owners starts a big advertising campaign in which he proclaims that his shop sells the best goods for the lowest price: nowhere cheaper in town! The other shop owner refrains from such a campaign because he knows that there is no difference between the shops. What is likely to happen? At least the advertising will attract some people. The advertising shop will sell more and expand, and the other will shrink if it does not change its strategy. Money, which is the equivalent of ATP in a cell, is the driving gradient of the economy.

Another example comes from politics. Imagine a country in which a 'strong man' comes into power with a small number of followers by means of a coup. As soon as possible, he and his followers will brainwash the population with propaganda about their ideas - if necessary, by force. The citizens who do not like the ideas will suffer opposition from the rulers and are likely to be inactivated, just like the sensitive macromolecules in the acidified cytoplasm. Besides the inactivation of the persons who do not 'fit' into the new system, the rulers will gradually increase their 'territory', and this may finally lead to invading neighboring countries: a phenomenon once again comparable to the expansion of the cell under the influence of increased osmotic concentration.

If one looks around one easily discovers many examples of this principle in daily life: I think that the GP-Triple S Principle may be a basic principle in sociology and in economics. I will come back to the issue of "volume changes" of compartments and self-selection in Chapter 15, which deals with the mechanisms of evolution.

A third major problem encountered by the first cell was that of motility. If a sort of primitive *contractile* cytoskeleton was not yet present at the moment when the building up of the first ionic gradient started, there is a problem. Indeed, in such case the only means of locomotion for such a cell would have been passive: either being carried along a stream of water or small movements resulting from osmotic swelling and shrinking. How such a cell devoid of any cytoskeletal framework could have prevented itself from bursting because of osmotic swelling and how it could have entered into the process of cell division which is necessary to extend life over more than one generation, is not clear at all.

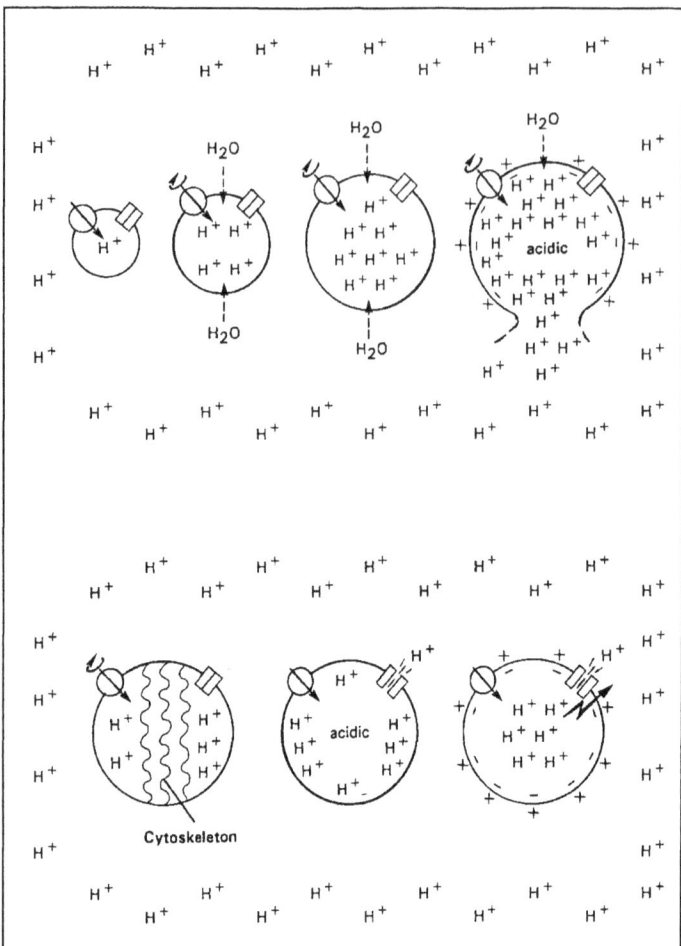

Figure 12.3 Schematic representation of the "GP-Triple S Principle". By the accumulation of H^+-ions in the interior of the cell, an acidification takes place and water enters by osmosis. If the cell cannot cope with this acidification it is likely to die. If it has no means to counteract the increase in pressure, it will burst. Possible mechanisms to cope with the increased pressure include a cytoskeleton, a strong cell wall, ion channels which are sensitive to pressure increases in the cytoplasm and/or ion pumps which stop pumping in time, etc.

12.8   Conclusions

From my definition of life it follows that the living state could only have come into existence after a long period of prebiotic evolution during which a certain type(s) of compartmentalized aggregate acquired the ability to communicate at its highest level of compartmental organization, namely its plasma membrane. Furthermore, I repeat that I think that for the living state to come into being, the formation of the primitive cytoskeleton was as important as that of the synthesis of nucleic acids.

Because the soft parts of organisms are not readily fossilized, the only way left to find out which biochemical pathways might have been present in the Progenote is to undertake comparative genome analysis of extremophile- and 'normal' bacteria (Archaea and

Eubacteria). The combination of genomics and geosciences may in due time yield a fairly good picture of the physical and chemical properties of the environment, as well as of the metabolic reactions that enabled life to come into existence (Banfield and Marshall, 2000).

ESSENTIALS

LIFE'S ORIGIN

1. Life probably originated some 3.7 to 4 billion years ago, thus relatively soon after the earth itself came into being about 4.5 billion years ago.
2. A number of molecules essential to the form of life that is now present on earth seem to be present outside the earth as well. Hence, some of the building blocks of living systems may have been extraterrestrial in origin.
3. The exact site of origin of the first cell on earth, called the Progenote, will remain forever a matter of speculation. To date, among the different possibilities, deep-sea vents are thought to be a likely site, despite the extreme conditions prevailing there.
4. Recent discoveries concerning a variety of enzymatic reactions that can be executed by RNA add plausibility to the view that an 'RNA world' preceded our current 'DNA world (the Cech hypothesis).
5. By way of analogy with the 'big bang' or 'big splash' that marked the origin of the physical universe, the term 'big hello' might be appropriate to denote the coming into existence of communication activity, and thus of life.
6. The first cell must have had all the essential properties of a communication system to be able to execute its essential function, which was problem solving.
7. The appearance of the ability to produce cellular electricity carried by ions (and not by electrons) was absolutely essential for life to come into existence.
8. The first cell must have had a sort of cytoskeleton and was probably polarized (= not spherically symmetrical, thus with 'an anterior' and 'a posterior') right from the beginning. All contemporary cells, both prokaryotes and eukaryotes, have actin or actin-like microfilaments. This actin, which has a double helix structure (useful for replication), has been very well conserved in evolution. Actin in the chromosomal skeleton should be as long-lived as the DNA for which it serves as an anchor. This strengthens the view that, in addition to its other functions, actin may have been - and might still be - part of the cognitive memory system.
9. The first cell was probably faced with the problem of osmosis, which made it either shrink or swell as a result of differences in concentrations of solutes inside and outside the cell. This mechanism of Gradient-Provoked Swelling or Shrinking and Self-Selection has been called the GP-Triple S Principle. It applies to all levels of compartmental organization. It is also a key principle in economy and sociology.

CHAPTER 13

REPRODUCTION

GENDER AND SEX

How 'it' may have given birth to 'him' and 'her'

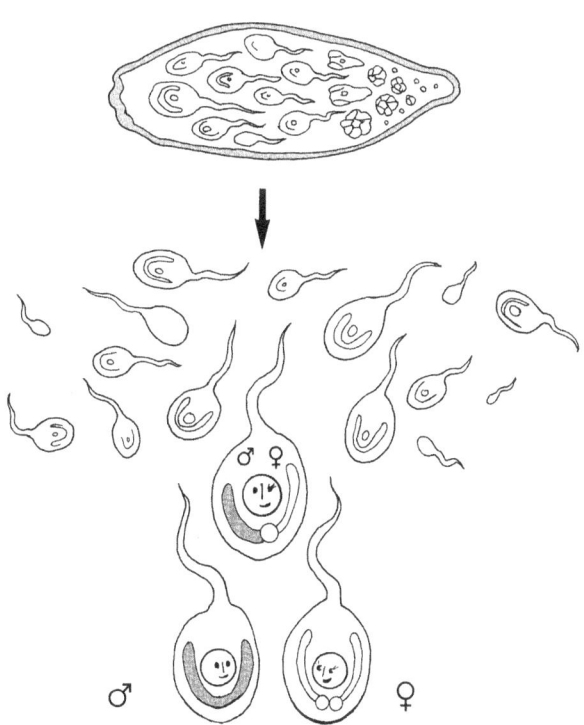

Males and females: a late 'invention' in evolution

Contents

13.1 Introduction
13.2 The Progenote: why did it engage in cell division? Immortality as a reward
13.3 Mitotic cell division in the Urkaryote: a problem of coordination
13.4 Asexual reproduction in multicellular organisms: by regeneration and from stem cells
13.5 The sexual mode of reproduction: by means of gametes resulting from meiotic cell division
13.6 Formation of the germ cell line. Solitarization of the presumptive germ cells in insects as a model
13.7 What made the primordial germ cells come into existence? An ancient bacterial infection?
13.8 How to increase the number of gametes? Rounds of mitosis precede meiosis in higher animals
13.9 How do bipotential germ cells get committed to either the oocyte or the sperm cell scenario? Sex steroids
13.10 Gamete formation in the gonads: a benign form of cancer? The 'fight' between somatoplasm and germ plasm
13.11 The 'idea' behind yolk accumulation in eggs: based upon an ancient innate immunological mechanism?
13.12 Oviposition and sperm ejaculation: again 'hostile' acts by the somatoplasm
13.13 Gender: why only two isoforms, males and females? Gender-determining genes
13.14 Cholesterol, sex hormones and gender. Again the bacterial infection route. Parthenogenesis and sex-reversal
13.15 Males: from superfluous to indispensable?
13.16 DNA methylation and imprinting
13.17 Death of the somatoplasm: the price to be paid for meiosis. Sexual pleasure in return. The progeny: goal or free surprise bonus?
13.18 Sperm competition. The choosing female
13.19 Conclusions: the origin of reproduction is not what you thought it was

Essentials

13.1 Introduction

What is and happens around us all the time does not make us wonder any more: it is considered normal and self-evident. We find it simply normal that organisms grow, that they reproduce one way or another, and that they finally die by aging or from an accident. Gender with two forms, male and female, is so ubiquitous and asexual reproduction is rather rare, particularly in animals, so it may look as if gender has existed right from the beginning of life on earth. This cannot have been the case. Gender is probably a relatively late 'invention' in the course of evolution.

This chapter does not focus on the well-known aspects of reproduction such as the mechanisms of mitosis and meiosis, oogenesis and spermatogenesis, the hormonal differences between males and females, etc. All these topics are sufficiently covered in common textbooks. Some very basic questions will be asked. The major one is: "How could such a strange phenomenon as sexual reproduction ever come into existence?"

If we want to reconstruct the evolutionary history of sexual reproduction, no other option is left than to dig into the known principles of physiology and cell biology in order to find plausible explanations for events that took place long ago. Because it is not possible to redo evolution, the scenario that will be forwarded cannot possibly be 'waterproof'.

As a reminder: the term 'asexual reproduction' applies to all modes of reproduction that do not make use of sex cells (egg and sperm cells, also called gametes). Sexual reproduction invariably involves the use of gametes. In the case of parthenogenesis, only egg cells are used, no sperm cells. An individual organism that produces both sperm and egg cells is called a hermaphrodite.

13.2 The Progenote: why did it engage in cell division? Immortality as a reward

Imagine that you are a being that lives completely on its own, that you have never seen any other being like yourself and that, as a consequence, reproduction is a meaningless term to you. An angel, named Hallelujah, descends from heaven and tells you that you have to make a choice. You may choose either to continue living the way you are living now, thus completely on your own, and finally die from old age or from some accident. Or, you may decide to divide yourself into two halves, each of which will be able to regenerate into two normal individuals. But a major string is attached to the second choice: no guarantee is given that you will survive the risky splitting-into-two operation. If not, instant death will occur. My guess is that there would be few candidates to take the risk. For the individual itself, there is no advantage to continue in duplicated form instead of singly. Furthermore, the division might be a far from pleasant experience.

However, the Progenote, the (primordial) ur-ancestral cell of all living beings on earth, did engage in the division process and successfully survived. The self-closing properties of the lipidic plasma membrane were probably crucial in this success. Why did the Progenote take the risk? Because of physiological necessity: it had no other choice. One plausible explanation could be that some 'toxic' metabolite, whatever its nature, had started to accumulate in the cytoplasm and that it made some contractile proteins constituting the primordial cytoskeleton contract with cell splitting as a result (Fig. 13.1).

For this to yield any advantage in the long run, the genome first had to be duplicated in such a way that it could be halved again in a symmetric way. After the splitting, regeneration took place and the original cellular morphology was regained. Toxin production and accumulation continued, resulting again in cell division, and so on and so forth. Cell division as a cyclic event had come into existence. The reward, in the long run, was immortality, at least if death from mechanical accidents is not taken into account. This strategy, that is still in use in the dominant life form on earth, the prokaryotes (bacteria), does not yet involve gender or gametes (sex cells).

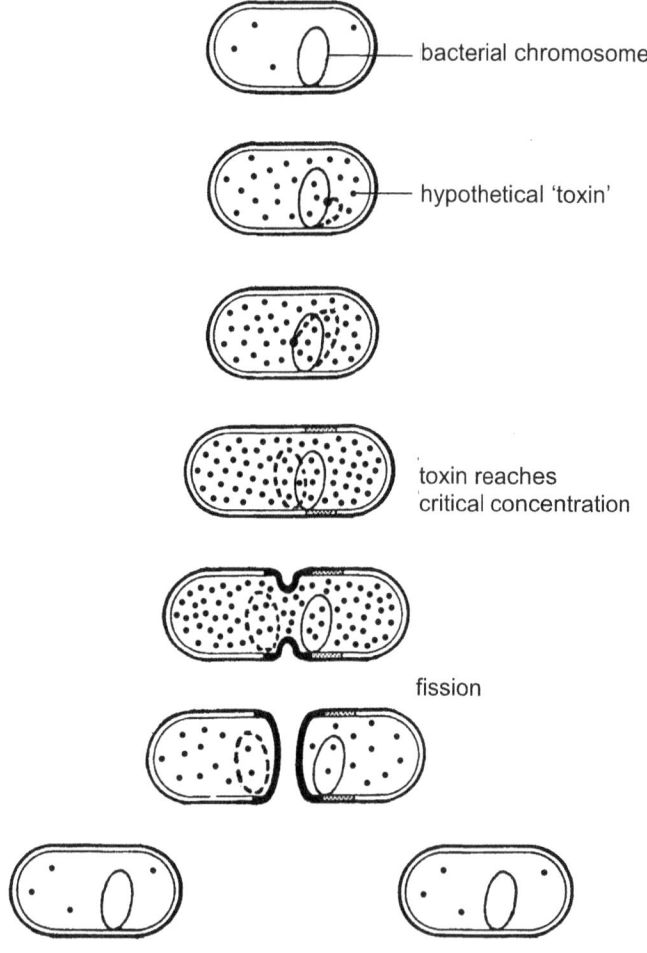

Figure 13.1 The triggering mechanism for initiating fission in the hypothetical first cell. The hypothesis is forwarded that some 'toxic substance', whatever its nature, started accumulating and triggered the fission. Upon subsequent growth of the daughter cells, the concentration of the toxin gets temporarily diluted. More toxin is produced with a next round of fission as result.

One could call this the indifferent or neutral mode of reproduction. In essence it is reproduction based upon regeneration.

In some present-day bacteria, a more complex system of reproduction is occasionally operational, namely conjugation. Two bacteria get joined by a cytoplasmic bridge through which a small circular chromosome, called a plasmid, can be transferred from one individual to the other.

13.3 Mitotic cell division in the Urkaryote: a problem of coordination

As already mentioned several times before, according to Lynn Margulis, the eukaryotic cell-type originated from symbiosis between at least two types (species) of bacteria, an unmatched masterpiece of natural genetic engineering. This hypothesis is strongly supported by the fact that cell organelles such as mitochondria and chloroplasts still have a small bacterial-type genome (DNA) of their own. A lot of their genes have been translocated to the nucleus and exert their activity there (Gray, 2000). In addition, these organelles divide inside cells just like regular bacteria would do. The first unit that successfully integrated the collaborating bacteria and succeeded in division is called the Urkaryote. Its descendants are called the eukaryotes.

The Urkaryote being a mixture of several organisms, successful cell division requires that the genomes of all constituent gene-carrying units (nucleus, mitochondria, chloroplasts…) somehow be transferred to the daughter cells in a conservative manner. To date we assume that cell division in eukaryotic cells is always conservative because we assume that all mitochondria and chloroplasts in a given cell have an identical genome. In the Urkaryote, the first organelle population might have been less homogenous. If, for example, two populations of mitochondria had been present in a cell, and if during cell division there had been a means to separate them into the two daughter cells, the progeny might have differentiated even though the nuclear genes would have remained identical. Later, a few examples will be discussed in which either mitochondria or regular bacteria play a role in reproduction and gender formation in animals.

The typical cell division of eukaryotes is called mitosis. First, the DNA in the nucleus is duplicated. In the next step, the doubled chromosomes are halved. The resulting daughter cells again have the same amount of DNA and number of chromosomes as the parent cell. Mitosis does not involve gender, it does not generate genetic variability and, under ideal conditions, it confers immortality. It is omnipresent in Protists and in the body cells (somatoplasm) of Fungi, plants and animals. Prokaryotes have binary cell division.

Some present day unicellular Protoctists display some forms of sexuality, such as the conjugation process in some ciliates, or the formation of micro- and macrogametocytes in *Plasmodium*, the Protist that causes malaria.

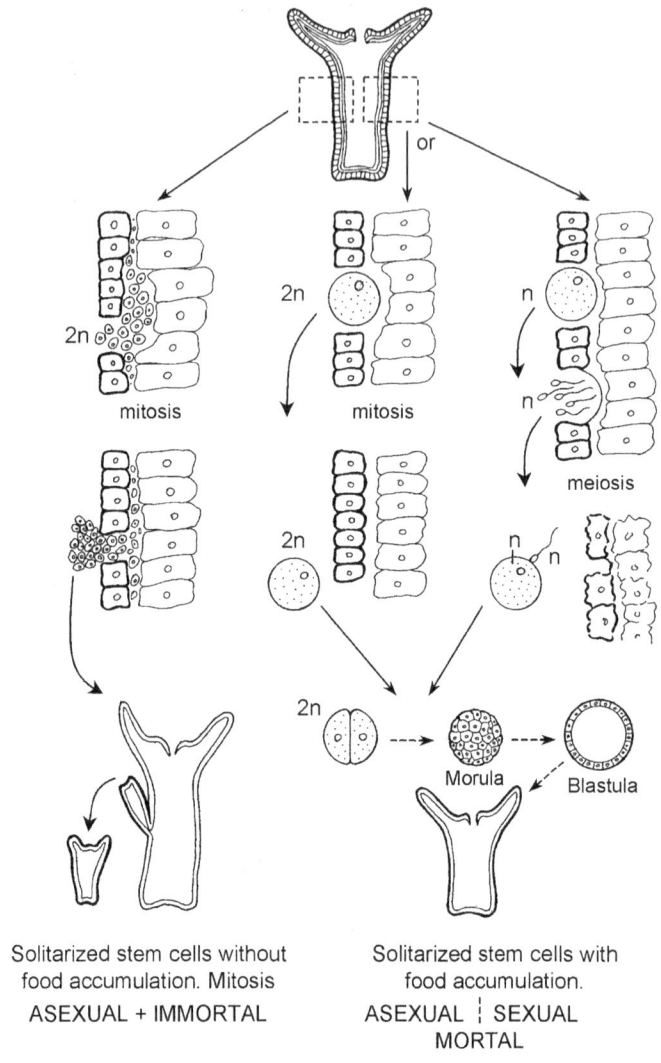

Figure 13.2 Early in evolution, asexual reproduction, which means reproduction without eggs or sperm cells, was the only possible way of multiplication. It uses the principles of regeneration. A new organism forms from undifferentiated stem cells that are not very rich in nutrient reserves. At some point in evolution, some stem cells may have appeared that accumulated a sort of yolk and that became solitarized. From such big cells, a new organism could perhaps be formed by regular mitotic cell divisions. For reasons that continue to be enigmatic, a special form of cell division, called 'meiosis', came into being. This resulted in the formation of haploid oocytes and sperm cells. To restore the normal diploid situation, fertilization became necessary. To illustrate all this, a hypothetical coelenterate (polyp) is used as an example.

13.4 Asexual reproduction in multicellular organisms: by regeneration and from stem cells

From broken parts

Everybody knows that many plant species can propagate asexually. For example, strawberry plants form stolones, from which new plants develop. In some grasses, small pieces of roots can give rise to new plants. The planting out of cuttings and meristem cultures is common practice in horticulture. If conditions are right, plants can grow from a single somatic cell.

From undifferentiated (stem) cells

In sponges and some coelenterates, such as the polyp *Hydra*, buds can be formed. Groups of undifferentiated cells (archeocytes) migrate to the presumptive bud area. Upon detachment of the buds, new individuals develop (Fig. 13.2).

Many freshwater and marine sponges disintegrate at the beginning of winter, leaving behind reduction bodies, compact masses of undifferentiated cells, often laden with food reserves. When climatic conditions become favorable again, a new individual is formed.

In some flatworms, asexual reproduction is part of a complex life cycle. The liver fluke *Fasciola hepatica* has a larval stage that is called a 'redia'. Such a larva can reproduce asexually. In the interior of the body are undifferentiated germ balls. In the winter, new redia-larvae can develop from them. In the summer, bodies termed cercaria larvae will be formed.

In mammals, stem cell research has boomed in recent years. Such cells are pluripotential and are important for tissue differentiation. It is easier to isolate stem cells from young embryos than from adult tissues. It should be emphasized that the differentiation of stem cells involves only mitotic cell division. Hence, gender does not exist in asexually reproducing organisms.

The primordial germ cells that are at the basis of sex cell formation are a sort of stem cell with restricted potency: they can only give rise to gametes. Here, mitosis will be followed by meiosis, as will be discussed later.

13.5 The sexual mode of reproduction: by means of gametes resulting from meiotic cell division

Even in animal phyla in which asexual reproduction is a possibility, as in sponges, coelenterates and flatworms, sexual reproduction also occurs and is often more important than the asexual mode.

The switch from asexual to sexual reproduction involves the introduction of a second type of cell division that is called meiosis (Fig. 13.3 and 13.4). The resulting cells have only half (= haploid) the number of chromosomes of normal diploid somatic cells. They cannot divide any further and they cannot survive on their own. This type of cell division only occurs in a very small part of the body, namely in two types of special stem cells that in higher animals are invariably located in the gonads, all other cell types of the body (the somatoplasm) maintaining mitosis as their basic strategy of division. The stem cells are called spermatogonia in males and the oogonia in females. The term 'germ (cell) line' denotes the totality of cells from which gametes are formed. All other cells of the body constitute the 'somatoplasm'.

The functional end result of meiosis is gender-specific. In females, meiosis yields only one functional gamete, the egg cell. The other three cells, the pole cells, are not functional and die. In males, all four cells become functional spermatozoa.

In brief, there are two major reasons for the evolutionary success of sexual reproduction. First, by means of meiosis, a larger number of offspring can be generated in a shorter time as compared to the relatively slow process of reproduction by regeneration. Second, meiosis generates genetic variability in the offspring, whereas asexual reproduction does not.

No wonder that in the course of time this form of reproduction could overtake asexual reproduction. The overtaking of asexual by sexual reproduction was probably not a sudden event. It is likely that this transition was gradual and that both mechanisms coexisted for quite some time in the same organisms. This duality is still present in sponges, coelenterates and plants as well. At some point, the end result of meiosis was refined enough to allow fertilization of an oocyte by a sperm cell followed by successful embryonic development. From then on, the asexual mode could be abandoned. In this view, the production of not yet functional gametes could have been a seemingly useless process and a waste of energy during long periods of evolution.

13.6 Formation of the germ cell line. Solitarization of the presumptive germ cells in insects as a model

Here, the key question is: in which properties do the presumptive germ cell precursors differ from the other cells of the body (the somatoplasm) so that they start to engage in meiosis, while the somatoplasm continues to rigorously stick to the mitosis strategy?

In my opinion, the key event in the process of delineating the germ cell line is the moment when the presumptive germ cell precursors are prevented from acquiring the epithelial status. In other words, for one reason or another they lack the ability to remain in tight contact with their neighboring cells by means of special formations of the cytoskeleton such as tight junctions. One could say that they are kept from leading a normal gregarious way of life and instead are forced into a solitary existence.

One could even say that this solitarization process alienates them from the gregarious somatoplastic cells and makes them look as if they were non-self or even foreign. This view, an unusual one in common thinking about reproductive biology, deserves to be taken into account. It makes it possible to fit together, in a logical way and in accordance with some known principles from immunology and cell biology, the cascade of events that finally lead to the ejection of gametes out of the body.

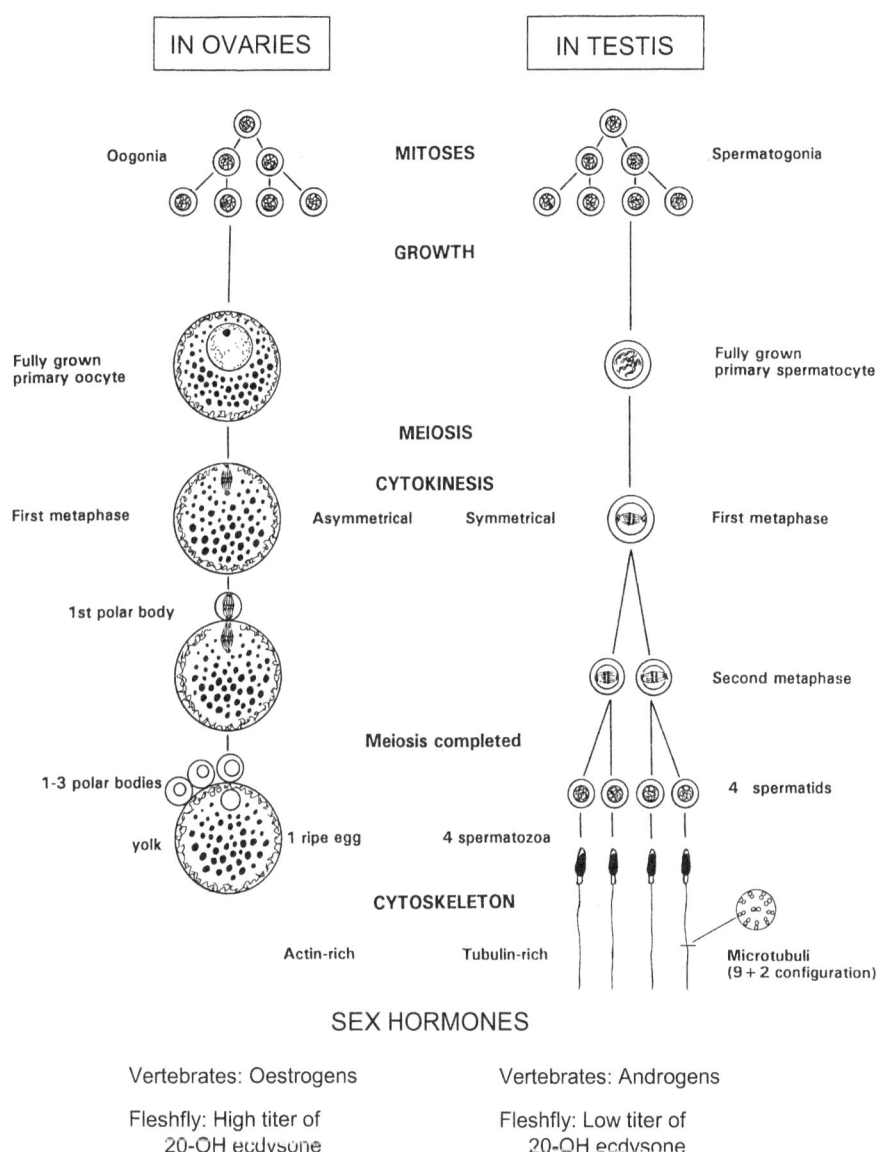

Figure 13.3 Major physiological differences between female and male gender which result from the genetic differences, whatever their nature, between the sexes. A first difference is that only the oocyte is capable of accumulating, by pinocytosis, yolk proteins (vitellogenins) from extra-ovarian origin. A second difference is that during meiosis the splitting of the cells (cytokinesis) is symmetrical in males while it is asymmetrical in females. This reflects a differential organization of the cytoskeleton in both sexes. A third difference is that the cytoskeleton of oocytes is actin-rich, while that of spermatozoa is rich in tubulin. Microtubules organized in the typical 9+2 configuration allow fast movement. Actin can bind a variety of different molecules, for example some maternal messenger RNA molecules and proteins (see also Chapter 14). A fourth difference concerns sex hormones, when present. In vertebrates the estrogen-androgen situation is well understood. In the fleshfly *Neobellieria bullata*, the best documented species in this respect, the molting hormone 20-OH-ecdysone induces vitellogenin synthesis. In most invertebrates, the nature of the sex hormones, if present, is not yet known.

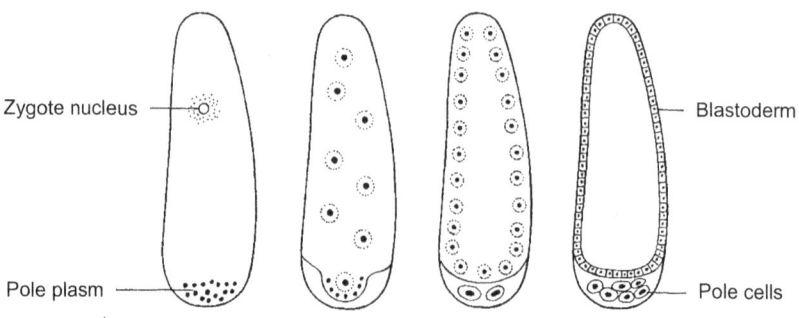

Figure 13.4 The definition of an animal is that it is an organism which develops from a blastula (De Loof, 1992). This means that the young embryo passes through a stage in which it organizes itself as an epithelium. At some early stage in development, some cells escape from being integrated in an epithelium and form the germ cell anlage. Insects are somewhat special with respect to the first divisions early in embryonic development because the central part of the egg contains so much yolk that nuclear divisions are not followed by plasma membrane formation. Plasma membranes are not formed earlier than in the blastula stage (blastoderm). In some insect species (*Drosophila* has been well studied in this respect), the segregation of the gonadal anlage happens already before the epithelium is formed. At one end of the egg, there is a group of particles with as yet incompletely characterized chemical composition (anyhow rich in RNA), called the polar granules. When a nucleus comes in the vicinity of these granules, segregation is induced. The isolated cell will undergo several rounds of mitosis, thereby forming the pole cells that represent the germ cell anlage. This anlage will give rise to oocytes or sperm cells later in development. From De Loof et al. (1998).

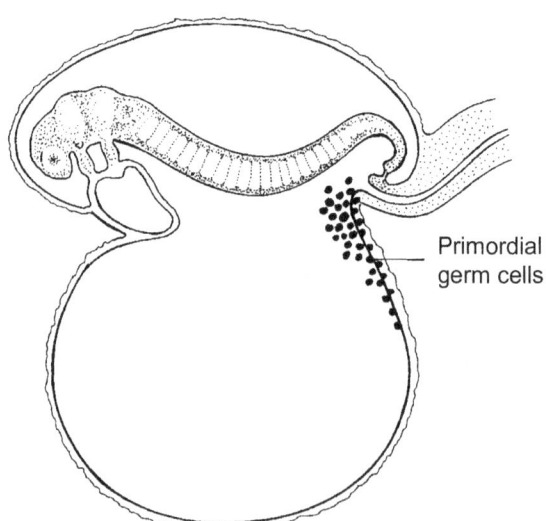

Figure 13.5 In a human embryo of "-weeks old, the primordial germ cells are found in the wall of the yolk sac, close to the site where the future umbilicord will be attached (redrawn after Langman, 1976). Here again, the primordial germ cells are not epithelially organized. By ameboid movements, the cells migrate towards the gonadal primordia, which they invade. From De Loof et al. (1998).

The solitarization process is clear in dipteran insects (flies and mosquitoes) (Fig. 13.4). In their egg, the zygote nucleus lies more or less in the middle of the dense yolk mass. Insects overcome the hindering effect of yolk on membrane formation by duplicating the nucleus without concomitant plasma membrane formation. Such nuclear divisions go on for some time. At a given moment, the majority of the nuclei migrate toward the periphery of the egg, into the thin cytoplasmic cortical layer that excludes yolk particles because of the dense cytoskeletal network which is present in this part of the structure. Within a few minutes, plasma membranes are formed around all these nuclei. A blastula with normal plasma membrane-bound epithelial cells is formed.

However, much sooner than blastula formation, a crucial event in the formation of the germ cell line takes place in the 8-nuclei stage. One of the 8 nuclei enters the pole of the egg where some RNA-rich granules called polar granules have accumulated. This nucleus enlarges upon entering into this special environment, and it becomes surrounded by cytoplasm and plasma membrane. Next, it undergoes a few extra rounds of mitotic cell division. These yield the stem cells of the presumptive germ cell line.

The fact that the polar cells are the precursors of the presumptive gametes was discovered in experiments in which the nuclei of the polar cells were destroyed by irradiation. Despite this damage and the inability to regenerate the destroyed cells, the embryos continue their development, at first sight without any harm. The damage done by the irradiation does not become apparent until in the adult stage. The phenotypic appearance of the animals is either male or female but they are sterile because there are no germ cells in the gonadal tissues.

The usual situation in animals is that the germ cells are not formed inside the gonads. In the example of *Drosophila*, after their sequestration at the caudal pole of the egg, they migrate by ameboid movements to the gonadal anlage inside the body. In the three-week-old human embryo, the germ cells can be found attached to the inner wall of the yolk sac (Fig. 13.5). Next they migrate towards the gonad anlage, in the vicinity of the kidneys. After their arrival in the anlage, male embryos will form testis cords, while females form cortical cords. In both cases, cells of the somatoplasm encapsulate the primordial cells. Encapsulation is one of the strategies of the immune system to isolate and inactivate non-self material. Later in this chapter, other aspects of immunological defense strategies against the germ cells will be dealt with.

13.7 What made the primordial germ cells come into existence? An ancient bacterial infection?

Upon a first confrontation, such a title may provoke hilarity and disbelief among genuine biologists. But if the symbiotic theory of Lynn Margulis is correct, (and I think it is), then the possible role of bacteria in gender formation might be worth considering.

As already mentioned, the symbiotic theory says that eukaryotic cells originated as the result of several bacterial species (at least two in animal cells, and three in plant cells) starting to live in symbiosis. In terms of their origin, mitochondria were originally a sort of bacterium that invaded another larger bacterium. It helped the larger one by coping with the damaging effects of O_2 in the atmosphere and it helped the host with its ATP energy production. Chloroplasts in plant cells are descendants from a photosynthesizing bacterium.

In his search for the molecules that make the fruit fly *Drosophila melanogaster* form its stem cells for the production of the future gametes (called the germ cell line), the team of Prof. Masukichi Okada from Japan made a remarkable discovery (Kobayashi and Okada, 1989; Kobayashi et al., 1993; Amikura et al., 1996). After long experimentation, his team found that the large ribosomal RNA of mitochondria present in the *Drosophila* egg was involved in germ line delineation. Remarkably, the mitochondrial rRNA is not taken up by the presumptive germ cells. At the time of the discovery, the action of mitochondria as causal agents in germ cell line formation looked more like an artifact without biological significance than like a genuine mechanism. Former students of Okada have substantiated and extended the data. They found that in addition to the 'large ribosomal RNA', 'smaller ribosomal RNAs' also play a role in germ cell line determination in *Drosophila*. If Okada's results are correct, (and there is no reason to doubt that they are), and taking into account the fact that in their evolutionary origin mitochondria were bacteria, the hypothesis arises that perhaps germ cell line formation is an indirect consequence of a bacterial infection that took place long ago. In the mean time, researchers have established that quite a number of other (this time 'genuine') *Drosophila* genes, such as *nanos*, are also involved in germ line formation.

It is generally assumed that there are no differences between the mitochondria present in males or females. All the mitochondria that a mammal has come from its mother, not from its father. This is due to the fact that at fertilization only the nucleus of the spermatozoon enters the egg, not its mitochondria. Accidentally penetrating mitochondria from the spermatozoon are quickly destroyed (by the so-called ubiquitin-system). This means that when mitochondria play a role in meiosis and gender formation, it is in aspects that are common to both sexes. Their role might be in contributing to solitarization, but not in specifying gender as such. Later we will briefly mention the role of mitochondria in sex hormone production and in the sex reversal that occurs in some animal species.

13.8 How to increase the number of gametes? Rounds of mitosis precede meiosis in higher animals

Sponges and some coelenterates do not produce numerous gametes as higher phyla do. Here, meiosis is an event in few stem cells. In animals with well-developed gonads, numerous gametes are formed. The primordial germ cells first undergo a number of rounds of mitotic divisions, thereby greatly increasing their numbers. Next, meiosis takes place.

Haeckel, a famous German zoologist, suggested the biogenetic law: "Ontogeny is the recapitulation of Phylogeny". In simple words, Haeckel thought that embryonic development reflects the evolutionary (pre)history of the species. Although this 'law' cannot be either proved or falsified, and although it should be interpreted in another way than Haeckel did (Gilbert, 1997), its basic idea is attractive. An example to illustrate the 'law': very young embryos of reptiles, birds and mammals have gill slits, like fishes. Later in development, these gill slits are overgrown and replaced by other structures. In Haeckel's view, the presence of gill slits reflects the situation at the time that our ancestors were still in the 'fish stage'.

An interpretation in accordance with Haeckel's law of the observation that rounds of mitosis precede meiosis in present-day gonads, would be that long ago the mitotic cells were the stem cells from which new organisms developed, without gender being involved. Meiosis was a later acquisition and, along with it, males and females came into existence and became functional entities.

The mechanism that causes the rounds of mitosis to precede meiosis has been partially elucidated in the worm *Caenorhabditis*. The decision whether the germ cells inside the gonad will undergo mitosis or meiosis is controlled by the secretion of a single non-dividing cell at the end of each gonad, the distal cell tip (Fig. 13.6). The germ cell precursors near this tip cell divide mitotically and form the pool of germ cells. As the cells get farther away from the distal tip cell, they undergo meiosis. When the tip cell is destroyed by a focused laser beam, there are no longer rounds of mitosis and all precursor germ cells enter meiosis directly. Apparently two factors with opposite effects are at work inside the gonad. A distal tip cell secretion product, when present above a certain threshold, promotes mitosis and prevents meiosis. A second factor might be present in the highest concentration at the opposite (proximal) end of the gonad. This factor is likely to promote meiosis. The nature of this second factor is unknown, not only in *Caenorhabditis*, but in any animal species. A plausible guess is that this factor belongs to the family of sex steroids (androgens and estrogens in vertebrates, ecdysteroids in insects). There is indeed a very clear positive correlation between the exclusive occurrence of meiosis inside of the gonads and the high concentrations of sex-steroids in the gonads. Perhaps the gonadal tissues themselves do not undergo meiosis because they retain their epithelial status while the germ cell precursors are solitarized.

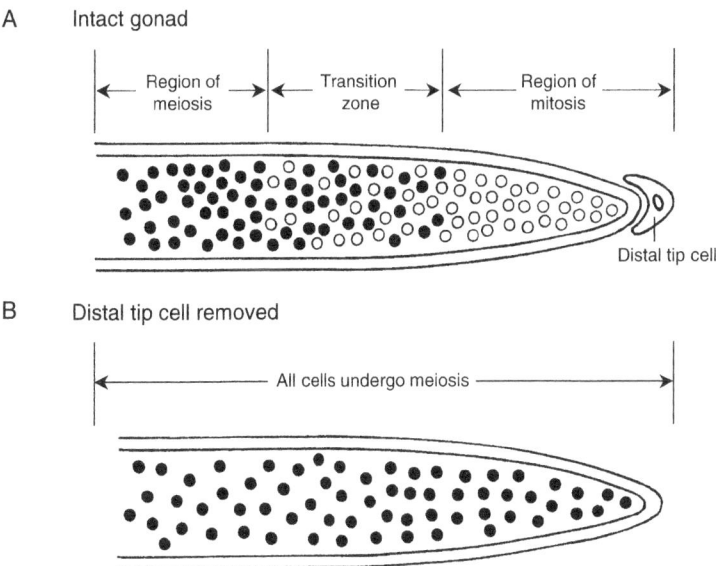

Figure 13.6 In the nematode worm *Caenorhabditis elegans*, a well-studied mode, the tip cell secretes a sort of growth factor that makes that in the gonad rounds of mitosis precede meiosis (redrawn after Gilbert, 1997).

Whatever explanation is given for the mitosis-meiosis sequence, it is clear that rounds of mitosis increase the number of gametes that can be generated in a lifetime. This, combined with the genetic variability that is generated by meiosis, turned meiosis and sexual reproduction into an evolutionary success story.

13.9 How do bipotential germ cells get committed to either the oocyte or the sperm cell scenario? Sex-steroids

In many species, the germ cells migrating into the gonad are bipotential and can differentiate into either sperm or ova, depending upon their gonadal development.

In the housefly and the mouse, which are the best-documented species, it is the gonad itself that is able to direct the differentiation of the germ cells. Thus in most organisms the sex of the gonads and its germ cells is the same. How gonadal tissue can direct meiosis into either sperm cell or oocyte differentiation is not fully understood. One possibility is that the differences in sex steroids present in the gonads, which (in vertebrates) are predominantly estrogens in the female and androgens in the male (see later), are important.

In hermaphrodites the situation is more complex. In the worm *Caenorhabditis*, the same gonad can produce sperm or ova. A cascade of genes controls this.

13.10 Gamete formation in the gonads: a benign form of cancer? The 'fight' between somatoplasm and germ plasm

Cancer cells have lost the mechanisms that prevent unlimited growth or division/multiplication rates. As a result, tumors arise and cancer cells start wandering around in the body to places where they do not belong. Our body has means for coping with disobedient cells and it eliminates them. Only when this defense system fails, do we get cancer.

Let us consider three aspects of the process these cells go through: growing big, dividing without restraint, and escaping from their normal place. Does this apply to gamete formation?

In order to give a plausible answer, a reminder of the definition of an 'animal' may not be superfluous. An animal is an organism that develops from a blastula, which is an epithelium (De Loof, 1992). Thus the basic cell type for animals is the epithelial cell type. This is characterized by its tight contacts with neighboring cells by specializations of the cytoskeleton and cell adhesion molecules. Growth and multiplication are restrained. Cancer cells are different because of their unrestrained multiplication rate and their abnormal growth and spreading.

Cells can grow abnormally in two ways. One possibility is that they can grow very large – much larger than the average size of a cell in the somatoplasm. Egg cells usually grow to a giant size. For example, the 'yolk' of a bird's egg represents one single cell. Even the whole egg is in fact a single cell because the egg white and the surrounding layers are acellular in nature. If a skin cell reached the size of a fully-grown oocyte, it would probably cause as much pain as an ulcer of similar size would cause. Developing eggs do not cause pain, perhaps because they secrete endogenous painkillers (endorphins).

The other possibility is that a cell looses its tight contacts with its neighbors, gets loose and starts to divide in an unrestrained way. This scenario applies to some extent to spermatogenesis.

If germ cells are considered to be a sort of cancer cell, the question then arises as to why the body does not get sick from the unrestrained germ cells and succumb? The answer is that the body does get sick, be it very slowly, and finally dies. The price to be paid for sexual reproduction is invariably death. Why can death be postponed for some time? Because the germ cells can be confined inside the gonadal tissues and finally ejected from the body after the body's arsenal of defensive strategies has been exhausted.

The most obvious mechanism for trying to silence the growing oocytes is yolk formation. This example of an anti-cancer strategy is dealt with next.

13.11 The 'idea' behind yolk accumulation in eggs: based upon an ancient innate immunological mechanism?

Our perception of the meaning of yolk is that it represents the nutrient store for the developing embryo. Females contribute to the generation of offspring by producing big, usually immobile, gametes, loaded with food. Sperm are specialized for swimming. They are slim and mobile.

Why did nature ever start to make this distinction? It cannot be stressed enough that nature does not deliberately plan for the future. In other words, there is no goal in nature. Thus, there has never been a deliberate strategy to replace asexual by sexual reproduction, to generate males and females, or yolk, or sperm cells that could fertilize egg cells. All this came into existence by accidental mutations. If a mutation increases the fitness of the organism(s) carrying the mutation, its frequency in the population will increase. There can be situations in which a mutation can be present in a population without any visible effect (silent mutations). When a second, third, etc. additional mutation takes place, at a given moment a novel communication pathway can appear. If this pathway provides a reproductive advantage, it may gradually replace the original system in the population.

Thus it was only after some organisms had started using preexisting cells loaded with yolk for reproduction, that this sexual form of reproduction could overtake the asexual form. But why did some cells engage in the accumulation of yolk precursors? The biochemical and endocrinological analysis of the nature of yolk and the way its formation is controlled in present day organisms allow us to gain insight into its evolutionary history (De Loof et al., 1998).

If one looks at a hen's ovary, with its numerous oocytes of increasing size, one might be inclined to think that the oocytes synthesize their own yolk. However, this is seldom the case. The common situation is that the storage proteins which are typical for yolk and which are called vitellogenins, are produced outside the ovaries. In vertebrates vitellogenins are produced by the liver, and in insects by the fat body in response to a hormonal signal that comes from the ovaries. Once released from the production site, the vitellogenins are transported by the blood to the surface of the oocytes. Here they bind to receptors. Next, by a process called pinocytosis (= drinking of the cell), the vitellogenin-receptor complexes are internalized by the oocyte. Small pinocytotic vesicles are pinched off from the membrane. Upon repeated fusion, the relatively large yolk platelets are formed. This process makes the oocyte turn into a giant cell. What we call 'the yolk' in a bird's egg is the single-celled oocyte. The 'egg white' is a watery solution of specific proteins that are produced by the oviduct when the oocyte descends into the uterus.

What type of proteins are vitellogenins? Do their properties shed some light upon the remarkable liver-gonad link? Vitellogenins belong to the protein family of the agglutinins and lectins (Stynen and De Loof, 1982). Lectins were first discovered in plants. They can be present in large amounts in the seeds of some plant species. They bind specifically to sugar moieties. This allows them to exert anti-microbial activity. Agglutinins are a similar type of protein, but they are present in animals. They too are part of our innate immune system. In vertebrates, the liver produces agglutinins. In the case of a mistaken blood transfusion, certain agglutinins in our blood plasma will agglutinate red blood cells originating from a donor with a non-compatible blood group. This agglutination ('sticking together') process allows the foreign blood cells to be destroyed: it is a system for protection against foreign intruders. Vitellogenins can also make red blood cells of certain types agglutinate, just as agglutinins and lectins do.

The following conclusion emerges: vitellogenins bind to the surface of the oocyte because it delivers the signal "I am foreign". It is likely that in the outer surface of the plasma membrane of oocytes, certain molecules are present that resemble certain ligands present in the cell wall of certain bacteria. Hence the 'hostile' reaction of the vitellogenins.

The next question becomes: what was the original role of estrogens in females and their counterparts in males, the androgens? Nowadays we say that estrogens allow the female to acquire its typical female phenotype and to become attractive to males. This is obviously true but this might not have been the original function of female sex hormones. In my opinion, sex hormones are substances whose original function was to warn the whole somatoplasm that a cancer had started developing at the site where the primordial germ cells were located. We must remember that after their confinement by follicle cells, the primordial germ cells remain dormant up to the end of larval life (time of puberty). Encapsulation by follicle cells is obviously an effective method of constraining the solitarized primordial germ cells for a period of time. At a given moment, for one reason or another, (e.g. because of decreasing production of some growth hormone), the follicle cells are no longer able to restrain the oocyte. It starts growing and is then recognized as abnormal or as non-self. The surrounding follicle cells emit an alarm signal in the form of a steroid hormone. That activates vitellogenin (agglutinin) production in the liver, but at the same time it also will affect all cell types that have receptors for this hormone. Thus other parts of the body, such as the integument and muscle tissue, can also be influenced. In this view, the acquisition of the phenotypes, which is typical for the two forms of gender, is a consequence of a defense reaction of the somatoplasm against the germ plasm. Perhaps the scenario I suggest explains why some sex steroids or their precursors can be used in cancer therapy: it was their original function.

One more question: why do only the gametes of females, except those of placental mammals, accumulate yolk? High levels of estrogens, the typical sex steroid hormones of female vertebrates, induce the synthesis of vitellogenins. Males do have the gene coding for vitellogenin. They do not activate this gene because they do not convert enough testosterone into estrogens. In insects, a similar mechanism is used but other hormones, namely ecdysteroids, are involved. Placental mammals do not produce yolk, probably because of a mutation in the vitellogenin gene.

13.12 Oviposition and sperm ejaculation: again 'hostile' acts by the somatoplasm

We are familiar with the fact that the body of vertebrates rejects grafted tissues originating from other individuals of the same population/species (xenotransplants). One says that an organism can discriminate between self and non-self. Non-self is usually rejected and eliminated. If one watches a bird laying an egg or a mammal giving birth to its young, the parallel with rejection of non-self is hard to deny. Normally, rejected material will not survive unless the environment offers an escape strategy.

Eggs are evidently self, but nevertheless they are ejected from the body as if they were non-self. Perhaps this ejection process is a normal physiological response of the female's body (somatoplasm) to the situation where the egg has grown to such an enormous size (for a cell) that it emits similar types of signaling molecules as if it were either non-self material, cancerous or an ulcer.

Since the genes of the offspring in mammals come for 50 % from the father and since many of these genes differ slightly from those of the mother, the offspring will have numerous proteins that are therefore non-self to the mother. For this reason, the immunological rejection of a fetus in the uterus of the mother would be as normal as the rejection of a piece of skin transplanted from the father to the mother. It is a remarkable achievement that the rejection in the uterus can be postponed so long that the embryo can develop into a viable being before the mother ejects it from her body.

From the point of view of cell biology, the ejection of sperm from the body of a male can hardly be seen as a friendly act. It will kill the sperm cells unless they succeed in finding an egg that has not yet been fertilized. But even then, the fertilizing sperm cells disintegrate: only their nucleus will survive.

The conclusion of all this is that the encapsulation of primordial germ cells by follicle cells and vitellogenin, the production of sex steroids and the ejection of gametes are all aspects of the hostile relationship between the somatoplasm and germ plasm.

13.13 Gender: why only two isoforms, males and females? Gender-determining genes

Because gender with only two forms, male and female, is the situation we see all around us in the animal world, it may appear that the living world was conceived that way. In the Bible it is said: "So God created man in his own image, in the image of God he created him; male and female he created them." He first made Adam, and from a rib of Adam he created Eve to be man's companion. Romantic but unlikely.

If the theory is correct that all living beings on earth are descendants from a single cell, then the first cell was by definition genderless. Indeed, gender implies the occurrence of two isoforms that are reproductively complementary. This situation may not have occurred earlier than many hundreds of millions of years after the Progenote. It could be that some mutation had a drastic effect upon steroid biosynthesis. This family of compounds plays a very important role in reproduction, at least in animals.

Our thinking about the genetics of gender is greatly influenced by the discovery of differences in gross structure of gender-linked chromosomes in mammals.

Female mammals have 2 X chromosomes and males have 1 X and 1 Y chromosome, in addition to their other chromosomes, the autosomes. If the Y chromosome were simply an X that has lost one its arms with the corresponding genes, the origin of male gender would be due to a reduction of the indifferent XX genome. The indifferent situation would then be the one with yolk accumulating gametes, on condition that such indifferent individuals could reproduce autonomously without males, in other words parthenogenetically. Mechanisms of parthenogenesis will be dealt with later.

To date, however, the mechanisms of sex determination both in mammalian and in insect model systems are much more complex than this simple theoretical model. The mammalian Y chromosome differs thoroughly from an X. In addition, one of the X-chromosomes becomes inactivated in a random way in all somatic cells of the body (Barr body), except in the germ cells. This means that female mammals, women inclusive, are mosaics of two cell populations that differ in their X-chromosomes.

The Y chromosome of mammals carries a gene called the SRY gene (sex-specific region of the Y chromosome), which upon translation into the SRY protein causes the formation of Leydig cells in the primitive gonads. These cells secrete testosterone. Through a cascade of molecular events (De Loof and Huybrechts, 1998), the male phenotype, physiology and behavior emerge. In animals with 2 X and no Y, the SRY protein is not produced. The resulting situation is the one we describe as female. Mammals are the only vertebrates with an SRY gene. The mechanisms of sex determination in all other classes of vertebrates are very poorly documented (Smith et al., 1999).

If gender, in its very origin, were only caused by one gene (product) that can either be present or absent without any intermediate possibilities, then no more than two different isoforms of gender would be possible. This way of thinking persisted until the discovery of additional sex determining genes that are located not on the X or Y chromosomes but on the autosomes. First the anti-testis gene DAX was discovered. Later DMRT was also found, not only in mammals but in birds and reptiles as well (Smith et al., 1999). Control of a given function by more than one gene is not at all unusual in nature. This fact, however, makes it difficult to clearly define the contributions of the respective genes.

Is it theoretically possible to have gender with more than two isoforms, i.e. males and females? The answer is positive. With only one variable, for example SRY present or absent, two isoforms can be produced, namely male and female. According to Mendel's genetic rules, with two variables, four combinations can be made, on condition that the genes act independently of each other. Thus, two sex determining genes could yield four gender isoforms, and three could yield 64 such forms. It could be that in nature there are already more than just males and females as sexual isoforms. Perhaps homosexual male or female, bisexual, confirmed bachelor, etc. are also isoforms. Like many traits, gender is also a multifactorial trait in which the importance of the different genes involved is not necessarily equal. This means that simple deterministic Mendelian genetics do not fully apply.

In *Drosophila*, three sex-determining genes are known, namely *sex-lethal*, *transformer* and *double-sex*. The RNA transcripts of these genes are different in males as compared to females (Chapter 2, De Loof and Huybrechts, 1998).

The structure of the sex chromosomes, if they occur at all in a given species, is not a good indicator of gender. In birds the sex chromosomes of females are ZW and those in males are ZZ in morphology. In bees the females (queen and workers) are diploid while the males (drones) are haploid. There are many other possibilities in chromosome configuration. However, it all comes down to gender determining gene products, no matter on which chromosome the genes themselves are located.

The next question is: how are sex-determining genes and sex steroids linked?

13.14 Cholesterol, sex hormones and gender. Again the bacterial infection route. Parthenogenesis and sex-reversal

The typical sexual phenotypic differences between male and female vertebrates are due to differences in sex steroids, which in turn are based on differences in genotype. Females have higher titers of estrogens than males. Males in turn have much more testosterone and dihydrotestosterone (=androgens) than females. In insects, the best candidate hormones to play a role in phenotypic gender formation are the ecdysteroids. The two major forms are ecdysone (E) and 20-OH-ecdysone (20E). The hypothesis has been forwarded that E functions as male sex steroid in some species and 20E as the female sex steroid (De Loof and Huybrechts, 1998).

All steroid hormones are synthesized from cholesterol. Many enzymatic steps are involved. Some of these enzymes are located inside mitochondria, others in membranes of the endoplasmic reticulum (Fig. 13.7). The cleavage of the side chain of cholesterol invariably happens inside the mitochondria.

But mitochondria are bacteria in their evolutionary origin. One could say that without mitochondria, there would be no sex steroids and hence no males or females as we know them to date.

Some fishes can undergo sex reversal. In one breeding season they are male, while in the next, they are female. The exact causes are not fully understood. In some arthropods, sex reversal can occur under the influence of a bacterial infection. In the isopod *Armadillidium*, the normal sex ratio is about 1/1. An androgenic hormone produced by androgenic glands, located in the vicinity of the testes brings about the male phenotype. Infection by *Wolbachia* bacteria leads to hypertrophy of the androgenic glands. If this happens in animals that are genetic females but have not yet differentiated, the functional gender and phenotype will become male.

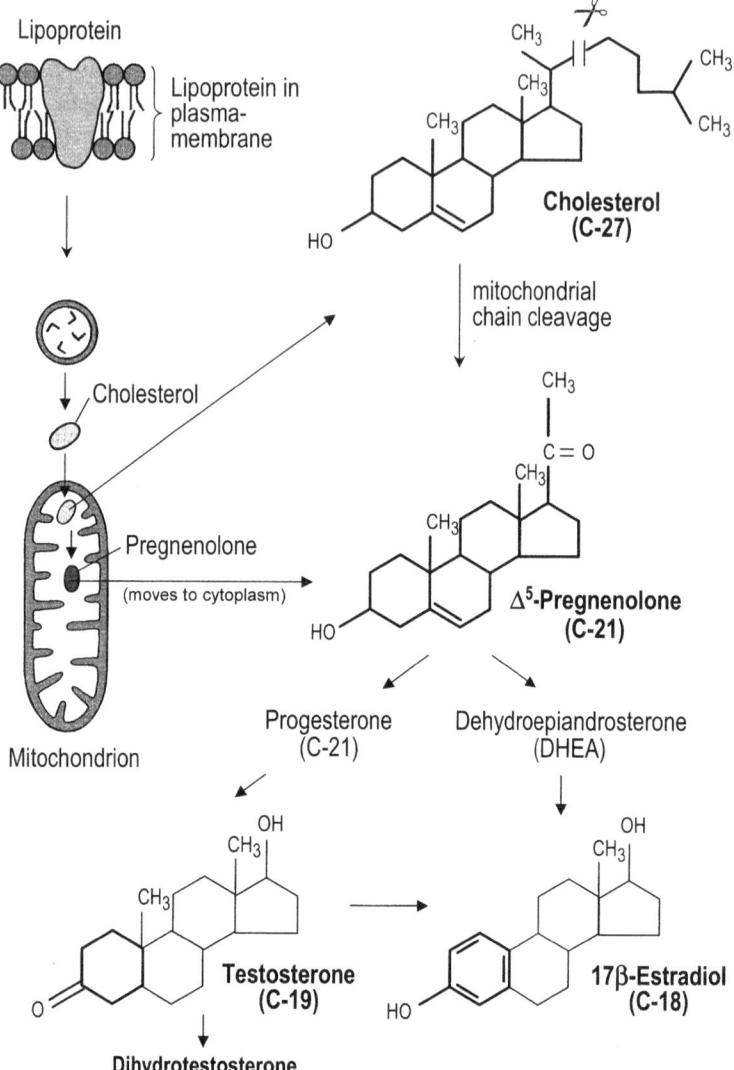

Figure 13.7 In vertebrates the so-called 'sex-steroids', are responsible for the typical male or female morphology. Males have higher concentrations of testosterone and of dihydrotestosterone while females have more estrogens. Cholesterol is their precursor. Part of the biosynthetic pathway takes place in the mitochondria. According to Margulis, these cell organelles are descendants from a bacterium that came to live in symbiosis with a bigger bacterium, thereby contributing to the formation of the eukaryotic cell type. Hence the deduction that sexual reproduction is made possible by an ancient bacterial infection.

In some other arthropod species, infection with *Wolbachia* causes feminization. In a wasp species, *Wolbachia* bacteria cause infected females to produce daughters without mating (Huigens et al., 2000). Thus, in this case parthenogenesis is a bacterial disease transmitted through the eggs, not through males. The exact mode of action of *Wolbachia* is unknown.

13.15 Males: from superfluous to indispensable?

It would have been an enormously lucky coincidence if the formation of both functional egg and sperm cells occurred at exactly the same moment in the course of evolution. It is more likely that there was a gap in between. The question then is: what came first, eggs or sperm? Gametes devoid of nutrients had no chance of making it into an embryo. Thus, eggs must have come first. Eve preceded Adam. But Eve's eggs were haploid and her daughters had to be diploid again, otherwise meiosis could not occur. Hence, the question arises as to how the haploid eggs could restore the diploid situation.

Perhaps, initially, oogonia underwent meiosis, yielding four equally sized cells. If such cells could fuse two-by-two, and if their nuclei were competent, then the diploid number of chromosomes would have been restored and two embryos would have resulted. However, in the course of time the symmetry in cytokinesis was abandoned in meiotic oogonia, but not in spermatogonia. Yolk accumulation became predominant in only one of the four oocytes. Perhaps, one of the three smaller cells could still fuse with the big one and give rise to subsequent embryonic development. Another possibility, still used by some aphids, is that one of the polar cells was retained inside the oocyte. This way, the diploid chromosome number is also retained. Finally, the asymmetry in cytokinesis became extreme with one very big oocyte and three small cells, the polar cells, which did not have enough 'vigor' left to engage in fusion. The problem of realizing the diploid situation had become acute.

If at this moment the male was capable of contributing a cell that could fuse with the big oocyte, then the diploid situation would be restored. The males thus became necessary, whereas before, in terms of the population, they had been mere 'decoration'; they had been producing sperm cells without any genetic function.

However, this hypothetical scenario does not automatically imply that right from the beginning the male's sperm also contributed in a functional way to genetics. This may appear to be contradictory to what is commonly stated in textbooks. However, the possibility that DNA can be rendered inactive by methylation must be considered as well.

13.16 DNA methylation and imprinting

Imprinting can occur in mammals. This means that a young child can resemble one of his parents more than would be expected on the basis of Mendelian genetics, which contends that the mother and the father make equal genetic contributions. This could be due to the selective methylation of genes.

One of the four bases in DNA, namely cytosine, is sometimes methylated, thereby forming 5-methyl-cytosine. When methylated, the gene is likely to be inactive, despite the fact that it can be transmitted to the zygote without any problem. About 5 percent of the cytosines in mammalian DNA are converted to 5-methyl-cytosine (Jones and Takai, 2001). In contrast, *Drosophila*, worms and probably most invertebrates do not (readily) methylate their DNA. After a long search, in *Drosophila* a few genes were found that are methylated during a short span of time in early development (Lyko et al., 2000). Again in mammals: some genes are methylated in the female, others in the male. Also, both the primordial germ cell nuclei are highly hypomethylated, but upon maturation, both the sperm and egg genes undergo extensive methylation. Perhaps this reflects a situation from long ago in the history of evolution.

One can only make guesses about the role of methylation at the moment an egg and sperm fused for the first time. Perhaps, there was no methylation at all and methylation was a later acquisition. Another possibility is that in one of the gametes, all the genes were methylated, thereby rendering them inactive. In the progeny, only the genes from the non-methylated parent were then active. That could have been either the male or the female. Later in evolution, perhaps the methylation was no longer affecting all genes, and perhaps it even got spread over the genes of both the male and female. It could be that the final outcome in the course of evolution was the situation we now see in most invertebrates: almost no DNA methylation, and equal contribution of males and females to the progeny.

For the females there is some consolation: maybe they have lost their full genetic dominance in the course of evolution, but they will never become indispensable. Males should not be so sure about that.

13.17 Death of the somatoplasm: the price to be paid for meiosis. Sexual pleasure in return. The progeny: goal or free surprise bonus?

From the point of view of longevity, one could wonder whether the introduction of meiosis was beneficial to the organisms that started using it. Ultimately it was not, because it led to their death. As long as organisms could use both the asexual form of reproduction (e.g. budding) and gamete formation, immortality was still prevailing. From the moment, however, that the asexual strategy no longer worked and organisms completely switched to meiosis and gamete production, the death of the somatoplasm was the price to be paid. I favor the idea that meiosis was an unfortunate accident for the individuals that suffered from it, but a blessing for the population. If we look around us, we see that the kingdom of the animals suffers more from death than any other kingdom. Indeed, the plants, the Fungi, the Protists and the Prokaryotes have retained more possibilities for asexual reproduction than the majority of animal species have.

Despite the fact that death is the dramatic fate of sexually reproducing animals, sexual reproduction not only vigorously persists but is even gaining in importance. This anomaly can probably be best explained by considering the compensating effect of the feelings of pleasure and happiness - most probably not only in humans - associated with sexual reproduction. If the feelings of well being associated with sexual reproduction (e.g. orgasm) are intensive enough, death will become acceptable even in those species that know what death means. Both the nervous and the hormonal systems play a dominant role in eliciting orgasmic feelings.

With respect to sex in the broad sense, most organisms are completely ignorant about its mechanisms and final outcome. They have no means for acquiring 'insight'.

First, the most common situation in nature is that - the studax *Homo sapiens* not taken into account - females and males do not know that external or internal fertilization involves the transfer of spermatozoa. Nor are they aware of the role of spermatozoa or of the principles of embryonic development. And neither do they know that there is a causal relationship between mating/copulation and offspring.

Secondly, in the majority of species the parents, and in particular the males, never see their offspring. Typical examples here include many species of corals, worms, fishes and reptiles. The number of species that practice brood care involving one or both parents is very low. In many insect species, the parents are already dead by the time the young larvae emerge from the eggs.

From all this, it follows that I do not like the jargon of evolutionary biologists who speak about the male or female 'drive' to produce offspring and become successful as though it were so obvious. The idea that males aim to produce offspring, and for this reason engage in a competitive struggle with other males is largely a myth. Males are not so clever in general, and certainly not clever enough to be interested in females because of potential progeny/offspring. This offspring-obsessed point of view concerning males that is widely being acclaimed in sociobiology these days (e.g. to 'justify' rapist behavior) overlooks some basic facts. If males and females engage in mating/copulation (external or internal fertilization), it is due to the sexual arousal and pleasure that partners of opposite gender can experience. Or, in the case of rape, it is rather for power than for pleasure. In the case of internal fertilization, males eject their sperm into the body of a female because this act rewards them with orgasmic feelings. Females accept males for similar reasons.

Thus, whether we like the idea or not, the progeny produced from the gametes is a free surprise bonus awarded along with the sex. It is a byproduct, not a goal.

13.18   Sperm competition. The choosing female

Males usually produce large numbers of sperm. In man, one milliliter of semen contains some 100 million spermatozoa, sometimes more, sometimes less. But in the end, only one spermatozoon will fertilize an egg. Hence, this overproduction of sperm may look like an enormous waste of energy. Sperm cells encounter many obstacles on their way to the egg cells and many are thus eliminated. Furthermore, many sperm cells are of bad quality. Motility is very important. Sperm with low motility has little chance in winning the race. Therefore, one could say that well functioning mitochondria in sperm cells are very important in sperm competition. Mitochondria are responsible for energy production in the form of ATP. Here, the bacterial route shows up again: mitochondria contribute not only to sex steroid production, but to sperm motility as well.

When sperm cells from different males are mixed, 'hostile' reactions may be elicited among them.

The picture of the passive female that just undergoes insemination by a male is loosing ground. In birds, the estimation is that in about 70 percent of all species, females actively seek copulation with a more dominant male than their regular partner. One could say that a female chooses the male by whom she wants to be chosen. A hen can eject the sperm of a rooster that she does not like (e.g. a young one) when the older dominant rooster is still around. Thus, the ability of females to selectively accept sperm is a means of influencing reproductive success. Basically, the preference (if existing) of females for strong, dominant males may reflect an unconscious search for well functioning mitochondria.

In bees, the queen mates with several partners during the nuptial flight – and only then. She will retain the sperm in the spermatheca in her body for the rest of her life. She can determine which eggs are fertilized and which ones not. Fertilized eggs give rise to females (workers and new queens), unfertilized eggs yield males (drones).

13.19 Conclusions: the origin of reproduction is not what you thought it was

In attempting to reconstruct the evolutionary history of sexual reproduction, one inevitably has to rely upon known principles of cell biology and physiology in order to find plausible explanations for phenomena that took place long ago. Because it is impossible to recapitulate history, and because important links may have completely vanished in the course of time, the proposed scenario is inevitably incomplete.

Most textbooks on general and on cell biology focus on the 'mechanics' of different aspects of reproduction. How do mitosis, meiosis, gamete production, fertilization, etc. work? The molecular-genetic approach of recent years has made the explanations quite complicated, even for specialists in the field. My approach aims at complementing the mechanistic approach by addressing the key question, namely why it all come into existence.

Anyone may occasionally wonder why nature evolved the way it did. Is there a goal in evolution? Or did it all evolve by accident? In the known laws of biology there is no ground whatsoever for goal-oriented evolution. Nature does not aim at anything. With its ceaseless attempts in an endless series of possible genetic combinations, it is inevitable that a few good hits should emerge from time to time.

Nature did not aim to develop meiosis in order to use it for sexual reproduction. It did so because of physiological necessity. At a given moment, the conditions inside the gonad anlage must have been such that the cells present there had no other choice than to engage in meiosis, even if there was not yet a function for the resulting cells. The exact nature of that ancient 'physiological necessity' remains to be determined.

In essence, sexual reproduction is the story a special sort of cancer, a germ cell cancer, that remains confined to the gonads because of an array of defense strategies exerted by the somatoplasm. Some of these strategies, however, have profound side effects on the physiology of the somatoplasm itself and on behavior. This 'germ cell cancer' differs from all other forms of cancer because it bestows moments of intense pleasure and feelings of well being upon animals that suffer from it. The intensity of these feelings is so high that death became an acceptable option in return for sexual reproduction – at least for humans.

Man likes to see 'love' as something almost supernatural, divine. The physiological approach that describes 'love' as one of the defense strategies of our immunel system against non-self may be experienced as completely opposite to what we intuitively feel. However, our intuition can mislead us. Consider the fact that our brain makes us believe that the sun turns around the earth.

The hypothesis that in due time, or in any case after a long delay, an ancient bacterial infection caused the bodies of the infected organisms to display gender in two isoforms that we call male and female may also look far-fetched. However, there are good arguments in favor of the proposition. Most humans are insufficiently aware of man's descent from a bacterial ancestor. Even more, all eukaryotic organisms are still symbiotic units of ancient bacteria. Thus, asserting that a bacterial contaminant in our cells or in one of our cell organelles, the mitochondria, made us acquire gender is just a further extension of the same line of thought.

ESSENTIALS

REPRODUCTION. GENDER. SEX

1. Why the Progenote engaged in the life-threatening undertaking of splitting itself into two parts will remain forever a matter of speculation. One possible explanation is that some 'toxic' factor accumulated above a given threshold, perhaps only locally, thereby initiating a chain of reactions that resulted in fission.
2. Asexual reproduction does not involve egg or sperm cells (gametes). It uses the mechanisms of regeneration. It involves only mitotic cell division and hence it does not generate genetic variability. In principle, asexually reproducing organisms are immortal.
3. Sexual reproduction requires at least egg cells (parthenogenesis), but usually sperm cells as well. Such gametes are formed by meiosis, a complicated form of cell division that only occurs in gonads (ovaries and testes). All resulting gametes are genetically different.
4. The price to be paid for meiosis is the death of the somatoplasm.
5. In animals the primordial germ cells are already formed very early in embryonic development, even before the gonad anlagen are formed. The primordial germ cells remain dormant, perhaps due to inhibition by the somatoplasm, until the end of larval life.
6. From the cell biological point of view, all phenomena accompanying gamete formation, including the formation of the primordial germ cells, sex steroid production, yolk protein formation and sperm ejaculation, fit the theory that gamete formation in the gonads is in fact a benign form of cancer.
7. The definition of an animal is important in this context. An animal is an organism that develops from a blastula. In histological terms, a blastula is a closed epithelium. Primordial germ cells escape from their original epithelial environment and would perhaps become unrestrained if the somatoplasm did not mobilize all kinds of mechanisms to prevent their abnormal growth and proliferation.
8. Given the complexity of the process, it is unlikely that gender with two isoforms, male and female, would have been invented more than once in evolution. However, a unifying principle to explain the master mechanism upon which gender is based is still missing.
9. Organisms do not aim to have progeny. Most of them do not even know that they have produced offspring. Hence, it must be concluded that, in general, individuals engage in mating primarily to increase their degree of contentment. The resulting offspring is a free surprise bonus.
10. Sperm cells originating from different males may compete with each other. In some species, the female is able to choose from which male she will accept or eject sperm (the choosing female concept).

CHAPTER 14

EMBRYONIC DEVELOPMENT

THE ACQUISITION OF HIGHER LEVELS OF COMPARTMENTALIZATION AND CORRESPONDING COMMUNICATION SYSTEMS IN THE COURSE OF THE LIFETIME

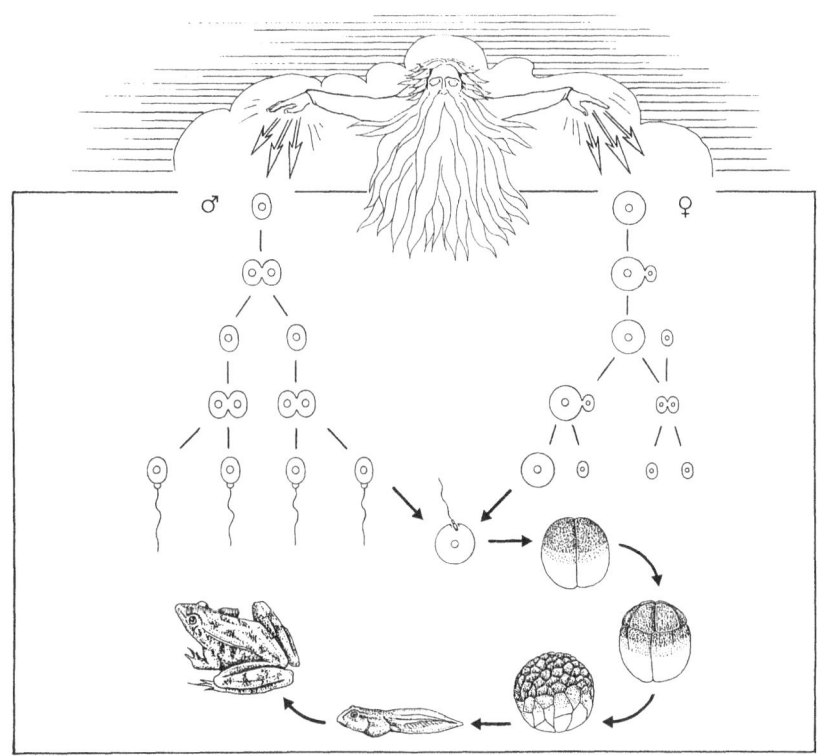

God said, "Let us use asymmetry again, this time to bring complex multicellular organisms into being."

Contents

14.1 The key questions in developmental biology
14.2 Model organisms for the study of development
14.3 The importance of asymmetry
14.4 Successive morphological stages in animal development: *Branchiostoma lanceolatum* as a model
14.5 The key morphological principle underlying embryonic development in animals is centered on epithelium formation
14.6 Mechanisms instrumental to the acquisition of ever higher levels of compartmentalization
14.7 The molecular biological approach in the study of development: *Drosophila* as a model.
14.8 *Drosophila* complemented with Dolly: the scientific importance of cloning
14.9 The role of the environment: the example of the identical twins again
14.10 The macromolecular environment around the genes: fine tuning of gene expression. Transcription factors
14.11 The ionic environment around the genes. The complexity of the 'internal ionic environment'. Six levels of control
14.12 Processes that can be controlled by inorganic ions
14.13 The universal principle upon which differentiation in animals is based
14.14 Self-selection during development
14.15 The 'vital force' of Hans Driesch
14.16 Changes in the level of compartmentalization during development. *In vitro* fertilization, cloning, abortion: ethical aspects

Essentials

14.1 The key questions in developmental biology

Just imagine if a long time ago, in the early days of microscopy, somebody would have succeeded in isolating one by one all the different cell types of our body and culturing them in artificial medium. Next imagine that this person played a trick on a colleague, a specialist in systematics of unicellular organisms, by asking him to determine these strange organisms. The story told by the cell biologist was that they had been accidentally found in a new environment that tasted a little bit less salty than regular seawater. There is a good chance that the systematics specialist, in good faith, would have come up with a list of 'new' species of organisms. Indeed, a nerve cell is so different from a red blood cell or from a muscle cell that, on morphological grounds, there would have been no reason to classify these cell types as isoforms of the same species.

Yet, all these cell types do have the same genome, despite their very different morphology and physiology. How this conclusion was reached will be explained later in this chapter.

In order to better understand evolution, in particular mega-evolution, a sound grasp of the principles of development is needed. Evolution and development are tightly interlinked. Evolution deals with the formation of new species out of existing ones. Development deals with the formation of new cell types out of existing ones. Changes in the embryonic development of particular individuals precede later changes in gross morphology in the population.

The major question in micro-evolution (of the hardware) is: "How can shifts in gene frequency in a given population take place so as to yield a new species?"

With respect to mega-evolution, the key question is: "How do additional levels of compartmental organization come into existence in the course of geological time?"

The corresponding questions, but now on the time scale of embryonic development are:

1. "How can the morphology and physiology of the descendants of a cell change while the genome remains unaltered?"
2. "Which mechanisms underlie the coming into existence of ever higher levels of compartmental organization?"

Excellent classical and more recent textbooks on developmental biology include those of Balinsky (1975), Langman (1976), Davidson (1976), Hopper and Hart (1980), Gilbert and Raunio (1997), Gilbert (1997), and Wolpert (1998). Most of the figures concerning *Branchiostoma* are redrawn after Balinsky (1975).

14.2 Model organisms for the study of development

The morphological aspects of development have been studied in many species, belonging to all classes of plants and animals. General rules have been outlined. Since the 1970s, researchers have been focusing more on the molecular mechanisms of development. For the time being, the best-documented animal models are the fruit fly *Drosophila melano-*

gaster, followed by the worm *Caenorahbditis elegans*, the mouse, and the zebra fish. The best-documented plant model is *Arabidopsis thaliana*.

14.3 The importance of asymmetry

If a fertilized egg were perfectly spherically symmetrical, and if it divided into two perfect halves, which in turn also divided perfectly symmetrically, etc., then there would be no differentiation into different cell types. The only outcome of such a mode of development would be balls of identical cells. The size of the balls would be dependent on the number of rounds of mitotic division the system had undergone. It would be restricted to asexual reproduction because without the introduction of some sort of asymmetry, no gametes could be produced.

This is obviously not what happens during development in sexually reproducing systems. Thus, there must be mechanisms that introduce functional asymmetry into the system. It may seem contradictory, but orderly systems, both living and non-living, are often based on the introduction of asymmetry.

In fact, embryonic development in animals, the systems to which this chapter is restricted, is based upon the introduction, very early in development, of asymmetry in two successive phases. After other aspects of development have been highlighted, this double asymmetry principle will be further explained and illustrated in Fig. 14.4. In brief, first the plasma membrane-cytoskeletal complex acquires a symmetry that is different from the spherical one. Next, some early cleavages are executed in such a way that daughter cells are formed which differ in some of their properties. This represents the second element of asymmetry.

14.4 Successive morphological stages in animal development: *Branchiostoma lanceolatum* as a model

It is easier to illustrate the morphological principles governing development with a relatively simple animal system. Consider the development of a lower chordate without vertebrae, *Branchiostoma lanceolatum*. This marine animal (Fig. 14.1) lives most of the time burrowed in the sand of shallow waters. The females produce small eggs with little yolk and release them into the seawater, where they may be fertilized. The early mitotic divisions of fertilized eggs are called cleavages. The resulting cells are blastomeres.

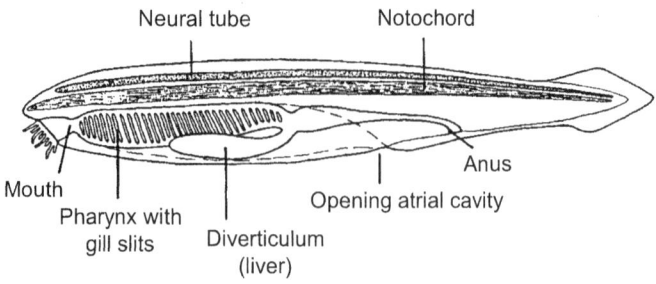

Figure 14.1 Basic body plan of *Branchiostoma lanceolatum*, a lower chordate.

Fertilization.

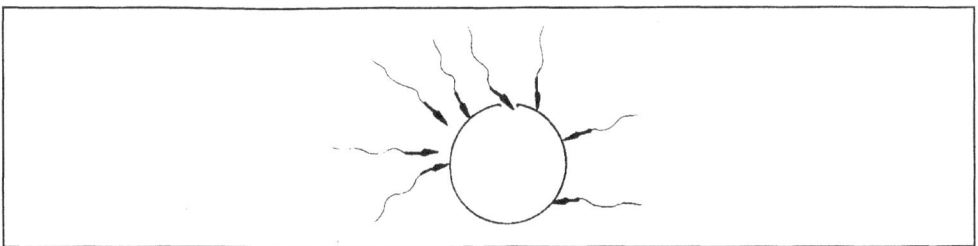

This is the first change in compartmentalization and one of the few instances where the number of compartments decreases with an increase in complexity as a result. The egg cell compartment and the spermatozoon compartment fuse, thereby forming a zygote and activating the 'developmental machinery'.

Cleavage.

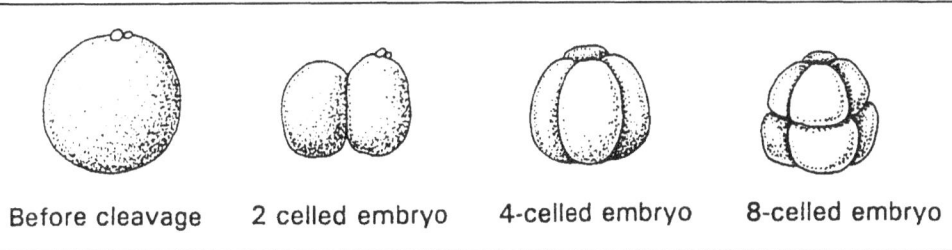

| Before cleavage | 2 celled embryo | 4-celled embryo | 8-celled embryo |

The fertilized egg (compartmentalization level 2 in Fig. 4.1) starts a series of mitotic cell divisions whereby the size of the cells decreases until it finally reaches that of typical somatic cells. The cells that result from these early mitotic divisions, referred to as cleavages in embryology, do not detach from each other at the end of the mitotic cycle but remain together. This novel level of compartmentalization can be achieved, for example, by the 'gluing' of the blastomeres together by cell adhesion molecules (level 3). As the blastomeres become smaller in successive cleavages, the embryo starts to look like a mulberry and is named a morula. In some animal species it has been shown that communication through gap junctions starts early in development. These events upgrade the embryo to compartmentalization level 4.

The blastula: mono-epithelial compartmentalization. Definition of 'animal'.

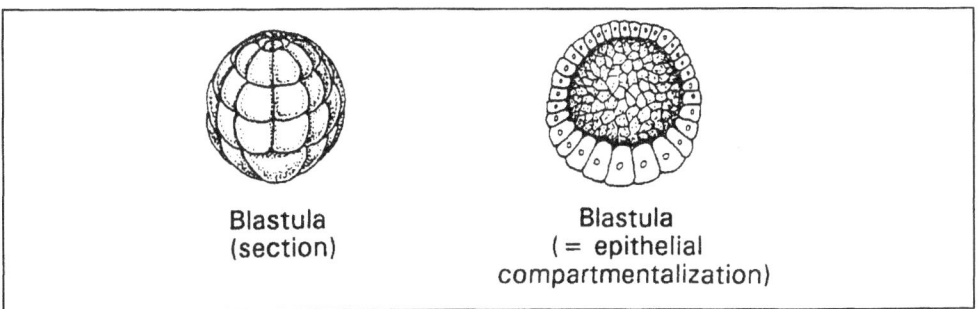

Blastula (section)

Blastula (= epithelial compartmentalization)

The cells of the morula align themselves in such a way that they form a simple epithelium that encloses a fluid-filled cavity (level 5 in Fig. 4.1). This is ***the*** crucial step in the development of any animal. From this event, the definition of what an animal is has been derived. In the Five Kingdoms classification system *an animal is defined as an organism that develops from a blastula or*, in other words, *that necessarily passes through the stage of organizing itself into an epithelium, with all the properties that are characteristic of such a tissue* (De Loof, 1992). The blastula represents the level of mono-epithelial compartmentalization (Level 5). The cells are kept together both by cell adhesion molecules and by specializations of the actin cytoskeleton. These allow neighboring cells to be attached to each other so tightly that nearly all the water and solute molecules which pass through the blastular epithelium are transported straight through the cells and not in between them. As soon as an epithelium is functioning, it can build up trans-epithelial gradients.

The formation of the primitive gut: gastrulation. The beginning of poly-epithelial compartmentalization..

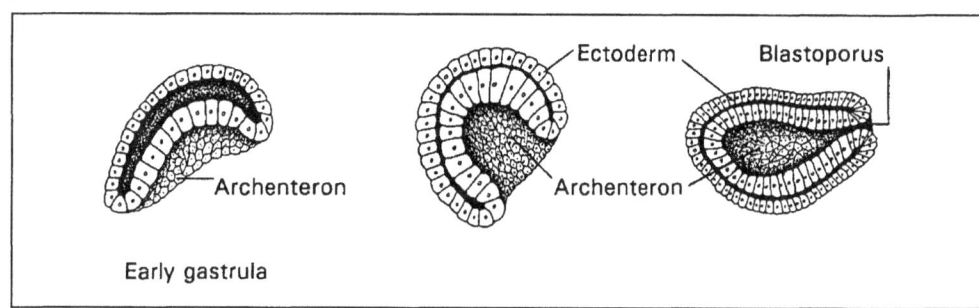

Gastrulation means the formation of a gastrula, and 'gaster' means 'stomach'. It happens this way. The blastular epithelium starts infolding at a distinct point, thereby forming an internal sac, the primitive gut or archenteron (compartmentalization level 6). This event is an example of the internalization of an epithelial compartment inside an outer epithelium. In this way some groups of cells that were far away from each other in the blastula come into each other's vicinity. Next, they can start communicating with one another in a chemical language, for example by releasing organic molecules (morphogens) that induce the activation or inactivation of specific genes in cells that have receptors for these morphogens. The presence of gradients of morphogens is well documented in some model animals.

Organogenesis and segmentation

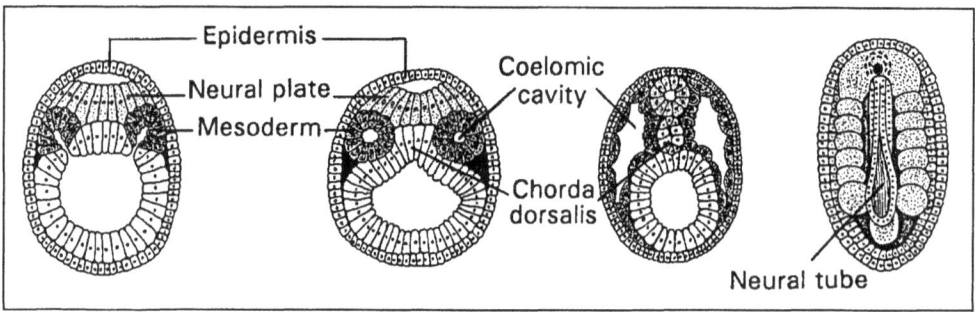

Following gastrulation more inner compartments are formed. At given locations, both the outer epithelium (= skin or ectoderm) and the inner epithelium (=gut lining or endoderm) infold. Some of the pouches that are formed this way are severed, thereby forming novel, epithelium-lined compartments. Two typical examples can illustrate this.

The coelomic cavity arises by the evagination of cells to form of two small pouches from the archenteron. The pouches then detach from their parental epithelium, enlarge and finally fuse.

The second example is the formation of the central nervous system. A small longitudinal ribbon of the ectoderm, located at the dorsal side of the embryo, starts infolding under the influence of an underlying rod-like structure, the chorda dorsalis. The invagination proceeds so far that the groove gets closed and the resulting entity becomes completely internalized as a hollow tube, called the neural tube.

Later in development some cells escape from epithelial structures and form non-epithelially organized structures such as muscles, the skeleton and blood. At a given moment, the embryo gets subdivided into a linear series of entities, the segments, thereby reaching level of compartmentalization 7.

14.5 The key morphological principle underlying embryonic development in animals is centered on epithelium formation

From this brief description it may have become clear that the basic morphological principle underlying *animal* development is *epithelial compartmentalization*. First an epithelium is formed, and then the embryo becomes more or less spherical. Next comes invagination. Then the severing of pouches and the folding of epithelia complete the job.

At the time of the writing of this book, numerous key genes which code for the signaling molecules that control a variety of important developmental events have been identified (Gilbert, 1997; Wolpert, 1998). However, little attention has been paid to the genes that are essential for the formation and transformation of epithelia.
It is evident that any time a higher level of compartmentalization is acquired, additional communication systems must be activated. According to my definition of life, life comes into existence when a given compartment acquires the ability to communicate at its highest level of compartmental organization. A fertilized egg corresponds to level 2 of compartmental organization, a segmented animal to level 7: thus, levels 3, 4, 5, 6 and 7 *are all different forms of life of the same organism*. This also means that 'life' changes in leaps during early embryonic development.

14.6 Mechanisms instrumental to the acquisition of ever higher levels of compartmentalization as relevant to mega-development

These mechanisms have already been outlined in Chapter 4 and are summarized in Fig. 4.1.

The mechanisms that are used to reach level 6 or 7 are:
- gluing together by means of cell adhesion molecules and by specializations of the cytoskeleton;
- formation of cytoplasmic contacts (gap junctions and cytoplasmic bridges);
- epithelium formation and transformation;
- internalization of epithelial compartments;
- formation of a nervous system for the headquarter function in animals.

It is likely that the search for the genes that govern these phenomena in different taxonomic groups may result in lists of genes that, in combination, yield similar morphological and/or physiological effects. This topic is interesting for a better understanding of what is called 'convergent evolution'.

14.7   The molecular biological approach in development: *Drosophila* as a model

Progress in unraveling the genetic program underlying animal and plant development has been extremely rapid in recent decades. The fruit fly *Drosophila melanogaster* (Fig. 14.2) has served as an ideal animal model to unravel the genetics of development. There were good reasons to choose this animal. *Drosophila* is small and it can be easily reared in the laboratory. Its developmental cycle is short. It has only 4 chromosomes. The ones that are present in the salivary glands are very large and visible (under the microscope) at all times of the cell cycle. The genetics of *Drosophila* are very well studied. Thousands of mutants have been characterized. The animal can be modified relatively easily by molecular biological approaches. Finally, the (nearly) full genome has been sequenced and was published in 2000 by Adams *et al.*

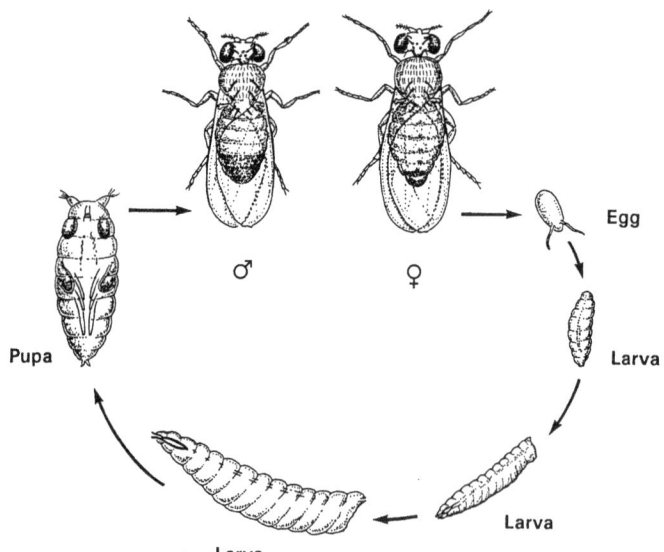

Figure 14.2. The life cycle of *Drosophila melanogaster*. There are three larval stages. The fully-grown third instar larva leaves the food and pupates. A few days later, the adult fly emerges.

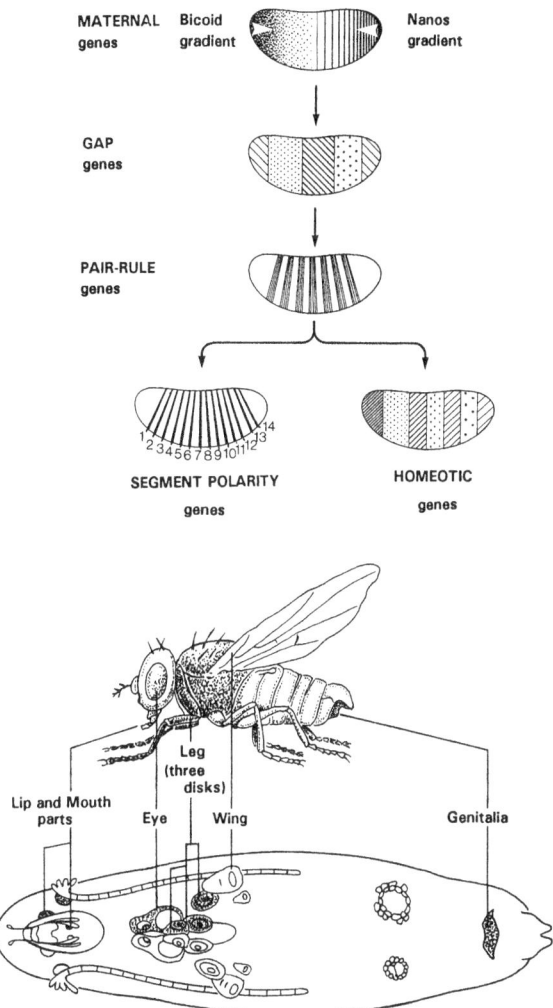

Figure 14.3. Development of the fruit fly *Drosophila melanogaster*, the best documented model system with respect to the molecular biology of pattern formation in a segmented animal. Although the morphology of vertebrates is quite different from that of insects, the basic molecular processes underlying the embryonic development of both groups are much less dissimilar. The essence is that a fertilized egg, which corresponds to compartmentalization level 2 (Fig. 4.1), is progressively subdivided into more and more subcompartments, resulting in the appearance of the adult form, which corresponds to compartmentalization level 7.

Drosophila has about 13,300 genes. A relatively large percentage of these genes display similarities with genes known to exist in other organisms, including man. Therefore it is not surprising that the basic principles that govern development are similar throughout the animal kingdom.

The genes that are needed to make a head, a rear end, a dorsal and a ventral side, segmentation, legs, antennae, eyes, a male or female, etc. are known. Ed Wilson, Christiane Nüsslein-Volhard and Eric Wieschaus received the Nobel Prize for medicine in 1995 for their pioneering work in this domain. For more information on this exciting topic, I refer to the review paper by St Johnston and Nüsslein-Volhard (1992), the textbooks on Developmental Biology by Gilbert (1997) or Wolpert (1998), or the book *The development of Drosophila melanogaster* edited by Bate and Martinez Arias (1993).

The major groups of genes that govern development in *Drosophila* are represented in Fig. 14.3.

The story begins with the adult female, which lays eggs in which two *gradients* of regulatory proteins are present. These proteins are called 'bicoid' and 'nanos', and they are involved in anterior-posterior axis formation. These proteins are not evenly distributed throughout the entire yolk mass, but rather are present in the form of gradients. The highest concentration of the bicoid protein is found at the anterior pole of the egg, the lowest one at the posterior pole (= first gradient).

The nanos protein is most concentrated at the posterior pole. Its concentration diminishes towards the anterior pole (= second gradient). Thus, at the moment that the egg is still present in the uterus of mother fly, the determination of what will become the anterior and posterior parts is already accomplished. The genes that code for 'bicoid' and 'nanos' are called **maternal genes** because they are already expressed when the egg is still inside the female. In concert, these two gradients cause the formation of a novel gradient, namely that of the 'hunchback' protein, which occurs in the highest concentration in the anterior part of the egg. In its turn, this protein gradient controls the expression of the **gap genes**, which specify broad areas in the embryo. The name gap genes is derived from the fact that mutations in these genes result in gaps of groups of segments (areas without a cuticle at the ventral side). In their turn, the gap gene proteins cause the expression of the **pair-rule genes** in the primordia of each alternating segment. These pair-rule genes control the expression of **segment-polarity genes** in each individual segment. These genes are responsible for the phenomenon that the two-segment primordia are divided into an anterior and a posterior part. Hence, the embryo is subdivided into fourteen segment primordia (three for the head, three for the thorax and eight for the abdomen). Now, the proteins of the gap genes, the pair-rule genes and the segment polarity genes interact to regulate the expression of another class of genes, the so called homeotic genes. The proteins from the homeotic genes make each segment acquire its final identity.

A fly larva has no legs and no antennae or other appendages. This contrasts with the situation in most other insects (e.g. locusts), where the newly emerged larvae have appendages. In insects undergoing complete metamorphosis, the different appendages (legs, wings, etc.) are formed out of what are called *imaginal discs*. These discs differentiate during metamorphosis under the influence of hormones.

14.8 *Drosophila* complemented with Dolly: the scientific importance of cloning

Nevertheless, the molecular approach, which yielded such a wealth of data about the numerous mechanisms instrumental in development, was insufficient for solving the following major problem in developmental biology.

All the different cell types of an adult organism are the progeny of the same single cell, namely the fertilized egg. They have been generated by *mitotic division*, which, by definition, should *yield cells with identical genomes*. If the genomes of all somatic cells of an organism are indeed identical, how can it be explained that the cells themselves nevertheless acquire different morphologies and engage in different functions?

The 'determinant hypothesis' of August Weismann, which dates from the early 20th century, postulated that the appearance of ever new cell types is due to the progressive, selective elimination of various sets of determinants. As a result, each differentiated cell type ends up with only its own specific sets of determinants (genes). According to Weismann's hypothesis, each fully differentiated cell type no longer has the full genome that is typical for the species, but only a subset of this genome. Since Weismann, the term 'determinant' has been abandoned. Nowadays, the unit of heredity is called the gene.

One cannot deny that there is good logic in Weismann's hypothesis. Yet it turned out to be wrong. The right explanation is that differentiated cells still have the full genome, but that they use only part of it, in a cell-type specific way. It took quite an effort to disprove Weismann's hypothesis with reliable experiments. Within this context, the cloning of the sheep Dolly was the crown (in terms of publicity) on a century-long endeavor involving great amounts of preparatory research.

What is cloning? First, an unfertilized egg cell is harvested from a (hormonally treated) ovulating female. By means of a micropipette, the nucleus is removed from the egg cell. With another micropipette, a nucleus is sucked up from a differentiated cell of the body. In Dolly's case, this was the nucleus of an udder cell. This nucleus, which is diploid, is injected into the enucleated egg cell. Fertilization is no longer required because the implanted nucleus is already diploid. If an embryo develops, its genome will be 100 percent identical to that of the mother. Furthermore, the formation of a normal adult proved that the differentiated udder cell still contained the full genome. Dolly is indeed a normal sheep, which does not mean that every baby animal obtained by cloning is normal. The success rate of the cloning procedure is still very low (a few percent) and there are many malformations and early deaths.

It is an illusion to think that in mammals a cloned baby will be an exact copy of its mother. As mentioned before, one of the two X-chromosomes gets inactivated in all somatic cells. This happens in a random manner. As a consequence, any mammalian female is a mosaic of two genetically different cell populations. This applies to the clone as well: any clone will be different from its mother.

In the popular news reporting on this topic, the impression was given that Dolly was the first success in the cloning of a vertebrate. This was not true. Years before, the team of Gurdon (1977) in the UK had done many such cloning experiments with amphibians. From Gurdon's experiments, from Dolly and from other successes in cloning mammals, the conclusion could be drawn that Weismann's hypothesis is wrong. Differentiated cells still have the full genome. Hence, it must be concluded that they use only a part of their genome, in a differential manner.

The next question is: What is it that changes during the differentiation from egg to somatic cells if it is not the genome that changes?

14.9 The role of the environment: the example of the identical twins again

By deduction, the conclusion is reached that if the genome remains unaltered, then the only possibility left is to introduce changes in the 'environment' around the genes.

The following example may help to illustrate the problem. Let us reconsider the theoretical example of identical twins born in Leuven, the Dutch-speaking city where I live. They are separated in such a way that one will grow up in Leuven and the other in Tokyo. The one raised in Leuven will speak Dutch while the one in Japan will speak Japanese. This illustrates that the genome itself is only part of the story. It contains enough information for any of the twins or any child to learn any language in the world and to behave according to any local cultural standards wherever on earth, on condition that they grow up in an appropriate environment. *Thus, it is the environment that tells the twins how to use that part of their genome that is used for this type of 'communication'*. This example illustrates that one should not exclusively focus on the genome: the environment is also very important.

A similar situation prevails in the different cell types of a differentiated organism. As has been demonstrated by cloning experiments and other research, all somatic cells do indeed have the same genome, (with the exception, for example, of the antibody-producing cells and the fact that in all somatic cells of female mammals one of the X-chromosomes gets inactivated to form a Barr body).

The question emerges as to what the term 'environment' means in relation to genes. In other words, what is the environment around the DNA double helix? This environment has a macromolecular component consisting mainly of proteins that can bind, whether alone or in the form of complexes, to the regulatory sites of genes. It also has another component consisting of the inorganic ions present at those sites in the cell that are relevant for protein synthesis. Both components can influence the expression of specific (sets of) genes.

14.10 The macromolecular environment around the genes: fine tuning of gene expression. Transcription factors

A gene has a promoter region, a start codon, a coding region (exons and introns, depending on the gene), a stop codon and, usually, a signal sequence for making a polyA tail (Chapter 2). The expression of many genes requires that a variety of regulatory proteins (enzymes and others) bind, often in the form of complexes, to specific regions of the promoter sequence, or sometimes even to sequences far away from the promoter, or even in introns.

Trans-regulators is the general name for soluble macromolecules (including proteins and RNAs) that are made by one gene and interact with genes on the same or different chromosomes. The term 'transcription factors' is used for proteins that specifically bind to regulatory regions of a gene that are called 'enhancers' or 'silencers'. By doing so, they contribute to controlling whether the expression of a gene will be stimulated or suppressed. Numerous transcription factors controlling the expression of key genes for development have already been identified.

It is an acceptable approximation to state that the macromolecular environment around the genes is responsible for the fine tuning of gene expression.

However, from the evolutionary point of view the problem arises as to "What came first, the chicken or the egg?". It should be remembered that any protein, whether regulatory or not, is itself an end product of gene expression. How was the expression of the very first transcription factor regulated? Logically, one should say that it was not regulated by a proteinaceous transcription factor. But was there any alternative?

There probably was, in the form of changes in ionic concentrations, in particular of H^+ (changes in pH) and in Ca^{2+} (explosions and oscillations), the two major regulatory ions even up to the present time.

14.11 The ionic environment around the genes. The complexity of the 'internal ionic environment'. Six levels of control

Living matter makes use of only a few inorganic ionic species, the major ones being K^+, Na^+, H^+, Ca^{2+}, Mg^{2+}, Cl^- and HCO_3^-. Such ions are present in all cells, no matter whether they are prokaryotic or eukaryotic, animal, plant or fungal in nature. One might be inclined to deduce from this universal occurrence that such 'simple' ions can hardly be used for control of specific physiological functions and, therefore, their role should not be more than marginal. Such is not the case.

In Chapter 9, I mentioned that impulse conduction in nerve cells largely depends upon fluxes of Na^+, K^+ and, to a lesser extent, of Ca^{2+}. The generation of an action potential and impulse conduction are certainly not the only processes in which inorganic ions play a key role: muscle contraction, gene expression, osmoregulation and transepithelial transport should be included as well.

The 'ionic environment' in a given compartment comprises the total concentrations of the different inorganic and organic ionic species present. Ionic activities (= concentrations of free ions, thus excluding the ions, for example, that are bound to macromolecules), potential and ionic gradients, ionic currents, and secondary chemical gradients that can be formed as the result of a potential gradient, for example, in systems in which self-electrophoresis has been reported.

All cells invest a lot of energy in maintaining their own specific internal ionic environment. If there were no need to do so, it would be a tremendous waste of energy and mutations would probably have eliminated this useless energy expenditure long ago in the course of evolution. Obviously this did not happen. This suggests that good control of the ionic environment is crucial for the well functioning and survival of all cells.

If one analyses the system, six levels of control become apparent. However, It would lead us too far afield to elaborate in detail on their mechanisms. I will simply enumerate the six levels. More details and figures can be found elsewhere (e.g. De Loof, 1986).

First level of control: the plasma membrane and its different types of ion transporting molecules (ion pumps and channels), including the gating mechanisms of ion channels.

Second level: properties of the membranes that form the boundaries of cell organelles. In some cases, under the influence of certain hormones, oscillating Ca^{2+} explosions with a different intensity in the cytoplasm as compared to the nucleus have been observed. The nuclear envelope may thus be a controlling barrier for some types of ion transport.

Third level: anchoring of ion transporting proteins at specific locations in the plasma membrane. For example, epithelial functioning requires the spatial separation of ion pump and ion channel activity.

Fourth level: contacts between neighboring cells. In particular, gap junctions and cytoplasmic bridges can contribute to changes in the internal ionic environment of connected cells.

Fifth level: the specific location of cell organelles. From the cell physiological point of view, it may make a difference whether a given cell organelle, such as the nucleus, is situated inside or outside an electrogenic ion flux. In such situations, secondary gradients of regulatory factors can de generated. The site where the nucleus of the different muscle cell types is located is illustrative in this respect. In skeletal muscle, the nucleus lies right underneath the plasma membrane. In heart and in smooth muscle, the nucleus is far away from this membrane.

Sixth level: the exact location of the chromosomes/genes inside the nucleus. In cells in which the nucleus can create its own internal ionic environment, the exact location of a chromosome can be important. In *Drosophila* it has been proven that each chromosome has its own specific location. As a consequence, each gene is situated at a well-defined and constant position inside the nucleus (Agard and Sedat, 1983).

In a nutshell: cells have 'invented' a variety of possible mechanisms for making their internal ionic environment a much better ordered system than a simple salt solution. No two cells can have a fully identical internal ionic environment: there are just too many variables and regulatory mechanisms involved for this to be statistically probable. Nature favors variability, again and again.

14.12 Processes that can be controlled by inorganic ions

The fact is very often overlooked that once a protein involved in development has been synthesized, it is not automatically active. Proteins are molecules in which the primary structure is a linear chain of amino acids arranged in a very specific sequence. They are only active after they have somehow ended up in the place in the cell where they will exert their function, and most importantly, *after they have adopted their functional three-dimensional structure*. Phosphorylation-dephosphorylation is a widely used mechanism that changes the conformation of some proteins. However, it is also known that changes in concentrations of some inorganic ions, in particular of Ca^{2+} and H^+, can have drastic effects on protein conformation and/or activity. Both ions can be quite toxic to cells if their concentrations become too high.

The internal ionic environment in a cell can also have effects on the following: the conformation of chromatin; the dynamics of membrane-, cytoplasmic-, and nuclear skeleton formation and functioning; the association of chromatin with the nuclear matrix (perhaps); protein-protein and protein-nucleic acid interactions; the association of the different classes of RNA with the cytoskeleton; the selective transport of certain molecules across membranes; protein synthesis. This list is not exhaustive.

In short: the 'ionic environment' controls development by contributing to the correct conformation of many macromolecules, not only in the cytoplasm but in the nucleoplasm as well. I consider this as the 'coarse tuning' of gene expression. Inorganic ions are not the only players. For example, changes in chromatin structure can also be brought about by chemical modification (acetylation and phosphorylation) of the tail of histones, the proteins around which the DNA thread is wound.

14.13 The universal principle upon which differentiation in animals is based

On the basis of all these observations, data, and deductions, and on the basis of all the research done by successive generations of developmental biologists during the past 150 years, I think that the highly complex process of development and differentiation in animals can be crystallized into its essence as follows:

> Differentiation in animals, (whose basic body plan corresponds to level 6 or 7 of compartmentalization, namely that of a polyepithelium, either without (level 6) or with (level 7) segmentation, is the stepwise formation of cells or cell clusters that differ in their plasma membrane-cytoskeletal properties. This asymmetry in plasma membrane-cytoskeleton results from the 'double asymmetry rule', as illustrated in Fig. 14.4.
>
> These differences in plasma membrane-cytoskeletal complex properties *cause* the distinct emerging cell types to display a differential pattern of gene expression resulting in a cell type-specific morphology and physiology, notwithstanding the fact that they all share an identical genome and similar mechanisms of protein synthesis and processing.
>
> Differences in the plasma membrane-cytoskeletal complex, the structure that confers a differential three-dimensional molecular scaffold on the cells, are first realized. Differential protein synthesis and pattern formation follow (of necessity): ***form precedes function***.
>
> The major 'strategy' used in the differentiation of somatic cells seems to be: ***"Keep the genome constant but change again and again the ionic and/or macromolecular environment around the genes."***
>
> Changes in the ionic environment may play a role in the coarse tuning of gene expression/protein synthesis. This can be achieved by influencing the three-dimensional conformation of a variety of macromolecules. It can also be achieved by realizing a specific form of the cytoskeleton.
>
> Changes in the macromolecular environment (especially in *trans*-acting factors) may be more appropriate for the fine-tuning of the expression of specific genes.

This rather simple basic principle does not exclude the many different molecular mechanisms that can be invoked by cells to realize differentiation. (For an overview of all the possibilities, see Gilbert, 1997).

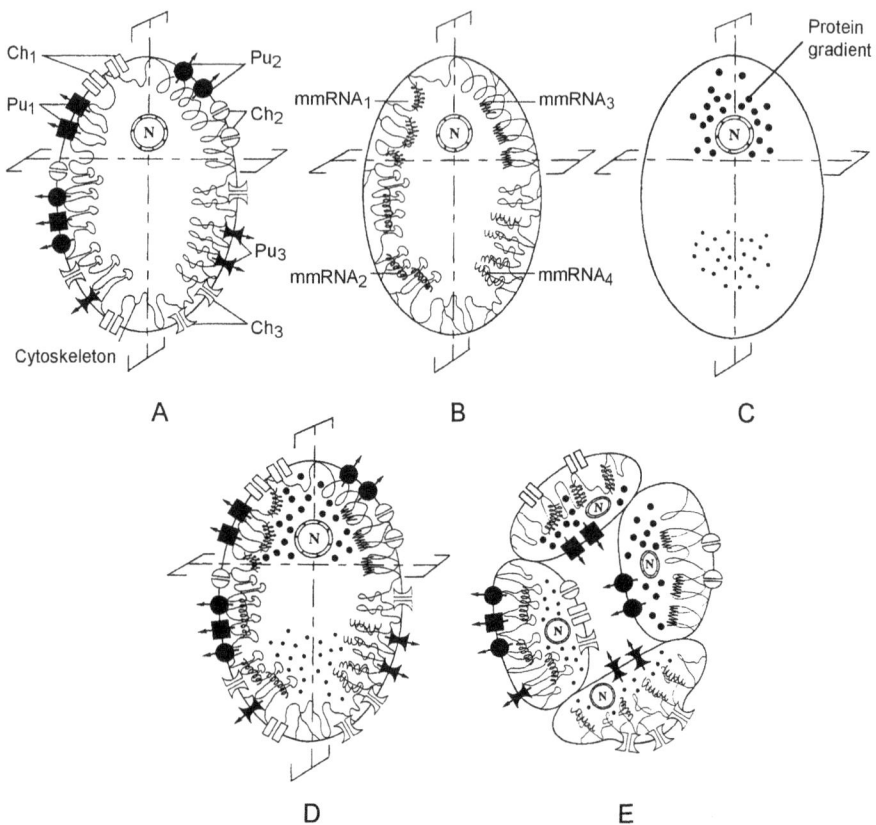

Figure 14.4 Schematic representation of the double asymmetry principle. Mechanisms instrumental in the generation of functional asymmetry in the stem cell (zygote) of a hypothetical 4-celled, epithelially organized organism (animal) (modified after De Loof, 1994). First, an asymmetrical distribution is realized (i) of the plasma membrane-cytoskeletal complex (A: Pu_1- Pu_3 = 3 types of ion pumps; Ch_1- Ch_3 = 3 types of ion channels); (ii) and/or of maternal messenger RNAs (B: $mmRNA_1$- $mmRNA_4$), some of which can be anchored to the cytoskeleton; (iii) and/or of gradients of certain proteins in the cytoplasm/yolk (C) (e.g. of the *bicoid* and *nanos* gradients in eggs of *Drosophila*, Fig. 14.3). Next, the egg is cleaved in an asymmetrical way. Sometimes this happens already during the first cleavage, sometimes in the second, but never later than during the third. In D, which represents the superposition of A, B, and C, it is the second cleavage which will give rise to four different cell types, all sharing the same identical genome.

14.14 Self-selection during development

When dealing with the origin of the first cell on earth (Chapter 12), I argued that life and self-selection go hand in hand, because life cannot exist without the buildup of an ionic gradient. Self-selection also operates in embryonic development. The following example illustrates this. In sea urchins and many other organisms, the penetration of a spermatozoon into the sea urchin's egg causes depolarization of the plasma membrane, pH changes and an 'explosion' of Ca^{2+} in the cytoplasm, all representing changes in gradients. Thus, fertilization causes changes in some gradients. These changes are thought to alter the conformation of certain macromolecules, thereby activating a number of enzyme systems such as those that repolarize the plasma membrane (= building up of a gradient) or intensify metabolism. The cytoskeleton also undergoes drastic changes. From its very first moment of coming into existence, the young zygote faces selection pressure. It is 'tested' to see whether or not it can survive the changes in its own internal gradients. If it cannot, it will be eliminated. If it can, it will develop further. Every time a higher level of compartmentalization is reached and every time novel communication systems start operating, the self-selection procedure is executed. The zygote immediately undergoes 'testing' to see whether it can survive all the changes in gradients: here again, from the very beginning, self-selection takes place, first at the molecular level and then at the cellular level, probably in similar fashion to the way the first cell that began life on earth was 'tested'. Self-selection goes on as long as novel compartments are being formed during development, when gap junctions between neighboring compartments become active, etc.

14.15 The 'vital force' of Hans Driesch

Anybody who has ever watched the development of an embryo is tempted to think that embryonic development requires more than just the execution of mechanical processes. Hans Driesch, the famous German embryologist and philosopher, well known for his 'vitalism' theory, thought that embryonic development was driven by a mysterious 'vital force'. Neither he, nor any of the others whose thinking tends in this direction, have ever been able to identify this 'force'. There is no evidence for its existence as such, though some now speak of a 'vital force' in another sense.

Nowadays we know a lot more about 'immaterial 'aspects of life. Driesch had no idea of the electrical phenomena that play such an important role in all living systems. In his book *A Study of Bioenergetics* (1986), Dr. Franklin Harold calls these electrochemical processes 'the vital force'. In addition, the analysis of the immaterial aspects inherent in communication activity dates from much later than the period Driesch was active in.

14.17 Changes in the level of compartmentalization during development. *In vitro* fertilization, cloning, abortion: ethical aspects

From my definition of life, which uses both qualitative (the nature of the acts of communication) and quantitative (the number of the different communication acts) parameters, it follows that life as an activity *increases* during development. The life of a zygote of a multicellular organism is only a fraction of that of an adult. The smaller an organism, the more likely it is that its life is 'smaller' than that of a bigger organism. However, the *quantitative* aspect of life cannot possibly be used as a measure of the 'value' of a given compartment: this depends mainly on the *quality* of the communication and on the 'interpretation' by a given compartment of what it considers to be 'good' or 'not good'. What is 'good' for compartment *a* can be 'bad' for compartment *b*. In my opinion 'good' could be defined as that which improves the communication of a given compartment, and 'bad' as that which worsens it. In Chapter 16 I will focus again on 'good' and 'bad'.

The power humans have acquired over 'life' by *in vitro* fertilization techniques, cloning, medical treatment, drugs affecting the nervous system, methods for euthanasia, etc. have raised many ethical questions. Answers that are acceptable to everyone, regardless of their religious and political convictions, are not possible.

A question often asked by students: When does a human embryo become a real human person? Possible answers: at fertilization of the egg, at blastula formation, or from the fourth month of intrauterine development onward, as suggested by the abortion laws in those countries which have legalized abortion through the third month of pregnancy, or perhaps as late as at the moment of birth.

My answer to this question is based on the combination of genetics with my definition of life, from which it follows that a given compartment starts to live from the moment that it acquires the faculty to communicate at its highest level of compartmental organization.

When a sperm cell touches the plasma membrane of an egg cell, a novel type of compartment is formed and a higher level of communication system is activated in a fraction of a second. The Ca^{2+} concentration in the cytoplasm increases dramatically and the pH changes as well: it can be stated that the 'life' of an embryo starts with a Ca^{2+} explosion. I think that this is the onset of the life of an embryo. Some centers for *in vitro* fertilization use the criterion that a human embryo comes into existence from the moment that the sperm and egg nuclei fuse. Others say that this happens only after several cleavages. I do not agree with these criteria.

Any mammalian somatic cell has the full genome of either a male or a female. Is such a cell a full human being? No, because it has no longer the ability to realize the compartmental levels 3 to 7 (Chapter 4).

I think that a human *being* comes into existence from the very moment of the fusion of the sperm cell with the egg, which results in the first higher-level gradient, the Ca^{2+} explosion. From genetics, a similar conclusion ensues. From the moment that the sperm nucleus has entered the egg, all genes that make up a human being are present. The fertilized human egg is a full human being, alive and well. All developmental changes that happen later are only quantitative and qualitative changes in levels of compartmentalization and accompanying acts of communication that do not change the genome. This means that there is no biological basis for denying an unborn baby, whatever its age, the status of being a full human being. It is not so that a fertilized egg is not yet a human being because it is very small. A new-born baby is also small compared to an adult, but nobody will say that a baby is not a full human being. The quantitative criterion is not a good one.

Let me be clear that I do not deny any person the right to decide for himself whether abortion or euthanasia are morally acceptable or not. Decisions about life and death are very personal and usually influenced by the pain of very difficult living conditions. When governments legalize abortion, they invoke criteria other than biological ones and these criteria are not a tribute to human dignity. Nowadays, there are societies where some animal species are better protected and have more rights than some humans. Animal rights activists would most surely be opposed to the idea that abortion methods routinely applied to women should be applied to pet, farm, or laboratory mammals. There are countries where abortion can be carried out up through the fourth month when a pregnant woman says that the pregnancy interferes with her plans for a holiday. There are abortion centers that call themselves 'Help Centers'. There is a television station that gave a woman the opportunity to declare in public that she had two abortions "because she loved her unborn babies so much": this happened in the Netherlands in 1995. There is something very wrong in such societies. In Chapter 16 I shall explain that life is hierarchically organized and that the interests of the higher compartment prevail over those of the lower. The motto "Master of my own uterus" is not compatible with this principle. In fact, *reproduction is regeneration of the population* (De Loof and Vanden Broeck, 1995). Thus the process of reproduction goes beyond the interests of the individual.

With respect to cloning, there can be no doubt that some day some people will offer such procedures for cloning humans. Neither can there be any doubt that some people will be willing to pay lots of money to be cloned. But why should a human society as a whole engage in such practices while often in the very same clinics, numerous perfectly healthy unborn babies are being eliminated? Many couples are on waiting lists to adopt a baby, often in vain.

ESSENTIALS

EMBRYONIC DEVELOPMENT

1. Long ago, Ernst Haeckel, a famous German embryologist, formulated the biogenetic law: "Ontogeny recapitulates phylogeny". He thought that a developing embryo passes very quickly through the stages that the species it belongs to passed through in geological time in the course of evolution.
2. Despite all the inadequacies in the arguments forwarded by Haeckel and others in favor of this hypothesis, it remains an interesting idea.
3. In animals, the fruit fly *Drosophila melanogaster* with its giant chromosomes in its salivary glands, its numerous mutants, and its advanced genetic transformation possibilities, etc., has become the most frequently used model for studying development. We now know which genes are responsible for the formation of the head and the rear part, of the dorsal and the ventral sides, of legs, eyes, segments, etc.
4. As already stated in Chapter 13, an animal is defined as an organism that during early development organizes its composing cells into a closed epithelium, which is called a blastula.
5. In principle, animal development always follows the same basic pattern: first, a closed epithelium is formed (the blastula); next, this epithelium folds and further folds upon itself, and then certain cells (groups) escape from their normal epithelial organization to form additional structures. Animals are largely constructed as folded epithelia.
6. The only type of cell division used in embryonic development is mitosis. Normally mitosis yields genetically identical daughter cells. As a consequence, all somatic cells of the body have the same genome, apart from a few exceptions.
7. Nevertheless a differentiating cell gradually acquires a different morphology and engages in different functions. This is possible because cells do not use all their genes. Numerous mechanisms exist to ensure that in the end only part of the genome is used in a given cell type, and that it is used in a differential manner.
8. In the early stages of development, animal embryos bypass some of the restrictions imposed by the mitotic mode of cell division upon the differential use of genes/proteins by exploiting the possibilities offered by asymmetric cleavage. This is known as the double asymmetry principle, a term referring to the two successive steps in the development of asymmetry in an organism: first the formation of a non-spherically symmetrical plasma membrane/cytoskeletal complex, and then the execution of at least two consecutive asymmetric cleavages.
9. The universal principle underlying differentiation seems to be: "Keep the genome constant, but change, again and again, the ionic/macromolecular environment around the genes."
10. In the course of development, ever-higher levels of compartmental organization come into existence. This means that the 'life' of a developing embryo progresses step by step as higher compartmental levels are formed.
11. Any embryo, including the human, becomes a full being at the moment of the first Ca^{2+} explosion that accompanies fertilization. Laws legalizing abortion are based not on biologically sound principles but rather on social or political considerations.

CHAPTER 15

EVOLUTION

THE *'HARDWARE-SOFTWARE'* OR *'DOUBLE CONTINUUM'* THEORY

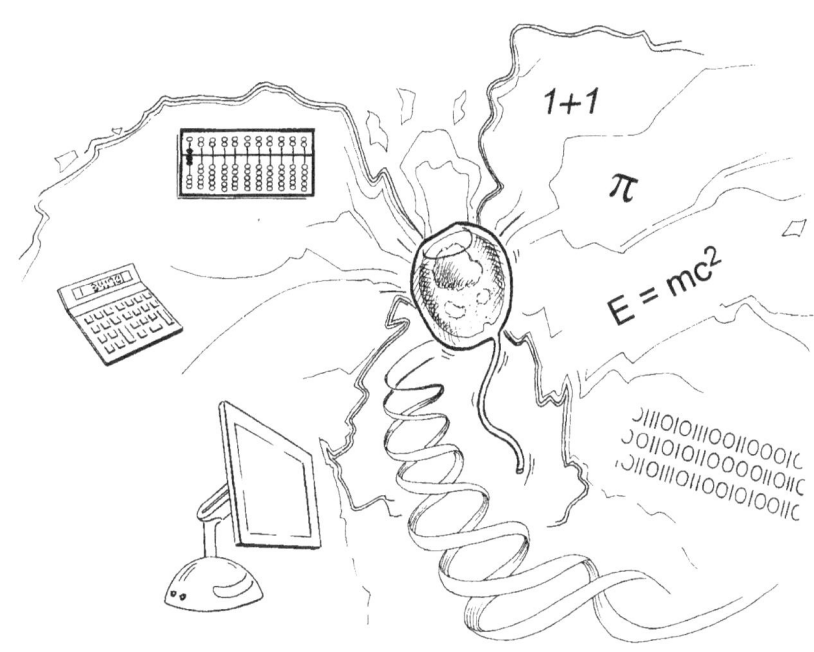

Evolution of 'life as a double continuum' began right away in the Progenote.
(After an idea by Frederik Lerouge.)

Contents

15.1 A sound theory of evolution should explain not only the formation of new species but also of 'life' at all its levels of compartmental organization
15.2 Micro-evolution. A brief overview of Darwinism
15.3 Elements in neo-Darwinism subject to improvement. Life is a double continuum
15.4 Thinking about evolution with insights from the computer era: not only the hardware evolves, but the software as well
15.5 Is evolution based merely upon software possible? The example of the identical twins for the third time
15.6 'Cultural evolution' is software-driven, being based on teaching-learning processes. Abstract thinking. Consciousness. The role of tools
15.7 Definitions of 'natural selection'
15.8 Natural selection: cause or result? The driving force(s) of evolution. Mutations and changes in the cognitive memory system
15.9 Summary of the essentials of the double continuum - or the hardware-software - theory of evolution
15.10 Input, through the software-aspect, of the organism's own input or free will in evolution? Creationism
15.11 Signal transduction pathways and the ongoing dispute between 'Phyletic gradualism' and 'Punctuated equilibrium'
15.12 An upgraded adage

Essentials

15.1 A sound theory of evolution should explain not only the formation of new species but also of 'life' at all its levels of compartmental organization

When Darwin and Wallace formulated the principles of what is currently known as 'the theory of evolution', the central issue was: "How can new species come into existence?" Without stating this explicitly, it was implicitly assumed that a theory that could explain the formation of new species should serve as well to explain the evolution of 'life' in its most general form. Because this was an unspoken assumption, the necessity to clearly define 'life' was pushed into the background. At the current time, this is still visible in nearly all theoretical studies on the theory of evolution: they do not even deal with the definition of life, while it should be normal practice to begin with it.

Darwinism is a splendid theory. Its greatest merit is that it created the conceptual framework for the common descent (with modification) of all contemporary organisms. The results of genomic analyses of increasing numbers of organisms undeniably demonstrate the validity of this concept. All organisms on earth are related to one another; they are 'family'. Their family tree goes back to one single ancestor, the bacterial Progenote that lived some 3.7 billion years ago. Although for the most part, Darwin is not considered to be a philosopher, he deserves a place of honor among them. For me, he is even the primus inter pares.

The theory of evolution can be regarded as definitely accepted as far its idea of evolution by descent with modification is concerned. Yet, the last word has not yet been said about the mechanisms governing evolution. Some convinced creationists are even likely to disagree with the first part of this paragraph.

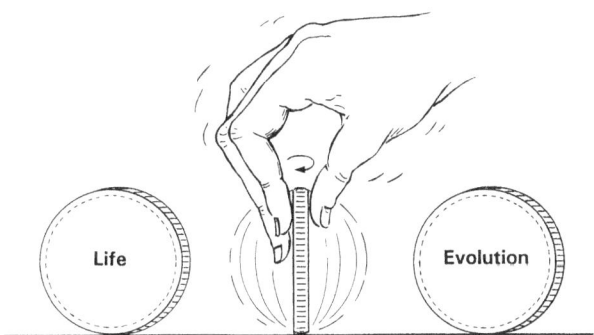

Figure 15.1 Life and Evolution are as closely interconnected as the two sides of a coin.

One important goal of this book is to demonstrate that the theory of evolution can be made even more encompassing and attractive by not restricting it to the evolution of population/species level but rather opening it up to the evolution of 'life'. This obviously requires that we first define 'life'. If my definition (in this book) of 'life' as 'communication activity' is correct, then the basic question in evolutionary biology becomes: "How can communication/problem solving activity change in the course of time, in particular of geological time?" In this perspective, the formation of new species (micro-evolution), no matter how important it may be, is only part of the broader topic. I will not focus on the genetic mechanisms of micro-evolution. This topic is amply covered in numerous textbooks on Evolution that have been published during the recent decade (e.g. Graur and Li, 1999; Gould, 2002).

15.2   Micro-evolution. A brief overview of Darwinism

When in 1859 Darwin published his theory on the origin of species, now known as The Theory of Evolution, the laws of genetics, of probability and of the thermodynamics of dissipative systems had not yet been formulated. The physics known at that time was formulated in terms of the dynamic models of Newton's mechanics, now referred to as 'linear Newtonian physics' (see next chapter).

The principal elements of Darwin's theory are:
1. Organisms have a very high reproductive potential. Many more individuals of each kind are born in each generation than can possibly obtain food, find shelter, survive and reproduce.
2. However, apart from exceptional circumstances, the size of a population either remains fairly constant or else fluctuates within moderate limits. This means that this high reproductive potential is not fully valorized. There is competition among individuals ('struggle for life') and selection of certain individuals that will finally survive.
3. Variability among individuals is a key issue in evolution: individuals are not identical in all their characteristics (identical twins and asexually reproducing organisms not taken into account).
4. Some individuals have certain characteristics which enable them to better survive and reproduce than individuals with other characteristics (the 20th century philosopher Spencer called this 'survival of the fittest').
5. Some of the characteristics that result in differential survival and reproduction are heritable.
6. Vast spans of time, measured in the geological time scale, have been available for change.

The final result is that populations adapt to changes in their environment, be it in a passive way.

Around 1900, the physicists Maxwell and Boltzman introduced the 'probability revolution' and this soon had consequences for the interpretation of Darwin's ideas. The 'Oxford school' of evolutionists emphasized genetic evolution and adaptive explanation while the 'American school' prompted a greater interest in drift and in population structure (Depew and Weber, 1994). The discovery of the double helix structure of DNA by Watson and Crick in 1953 was another milestone. The elucidation of the structure and role of DNA, RNA and proteins, the mechanisms controlling the expression of genes, and the different ways the genome can change (gene duplications, point mutations, inversions, deletions, etc), have greatly contributed to a better understanding of the genetic basis of the theory of evolution. All this led to the *neo-Darwinian synthesis, the centerpiece of which is the idea that natural selection, acting upon individual variations within a population to substitute one allele (= a variant of a gene at the same locus) for another, is the major force driving adaptive evolutionary change* (Kauffman, 1993).

However, not everybody fully agrees with this interpretation. First, there is the still ongoing debate between the theory of 'phyletic gradualism' which holds that evolution proceeds gradually, with progressive change in a given evolutionary line, and the theory of 'punctuated equilibrium'. The latter theory, of which Stephen J. Gould is the big promoter (see Gould, 2002) holds that evolution occurs in spurts, between which there are long periods in which there is little evolutionary change (Raven and Johnston, 1996). Furthermore, the 'neutralists' argue that much of evolution at the molecular level is selectively neutral.

The questions: Nature or nurture? Chance or necessity? are not yet unequivocally answered either. There are two major lines of thought. Richard Dawkins, author of the bestseller book *The Selfish Gene* (1976), and like-minded evolutionary biologists believe that natural selection is adequate to explain virtually every observation in evolutionary biology. Thus changes in genes and in their frequency in populations suffice to explain evolution. On the other hand Stephen Gould and his followers believe that natural selection is a very important force in evolution, but not the only one. As I will explain later in this chapter, I also hesitate very much to restrict natural selection to merely changes in genes: this is too simplistic.

At the end of the 20th century and about 25 years after Ilya Prigogine formulated the mathematics of how order can arise out of chaos in situations that are far from being in dynamic equilibrium, Darwinism entered a third phase. Models based on the dynamics of complex systems and the mathematics of chaos were introduced. Stuart Kauffman (1993) is an expert in this domain. In his view, evolution seems to work best when the effects of genetic units on fitness are complex and interconnected. The basis for this has become clear from the analysis of plant and animal genome projects, and from progress in molecular biology. Many genes code for more than one protein. Numerous phenotypic traits are coded by more than one gene. Physiological processes involve pathways of regulatory genes. Finally, insight has been gained into the interactions among different proteins and other macromolecules.

To summarize: Darwinism marked the advent of two key interlinking concepts: that of evolving species in branching phylogenies and that of natural selection as the force driving adaptation (cited from Kauffman, 1993).

15.3 Elements in neo-Darwinism subject to improvement. Life is a double continuum

In my opinion neo-Darwinism considers living matter too much as a single continuum, namely a hardware-continuum, in which the genes represent the only relevant source of variability. In my approach, living systems are double continua. As a consequence, in addition to genetic changes, learning processes can also play a role, particularly in free-living systems. This is inherent to the software aspect of communication.

Nature or Nurture? "Both". Nature for the hardware and some for the software, whereas Nurture is more via the software."
Neo-Darwinism would gain in comprehensiveness if certain imperfections were eliminated. Some of these imperfections and shortcomings have been discussed before by e.g. Kaufmann (1993). The ones I will focus on stem from the omission of the software as-

pect of living systems. Furthermore, neo-Darwinism has too narrow a view of several other aspects as well.

- The theory limits itself too much to evolution at the level of the population/species (micro-evolution). The key question in evolution should read: "How does *life* evolve at *all* its levels of organization?"
- It pays insufficient attention to the concept of evolution as a process involving *both* genetic and communicational alienation.
- Neo-Darwinism, in line with the emphasis that has been put on gene technology, gives too much the impression that living systems can be reduced to genes and gene products (proteins and peptides).
- The concept that many proteins act in concert with other proteins/molecules in what is called '*signal transduction pathways*' (problem solving networks) should be taken into account.
- The indispensable role of inorganic ions and water in the electric dimension of all cells and in the functioning of the nervous system, etc., is usually overlooked. Learning relies more on electric phenomena than on changes in genes. If it is assumed that genes regulate every process, then there can be no room for the organism's own input or for free will.
- The link between fitness and the number of fertile progeny is too restrictive.
- There should be a way to eliminate the tautology in "survival of the fittest" that says that "the survivors survive". Tautologies are correct by definition (they say the same) but they hardly have any explanatory value.
- Neo-Darwinism considers 'selection' as a cause of evolution. The possibility should be considered that selection is itself the result of something that happened before. If selection is a result, what is then the nature of the primary cause(s)? This question was already raised in 1927 by J.E. Wallin (cited in Margulis, 1981, p. 325):
  "Natural Selection, by itself, is not sufficient to determine the direction of organic evolution.... Natural selection can only deal with that which has been formed. It has no creative powers. Any directing influence that natural selection may have in organic evolution, must, in the nature of the process, be secondary to some other unknown factor."
- Gradients and gradient formation are hardly considered in neo-Darwinism. Hence, there is little room for gradient-driven self-selection. But living organisms are continuously confronted with internal and external gradients.
- Neo-Darwinism has to invoke 'cultural evolution' to explain the recent evolution of *Homo*. However, like all other species, *Homo* is a regular product of evolution. For his evolution, both organic and cultural, *Homo* cannot use any other mechanisms than the ones that are present in nature. Hence, there should be a general biological principle upon which cultural evolution is based.
- Finally, on the whole, neo-Darwinism says very little about the motivation that organisms need for engaging in problem solving, which is a major activity for all organisms.

In the next sections, a number of these imperfections will be focused on. It will be shown that by approaching evolution from the point of view of communication/problem activity instead of merely gene activity, the majority of the mentioned imperfections can be eliminated.

15.4 Thinking about evolution with insights from the computer era: not only the hardware evolves, but the software as well

It is evident that I have no problem whatsoever with the implementation of the achievements of genetics and of molecular biology in the further development of Darwin's theory of evolution. They contribute to solidifying its scientific basis. However, I think the time has come to implement the achievements of communication theory, as well.

The following question will probably ring a bell with many readers: "Is it possible to explain the evolution of computer technology in general, from let us say 1945 up to the present, by only taking into account the evolution of the hardware and omitting the evolution of the software?" The hardware evolved from very large machines to pocket calculators, sophisticated mainframes and powerful personal computers of the types XT, AT, 286, 386, 486, the Pentium series, the Apple system, etc.

Everybody will agree that limiting this historical overview to just the hardware is too restrictive. At least as much attention should be paid to the evolution of the software, with its hundreds of thousands of different programs. One need not be a computer specialist to know that the methods used to develop new hardware are different from the ones used by software developers. This is exactly the point in my criticism of neo-Darwinism. In fact, neo-Darwinism, with its emphasis on genes, is only interested in the evolution of the hardware of biological systems. It fails to incorporate the enormous importance of the evolution of the software.

In my opinion, one has to conclude that neo-Darwinism is becoming more and more incomplete as a theory, the more it is applied to species with ever increasing cognitive capacities (with their inherent capacity for learning and software evolution) and with greater mobility. This will be explained in more detail later.

15.5 Is evolution based merely upon software possible? The example of the identical twins for the third time

The following hypothetical example illustrates the possibilities of evolution based upon software only. In Chapters 10 and 14 I used the example of the genetically identical twins that were born in the Dutch speaking city of Leuven in Belgium to illustrate the influence of the environment on behavior. In this example one baby stayed in Leuven; the other was moved to Tokyo and raised there by a Japanese family. Despite their genetic identity, the twins would eventually speak totally different languages and behave each according to the cultural standards of their local environment. Mutations do not play any role in generating the differences because the instruments for communication, the bodies of the twins, do not change by being raised either in Belgium or in Japan. Nevertheless the communication produced by each of the twins is different due to the influence of teaching by the parents and other people, the environment, and learning.

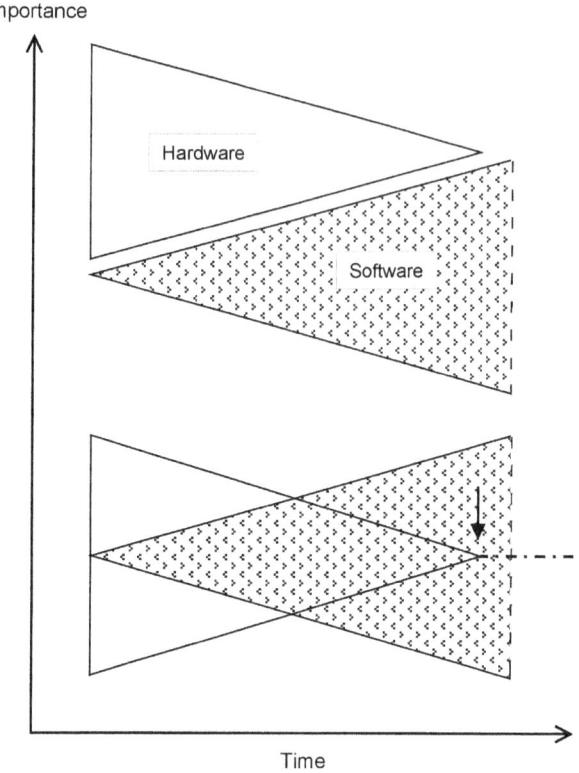

Figure 15.2 Early in evolution, additional changes in the hardware (mutations) have a greater impact than later, when the hardware is already highly complex. The opposite is true for the changes in software-based problem solving strategies. The species *Homo sapiens* has passed the point (arrow) where hardware evolution stops playing a role and 'software change' takes over as the sole mechanism of evolution.

The example of the twin babies obviously relates to the development of individuals and not to the evolution of a population. It could be extended to a population if we were to imagine that not one individual baby was transferred to a country with another culture, but rather one to two hundred babies, each having an identical twin, as could happen in a very special larger scale baby adoption program. Later on, when all the babies had grown up to become young adults, the ones in Japan would be brought together to form a community on their own. The same would be done with the ones in Belgium. The two newly formed populations, although having identical gene frequency distributions, would display such totally different behavior that they would no longer be considered as belonging even to the same parent population of which they are the progeny, let alone as being in fact identical. Both populations could even be taught to consider each other as enemies in order to ensure reproductive isolation if they were brought together again. Thus, under such conditions, a new population would have been generated without any mutation or change in gene frequencies in the subpopulation. This extreme example illustrates the important role of the environment (teaching and learning) in evolution. It also shows that evolution based on the software aspect of communication can proceed much faster than evolution based on mutations. It also suggests that species that develop teaching and learning in their evolution arrive at a turning point where software evolution through teaching and learning becomes more important than hardware evolution through mutation (Fig. 15.2). *Homo sapiens* apparently arrived at this turning point millions of years ago.

Imagine now that genetically identical twin lambs are treated the same way. The mèèèèèèè-language of the Japanese lamb, once grown up to a sheep, will not be that different from its counterpart in Leuven. The reason for this difference between humans and sheep is to a large extent due to the much higher number of possible bifurcation points in the brains of humans, which allow much greater flexibility in speech. The muscles involved in speech, which control the opening of the throat through which the outgoing air flows, as well as the muscles that are responsible for compressing the air before it flows out through the throat and the mouth, also play an important role. The mèèèèèèè language of the lamb corresponds to the crying language of a human baby. This type of speech is innate, simple, and not subject to much change. This situation could be compared to a computer, which is delivered with only an operating system (e.g. a DOS version), and only one very simple software program installed.

The complex language (and behavior in general) of humans is **acquired** by teaching and learning, and this teaching and learning occurs only minimally in sheep. In the computer analogy, one could say that a newborn human baby corresponds to a computer with an enormous hard disk-memory potential but which is delivered with only a minimal software package installed. Education is the introduction, again and again, of novel 'software' packages (e.g. all the different courses in school) in order to make the 'human computer' as versatile and powerful as possible and to increase the 'output'. Computers of the 'sheep brain type' have only little capacity for additional software to be installed on them.

Teaching and learning for the purpose of increasing people's capacities to solve problems are not processes that are instantly realized by mutations. Bifurcations are more important. Remember the example in Chapter 7 where different musicians produced totally different music with the same flute just by making use of the possibilities of the imaginary bifurcation points in between the holes in the recorder (Fig. 9.3)? Another analogy is that of two users of the same computer. They can make totally different images appear on the screen of the monitor simply by touching other combinations of keys on the keyboard.

15.6 'Cultural evolution' is software-driven, being based on teaching-learning processes. Abstract thinking. Consciousness. The role of tools

In classical biology, the rapid evolution of *Homo sapiens* is considered to be a special case of evolution because 'cultural evolution' is involved. The implicit assumption here is that no other species can undergo such 'cultural evolution'.

Let us briefly analyze what happened to the evolution of *Homo sapiens* in the 20th century. The behavior of *Homo*, and even to some extent his morphology (e.g. increase in body length) has undeniably evolved substantially in the 20th century. There are no indications that this is due to an increase in the mutation rate of specific genes in all human populations present on earth. Furthermore, if nevertheless such an increase had happened, the probability that the same genes (or allelic forms of these genes) would have mutated simultaneously in all human races in the same direction is nil. The changes in the *Homo* compartment are due in part to epigenetic factors (e.g. diet), but mostly to the drastic changes in 'toolization' (Fig. 4.1) and the degree of sophistication of mechanical communication systems. The use of highly efficient novel man-made mechanical systems of communication has resulted in a huge surge in information, which in turn has contributed to changes in 'life style' in the broad sense, etc. The speed of production of novel information has been so huge that genetics has not been able to keep up by generating larger and/or better functioning brains quickly enough to handle all this information. To cope with this problem, mechanical extensions of the brain (books, computers, compact discs, etc.) must be used to store and use all the additional information.

The relative importance of 'software-driven cultural evolution' as compared to 'gene-driven hardware evolution' increases as the capability of learning and teaching in a given population increases. At a given point, equilibrium between the two will be reached (Fig. 15.2). If teaching and learning continue to improve, then 'software-driven evolution' will become more important than genetic evolution. The more complex the software aspect in a given system, the more possibilities there are for further evolution because of the exponential increase in the number of bifurcation points which can be used for communication purposes any time the complexity increases. I think that this turning point is reached much faster than one usually thinks. In Chapter 7 I used the analogy of the whistle and the flute to illustrate how extremely fast the number of melodies that can be played on a flute increases each time one more bifurcation point is introduced. With an instrument as simple as a flute with four outlet holes (= the whistle situation plus only three 'mutations'), the number of different melodies that can be played is already infinite. One should always keep in mind that even the simplest cell type could experience an enormous number of different acts of communication. Kauffman (1993), using a totally different approach, also reaches the conclusion that evolution proceeds faster as the complexity of the system increases.

Whether there may be some form of teaching and forced learning in some other species is difficult to assess. It is likely that many populations which have the ability to learn and to teach (part of) what they have learnt in one way or another to other individuals, can undergo some sort of 'cultural' evolution. Examples include beavers, crows that manufacture a tool with a certain form and dimensions, birds that have adapted to life in cities, chimpanzees that use a stick to collect ants etc. Therefore I think that the term **'software evolution'** would be better than the well-established 'cultural evolution'. I realize, however, that it is very difficult to change any well-established terminology.

Not only in humans, but probably in all animals as well, there is a role for consciousness in evolution. For the state of the art in the field of consciousness, I would refer to the specialized journals and to the papers or books of Jaynes (1982), Crick and Koch (1992), Varela *et al.* (1992), Crick (1994), Penrose (1994) and Calvin and Ojemann (1994). In recent years there has been a growing tendency to study consciousness on the basis of its neural correlates.

With respect to evolution, the biologist Mark Nelissen posits five successive levels of consciousness in his book *Darwin's Glasses* (2000, written in Dutch).

First level: the presence of attention and the ability to purposely change the focus of attention.

Second level: the creation, processing and communicational expression of abstract ideas. For example, a frog must have some idea of a 'fly'; otherwise it would not capture it.

Third level: the possibility of evaluating in advance the meaning of a given action, as well as having expectations and making plans.

Fourth level: self-recognition (= It's me!) and recognition of other individuals. Sheep remember at least 50 other sheep faces for over 2 years (Kendrick et al., 2001). This example illustrates that one should not to easily categorize other beings as 'stupid' and devoid of consciousness.

Fifth level: the use of aesthetic and ethic values. This means being able to make the distinction between good and evil, beautiful and ugly.

The first four levels can be described in terms of objective parameters, and the fifth in terms of subjective parameters.

In recent years the methods for studying 'the neural correlates of consciousness' (a term introduced by Francis Crick) are continuing to be improved (Glynn, 2001). It can be expected that gradually an answer will be found to the questions as to why there is an association between pleasure and survival-promoting situations, as well as between discomfort and survival-threatening situations (Glynn, 2001).

15.7 Definitions of 'natural selection'

The general meaning of selection (*selectio*, Latin) is the act of choosing something or somebody for a certain characteristic. Here, selection has an *active* meaning. In *The Origin of Species by Means of Natural Selection* (1859), Darwin defined natural selection as follows: "*This preservation of favorable variation and the rejection of injurious variations, I call Natural Selection.*" Here, selection is something *passive*. The antagonism active-passive will be dealt with in the next section.

Darwin's definition is flexible enough to include variations resulting from changes in genes as well as from other sources. I can live with it. In neo-Darwinism, however, the meaning of this term has become much more restricted. Raven and Johnston (1996) define natural selection as the process whereby those individuals in a population that possess certain characteristics produce more surviving offspring than do those individuals that lack these characteristics. Only inheritable characteristics are assumed to be important, as is made clear in Stuart Kauffman's definition of Darwinism (1993), which was already cited earlier: " Natural selection, acting upon individual variations within a population to substitute one allele for another...." According to this strictly genetic interpretation of natural selection, evolution can only happen through changes in the gene pool.

If one were to stipulate that this restricted genetic interpretation is only valid for sessile organisms (plants, fungi, etc.), which display much less behavior (=total sum of all movements made by an organism), then perhaps this interpretation might still be acceptable.

However, in free-living organisms, which can use their cytoskeleton for substantial movement (e.g. by pseudopodia in amoebae, by cilia or by muscles), certain adaptive changes are not necessarily caused by genetic changes: the principles of communication (software aspect) already enable adaptation in a number of cases.

15.8 Natural selection: cause or result? The driving force(s) of evolution. Mutations and changes in the cognitive memory system

No doubt, selection is essential in evolution in both Darwinism and in my double continuum theory. However the two theories have a different view on the 'status' that should be attributed to selection. In neo-Darwinism selection is the *cause*, whereas in my approach selection, in its Darwinian formulation, is the *result* of something that happened before.

What is the 'force' driving evolution if selection is a result?

In this context 'force' does not have the same meaning as in physics, where force (F) is defined as mass x acceleration (F = ma).

Anything, not only mutations, that changes the existing genetic and/or cognitive signal transduction pathways with alterations in problem solving activity as a result, can in principle, contribute to evolutionary change. This is most obvious in micro-evolution but it also applies to macro- and mega-evolution.

Gradients

Living systems maintain themselves in environments that are not in equilibrium. This imbalance manifests itself in a variety of gradients (= higher-lower situations) such as: cold-warm; light-dark; fresh-salt water (osmosis); humid-dry; noisy-silent; oxygen rich-poor (aerobic-anaerobic); nutrient rich-poor, safe-unsafe (e.g. predators); toxin rich-poor; pathogen rich-poor; strong-weak radiation etc. Usually, gradients change in the course of time (e.g. diurnal, monthly, and seasonal rhythms).

Because of their internal ionic-electric and other gradients, living systems are far from thermodynamic **equilibria**. Hence, they need to invest energy to maintain this state of imbalance. Internal gradients in organisms can also change (e.g. quick changes in membrane potential, diurnal rhythms, developmental changes). Changes in gradients come first in the chain of events leading to evolutionary change.

Problem solving activity. 'Struggle for life'.

Individuals that cannot solve the problem(s) that are associated with changing gradients will not survive. On the other hand, the ones that can solve the problems will do well in the population. They will be rewarded with better growth and reproductive advantages than the non-problem solvers. If the improved problem solving capability can be transmitted from one generation to the next, either by genetic or cognitive means, a change in frequency of the signal transducing pathways involved is likely to occur, assuming that adequate time is provided.

The popular saying 'struggle for life' reflects our own experience that one has to make efforts all the time. Organisms do not 'struggle for life' in order to avoid death, they engage in problem solving in order to achieve equilibria (feel contented) in the short run.

Changes in the two memory systems

No doubt that changes in the architecture of genes and genomes (mutations) play a crucial role in evolution. However, the view that only mutations drive evolution is too simplistic, in particular in populations/species with a well developed cognitive system. One has to keep in mind that a gene is nothing else than a unit of genetic memory. A gene in itself does not do a thing: it is like a book sitting on a shelf in the library, or a chip plugged in a computer. It is the gene product(s), the corresponding protein(s) that can exert a function, seldom alone, usually **as part of a communication/signal transduction pathway consisting of a number of interacting proteins**. Some changes in gene-based signal transduction pathways can influence problem solving.

One can speculate that like the gene that is the basic unit of the genetic memory consisting of nucleotides as building blocks, there must also be a basic unit of the cognitive memory consisting of 'bytes' as building blocks (to use computer terminology). In analogy with 'gene', perhaps the term 'cogne' might be appropriate to denote the still hypothetical basic unit for the cognitive memory.

The combined action of the proteins in a genetic signal transduction pathway defines a genetic trait. The combined action of 'cognes' in a given cognitive signal transduction pathway defines a 'meme'. Both traits and memes can spread in a population.

Changes in energy

Problem solving is an activity. It requires energy. The best genes can do nothing if there is not enough ATP (food). Hence, shortage in energy supply will restrict the possibilities for evolutionary change.

Accidental changes

Another possibility is that the change in frequency of signal transduction systems is totally accidental, thus without adaptive value, as in the case of drift (see textbooks on Evolution).

From all this, it follows that in a given environment, the best communicators/problem solvers, both at the level of the hardware and at the level of the software, have a better chance to survive and to have more progeny, obviously on condition that they are not prematurely eliminated by accidental death. If the better problem solving activity is based upon mechanisms that ensure transmission to the next generations, changes in their frequency in the population are likely to occur.

In Fig. 15.3, I attempt to visualize the idea that different principles apply to the evolution of the hardware than to the evolution of the software.

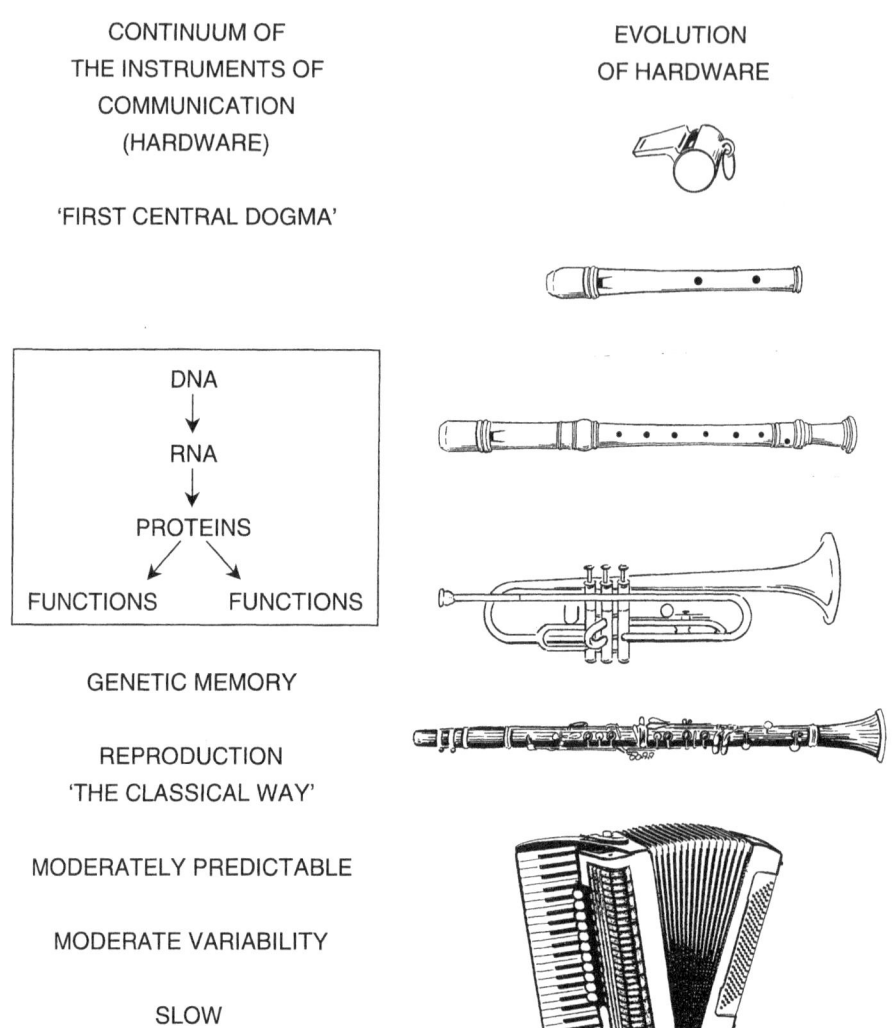

Figure 15.3 The essence of the double continuum theory. 'Hardware' and 'software' evolution use different mechanismes (see also Fig. 3.4 in Chapter 3).

EVOLUTION OF THE "SOFTWARE"-OUTCOME	CONTINUUM OF NON-GENETIC INFORMATION (SOFTWARE-BASED)
	'SECOND CENTRAL DOGMA'
	CENTRAL PRINCIPLE OF COMMUNICATION

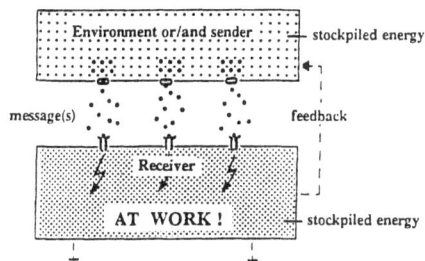

COGNITIVE MEMORY

LEARNING PROCESSES

UNPREDICTABLE

ENDLESS VARIABILITY

VERY FAST

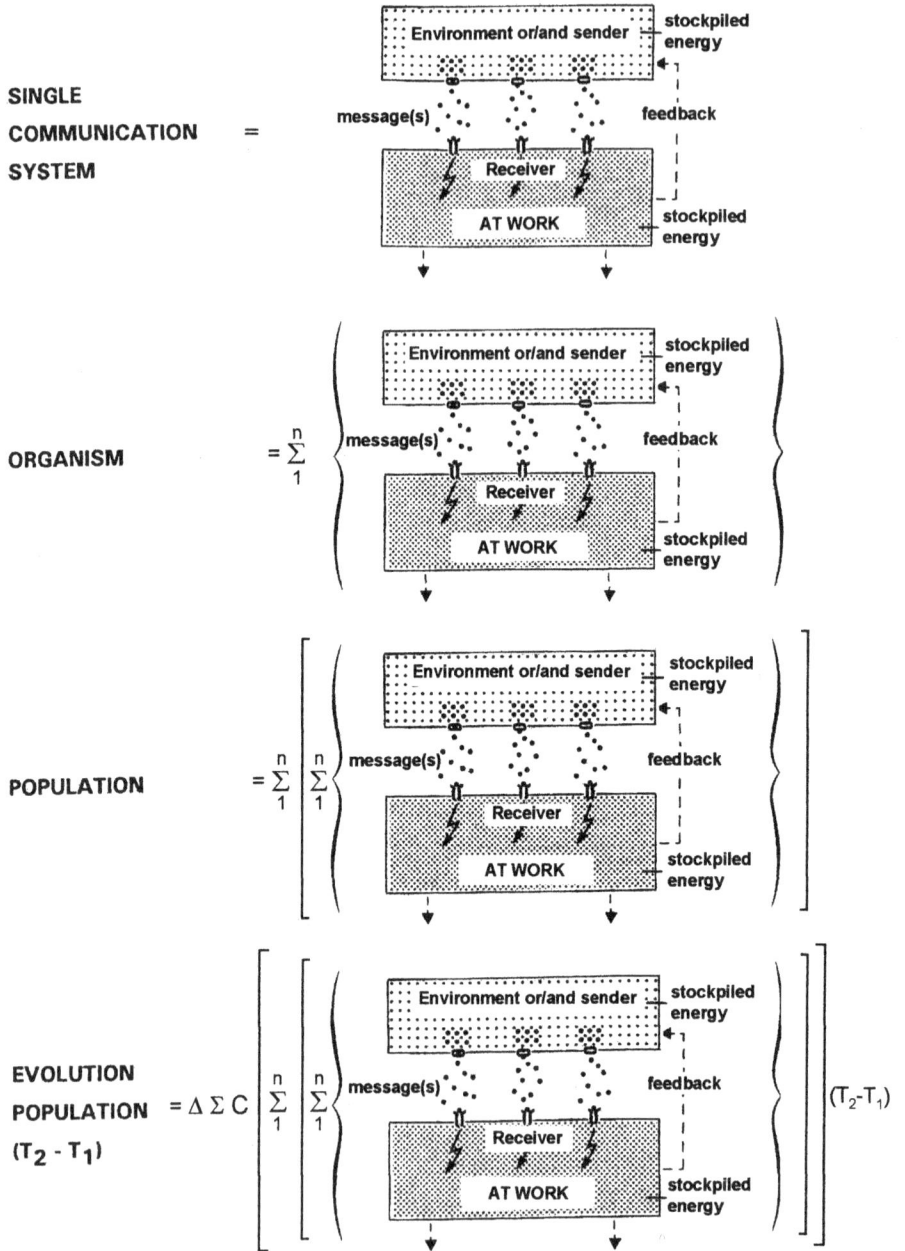

Figure 15.4 Summary diagram of the essence of evolution as seen from the point of view of communication. The basic building block of living matter is the communicating compartment. An organism consists of a given number of such compartments. Bacteria represent level 1 of compartmental organization and consist of only one communicating compartment. Eukaryotic cells are level 2, a segmented animal is level 7, a population is level 13 (see Chapter 4). Each communicating compartment is different from all the others and all are subject to change in terms of hardware, software, and energy mechanisms. This figure focuses on level 13, the population (micro-evolution). A population consists of a given number of individuals. In general, evolution of 'life' at a given compartmental level of organization concerns the changes in 'life', thus in communication activity over a given period of (geological) time (D($T_2 - T_1$)).

15.9 Summary of the essentials of the double continuum - or the hardware-software - theory of evolution

1. *The environment is unstable because of gradients*

Living systems must maintain themselves in an environment(s) that is (are) not in equilibrium. This imbalance manifests itself in a variety of gradients (= higher-lower situations) such as: cold-warm; light-dark; fresh-salty water (osmosis); humid-dry; noisy-silent; oxygen rich-poor (aerobic-anaerobic); nutrient rich-poor, safe-unsafe (e.g. predators); toxin rich-poor; pathogen rich-poor; strong-weak radiation etc. Usually, gradients change in the course of time (e.g. diurnal, monthly, seasonal, etc. rhythms).

2. *Organisms are also unstable because of their internal gradients*

Because of their internal ionic-electric and other gradients, living systems are also far from thermodynamic equilibrium. Hence they need to invest energy to maintain this state of imbalance. Internal gradients in organisms can also be subject to change, both in the short run (e.g. quick changes in membrane potential, diurnal rhythms, etc.) and during development.

3. *Hence, organisms must find an environment where a livable compromise is possible between the environmental gradients and their own internal gradients*

As long as no substantial changes in gradients occur, a relatively high degree of stability can be maintained. However, problems arise when the gradients start to change. In order to cope with changing gradients, organisms must first be informed about the changes.

4. *Organisms are constructed in such a way that they can sense (changes in) gradients*

This means that gradients and changes in gradients can be used used in a messenger system. In addition, the sensing of changes forms the physiological basis for some 'feelings', a term which is difficult to define in such a way that it applies to all types of organisms.

5. *Whatever their nature, feelings, and in particular feelings of 'contentment', are of vital importance to organisms*

Through their cellular signal transduction pathways, in which receptors play a key role, organisms are able to sum up the effects of (changing) gradients. The result yields the answer to the question: "Do I feel contented with my situation?" If the answer is yes, no new actions need to be undertaken. Organisms will proceed as usual. If the answer is no, then their problem solving machinery becomes activated: "**What is the problem and how is it to be coped with it?**"

6. *Organisms do not 'struggle for life'; rather, they either solve problems or they don't*

We cannot say that organisms 'struggle for life', since this would imply that they are aware of the meaning of 'death', which is most probably not the case, except in *Homo*. In order to stay alive, organisms have to solve problems on a continuous basis. If they fail to do so, they will accumulate disadvantages as compared to other (competing) organisms with better problem solving abilities. This is the very heart of evolution.

7. *Problem solving requires the proper tools. The architecture of a communication system*

The solving of a physiological problem is a multistep process that is initiated after the accumulation of incoming messages warns that something is going wrong. A whole mechanism called 'a communication system' is required in order to solve a problem. A typical communication system consists of a sender that emits a message in coded form, and a communication channel through which the message is transported to a competent receiver that decodes, amplifies and responds to it, thereby mobilizing part of its stored energy. This principle forms the basis for problem solving. By far the majority of all problems organisms are confrontented with are unconsciously solved. This makes us think that not all communication activity concerns problem solving. Yet, anytime a receiver has to decode an incoming message, it has to solve a problem. Thus communication and problem solving largely overlap. Communication usually involves motivation, decision-making and feedback.

8. *Problem solving is a task for each level of compartmental organization*

Living matter is organized into a series of hierarchically organized levels of communicating compartments, totaling 16 in the classification system proposed in this book. From level 2 on, all compartments contain subcompartments. There are subcompartments restricted to the same individual organism. There are also compartments that involve more than one individual of the same species. Finally there are also compartments that comprise individuals belonging to different species. As in human society, problem solving usually occurs at the level where the problem applies and may include feedback mechanisms at higher levels. The interest of the higher levels outweighs that of the lower.

9. *Problem solving by individual organisms is purposive, the purpose being to increase the quality of their life*

Individual organisms solve problems (an energy demanding activity) for the purpose of lowering their feelings of discomfort and/or of obtaining a higher degree of contentment. When an organism solves a problem of feeling hunger by eating, then it is rewarded by satisfaction. When an organism solves the problem of discomfort due to the retention of sperm by ejaculation, it is rewarded by the feelings that accompany an orgasm, etc. Organisms are comfort seekers that endeavor to feel contented ('enjoy life') as much as possible within a given environment that imposes its restrictions.

10. *All communication systems consist of hardware and software. Hence their reproduction and evolution relate to a double continuum. Nature or nurture?*

The hardware of organisms is their body of 'flesh and blood'. It is reproduced according to the rules of the first central dogma. Software programs are a special sort of memory. They are immaterial but often need a material carrier. Here the rules of the cognitive memory (second central dogma?) are relevant. The 'life as a hardware-software double continuum' may perhaps contribute to solving the question: Nature or Nurture? The answer is likely to be: Nature AND Nurture. Communication activity invariably requires that energy be stored beforehand in a suitable form in the receiver.

11. *Because all elements of a communication system are subject to change, an enormous variability is possible*

The evolution of the hardware makes use of a different set of rules than the evolution of the software, and it does not necessarily proceed at the same pace. The evolution of the hardware utilizes DNA as the carrier of the genetic information (genetic memory) and the first central dogma (DNA→RNA→Protein(s)) to do the work. Mutations are the key. The evolution of the software aspect uses the non-genetic (cognitive) memory, the molecular carrier of which is still ill defined. Cellular electricity plays a very important role. Decision-making presupposes the existence of imaginary bifurcation points. It can be expected that in due time a second central dogma will be formulated, explaining the full mode of action of the cognitive memory.

12. *Evolution is 'far-sighted blind', which means that it is not purposive but it does nevertheless anticipate future problems*

Organisms do not know in advance what changes in environmental and internal gradients will take place in the future. Hence, the only option left is to blindly generate changes in the genome and in cognitive problem-solving tools, and to 'hope' that at least one of the mutations or new 'versions' of software will enable the organism to solve the new problem(s). Thus, with respect to problem solving, evolution generates solutions in advance for problems that do not yet exist.

With respect to problem solving that requires changes in the hardware, one possible strategy is as follows. First, during the formation of a given gamete (or, in the case of asexual reproduction, of a reproductive stem cell), a gene or set of genes is duplicated without disturbing the normal physiology of the individual that will develop from it after fertilization. In later generations, one of the duplicated gene (sets) can undergo additional mutations. If this happens frequently enough, there is a statistical chance that, in due time, a *novel signaling pathway* can come into existence, along with the existing pathways that continue to operate. There may be conditions in which the novel putative pathway remains silent as long as there are no proper cues in the environment to activate it. When the proper trigger, (e.g. a given change in a particular gradient), makes its entry on the scene, the novel pathway is activated. This will be favorable to the organism(s) that have this pathway. The frequency in the population of the genes coding for this pathway is likely to increase over the generations.

In the case of problems that have to be solved by the software, a similar strategy applies, but this time based upon the principles of the cognitive memory. The best example in this respect is the education system in the human world. Children learn the principles of mathematics as a tool to help them solve future mathematical problems. They learn the grammar of a foreign language in order to speak that language in future situations that cannot be anticipated at the moment of learning the language.

13. *Evolution alienates*

Each time a gene mutates in a gamete, the degree of kinship with its producer decrease. The more mutations accumulate during successive generations, the lesser the degree of kinship becomes. At a given moment, populations have diverged so far that they have become new species.

Languages, whatever their nature, are also subject to change. Dialects may in due time become so different that they have become new languages. This is a non-genetic way for subpopulations to become isolated. It offers a basis for the divergence of a population into two different species while all organisms continue to live in each other's neighborhood (sympatric evolution).

Thus, evolution alienates both genetically and communicationally.

14. *The best problem-solvers have a greater potential for a faster growth and for a larger progeny than the competitors*

The best problem-solvers are those whose genetic or/and cognitive memories contain the information needed to solve the problem. Individuals that are better in problem solving have better opportunities for mobilizing more energy for their growth than competitors. Because gamete production starts only after a given developmental stage/body size has been reached, namely after the inhibitors present during larval life have disappeared, the faster growers will engage sooner in gamete production than the slower ones. In addition, faster growers have more chances for producing more gametes during their lifespan. The more gametes, the higher the chance that some of them may carry mutations that are beneficial for problem solving in later generations. Likewise, with respect to transmitting information from individual to individual in the software manner, namely by teaching-learning, the best problem-solvers are likely to have larger numbers of followers than less inventive individuals. Accidental elimination of good problem solvers by whatever cause (the Sewall- Wright effect) may lead to deviations in the sketched outline.

15. *Fitness*

The short-term success of an organism depends on its capacity to solve the different types of problems it is confronted with in its environment. It has to survive until adulthood. Next, it must reproduce, whether asexually or sexually, and produce fertile offspring, or acquire numerous disciples/followers in the case of the software mode of reproduction in the studax *Homo*.

Long-term success depends on the presence, in silent form, of mechanisms for solving problems that do not yet exist but which will show up in the future. In the long run, the gametes that are fitter than others are those that (1) solve the problem of getting united with the best partner gamete (sperm or egg); and that (2) also have the (silent) genes for future problem solving.

The fittest individual is the one that produces the highest number of fertile offspring with the best problem-solving abilities in the short- and long run, and, where relevant, also establishes the highest number of disciples. In this view Einstein, who had only two children, was nevertheless a very 'fit' individual.

16. *Adaptation is an active process in the short run, but not in the long run*

In the short run, free-living organisms can move away from harsh environmental conditions making use of the possibilities of their cognitive memory. Or they can activate escape mechanisms that pre-exist in their genes, if present. Sessile organisms can only dig into their repertoire of pre-existing mechanisms for coping with problems. The harsher the conditions, the more important it becomes for them to better disperse their gametes to improve their chances of ending up in a less harsh environment.

Because evolution is not purposive, adaptation in the long run is always a passive process. Long-term success is based on the possibility of the genome undergoing mutations that can remain silent in the genome until they are needed later when conditions get harsher or when new habitats are colonized.

17. *Selection is a result, never a cause. The GP-TripleS principle*

As soon as a new communicating compartment comes into existence, it is immediately subjected to the effects of its own gradients: the Gradient-Provoked Swellig/Shrinking and Self-Selection Principle (GP-Triple S Principle). If the compartment can cope with its own gradients, it will next be subjected to the gradients of the environment. Organisms that cannot cope with the mix of gradients will have a hard time doing as well in growth and reproduction as those that can. Changes in gradients come first, selection follows. Thus, selection is a passive process with a double face, namely a positive outcome for the better problem-solvers and a negative outcome for the weaker ones. It starts right from the moment that a compartment starts to live.

18. *Life and the driving force of its evolution. A spiral feedback mechanism?*

In my opinion, the best candidate for such a driving force is anything that changes communication/problem solving activity as already mentioned earlier in this chapter. But communication activity represents the very nature of life itself.

This leads to the remarkable conclusion that the evolution of life involves a feedback mechanism. Indeed:

> Life is communication activity and communication activity includes problem solving.
> Changes in communication/problem solving activity drive evolution
> Hence, changes in life (as an activity) act as the driving force of life's further evolution

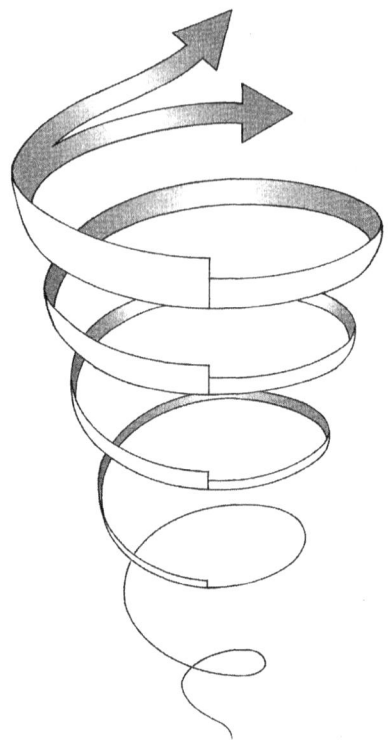

Figure 15.5 Evolution as seen from the viewpoint of communication (this book) is a spiral- or corkscrew-like process. This is largely due to the fact that communication involves feedback. At the end of any given problem solving cycle, the conditions are no longer the same as at the onset. Once a given problem has been solved, the compartment can engage in solving another problem of a higher degree of difficulty, from the moment that it acquires the necessary tools. This holds true for any additional problem solving cycle. Such activity is likely to generate ever-increasing complexity at ever-greater speed. The overall picture of an evolutionary process that is continuing to accelerate is correct. The end product of a given problem solving cycle may yield the tools for solving the subsequent problem in two (or more) different ways. At bifurcation points, side branches come into existence in the evolutionary tree.

One consequence of this spiral principle is that life cannot do anything else than continuously change and evolve. Furthermore, the more complex communication systems are, the greater opportunities they have to evolve, and this at an ever-faster pace. Indeed, if one problem-solving strategy fails, an alternative pre-existing strategy may be invoked. In addition, the progeny of a complex system have more opportunities for radiating in different directions in future generations. Finally the circular nature of the principle may explain why it is so difficult to unequivocally point at one specific primary cause of evolutionary change and why the question: "What is cause and what is result in evolution?" is often difficult to answer.

15.10 Input, through the software-aspect, of the organism's own input or free will in evolution? Creationism

Bionieuws [Bionews], a Dutch life sciences and biotechnology magazine, published a series of articles in the period 1994-1995 featuring different points of view in the controversy as to whether the theory of evolution should be included in the high school curricula and, if so, in what way. This issue is still relevant in some states of the US. F. Bretschneider (1995) raised the question as to why it is, again and again, just precisely the theory of evolution that acts as a red flag on a bull and who will finally resolve the issue?

One could ask the question as well this way: "Isn't there a way to end the sterile dispute between Darwinists and Creationists without the participants loosing face on either side?"

I think there is but let us first try to reduce the controversy to its essence. First, what happened in the past cannot be redone. Thus historical insights always involve some degree of 'believing'. Second, religions do not intend to provide insights into scientific issues; rather, they are about the relationship between man and God. Third - and this is the most important aspect from my point of view - a theory of evolution should not reduce any species, *Homo* inclusive, simply to sets of genes. Fourth, there is absolutely nothing wrong with *Homo* being the descendant of some primate, a reptile, a fish, or (in the very beginning) of the bacterial Progenote.

In my opinion, the strict genetic interpretation that neo-Darwinists have given to the theory of evolution, in accordance with the intellectual fashion of the epoch in which it was formulated, is to a large extent responsible for the resistance that still prevails in many people against Darwinism. If what we are and what we do can all be reduced to the action of our genes, then we have a very miserable portrayal of mankind. Because we cannot control our genes, we have no choice but to passively undergo what our genes command us to do. We have to await what chance brings us and act as automatons. As long as this applies to sessile organisms such as plants, we can accept it. But when applied to *Homo*, it seems to reduce 'life' to meaninglessness. Humans viewed through neo-Darwinian glasses are just a bunch of automata like plants. Everyone knows, of course, that we are not just mere automata, but that we have a certain degree of free will. There can be no doubt that we can make any number of decisions that can influence our own evolution. Evolutionary biologists will immediately argue that *Homo* is an exception because he can undergo cultural evolution. But the question as to how intellect and free will can play a role is usually left unanswered.

In my opinion, the solution is simple. We must no longer define micro-evolution as the strictly genetic evolution of populations but rather as the evolution of the population as being a particular level of compartmentalization, namely level 13 in my classification system. So doing, we will no longer need to invoke cultural evolution alongside the hardware evolution. The software aspect will automatically be dealt with in this approach. Because communication automatically implies the making of choices (Chapter 8), intellect and free will automatically come into play, where relevant. This way, a much more balanced view of evolution emerges, in which it is no longer needed to discriminate between chemical-biological evolution on one hand and cultural evolution on the other. The two are inextricably bound up with one another, complementary and universal. From this approach it follows that species that have a well-developed cognitive memory and are free-living can, to a certain extent, influence their own evolution. *Homo* does this already.

15.11 Signal transduction pathways and the ongoing dispute between 'Phyletic gradualism' and 'Punctuated equilibrium'

In my opinion, which contrasts with that of Stephen J. Gould (2002), there is no contradiction between the theories of 'phyletic gradualism' and that of 'punctuated equilibria'. In very simple systems, the effects of a mutation will be readily visible, while in complex systems the effect of a mutation may remain hidden until the time that the new gene fits into a new signal transduction pathway or contributes to the formation of a higher level of compartmentalization. The acquisition of higher levels of compartmentalization in Mega-evolution (Fig. 4.1) probably required the accumulation of many mutations over vast spans of time, until a final change in the genome resulted in the fitting together of the proteins resulting from the mutations into a novel, functioning higher-order communicating compartment or into a novel signal transduction pathway. At that moment, all the different parts of the puzzle fall into place and an intelligible picture becomes apparent.

15.12 An upgraded adage

By focusing on the possibility of personal input in problem solving, the double continuum theory offers a more positive outlook on life and its evolution than neo-Darwinism. The double continuum theory does not primarily focus on changes in genes and genomes as such, but rather on the mechanisms that make it possible for ever more difficult problems to be solved. Micro-evolution involves relatively simple signaling pathways for problem solving, whereas macro- and mega-evolution require more complex novel strategies.

Modern molecular biological techniques make it possible to determine which proteins interact with each other, thereby forming functional complexes. There can be no doubt that this approach, in combination with the possibilities of comparative genome analysis, will substantially contribute to a better understanding not only of development but also of the history and mechanisms of evolution.

Paraphrasing Kauffman's citation in 15.2, one could briefly say that: *Darwinism marked the advent of two key interlinking concepts: that of evolving species in branching phylogenies and that of acquiring better problem solving strategies as the force driving adaptation.*

ESSENTIALS

THE DOUBLE CONTINUUM THEORY OF EVOLUTION

1. A sound theory of evolution should explain not only the formation of new species but also the evolution of 'life' at all its levels of compartmental organization.
2. All beings on earth are the progeny of the Progenote. Hence they are all family. According to Margulis' symbiont theory, the eukaryotic cell type originated when several species of bacteria started to live as a single functional unit. This means that in fact there is only one basic form of life on earth: the bacteria. This life form is capable of producing an endless variety of morphological adaptations.
3. Evolution is far-sighted but blind. Organisms accumulate mutations and problem solving circuits more or less randomly without knowing which problems will manifest themselves in the distant future. Thus the tools for coping with future problems must be preformed.
4. Evolution unintentionally alienates, both genetically and communicationally, thereby yielding the enormous wealth of forms and species that make up the biosphere. In other words, by generating variability, it creates strangers.
5. Had Darwin lived in the computer era, he might not have overlooked the importance of 'software evolution', a weak point in neo-Darwinism.
6. To account for both the hardware and the software aspects of evolution, both feelings and energy must be taken into account.
7. Selection is not a cause but rather the result of something that previously acted or happened. Start with changes in gradients, and the rest follows of necessity.
8. The Darwinian concepts of "Struggle for life" and "Survival of the fittest" as applicable to micro-evolution can perhaps be made less harsh and more all-encompassing when viewed in the following perspective:
   - **Organisms live in changing environments.**
   - **They have means for sensing changes in gradients of temperature, humidity, salinity, etc.**
   - **Organisms are comfort-lovers.**
   - **Their basic feeling is of contentment.**
   - **All sorts of influences can disturb this state of contentment.**
   - **Hence, organisms must solve problems in order to regain their state of contentment and/or improve their quality of life.**
   - **Their basic architecture as communicating compartments with their hardware, software and energy aspects allows endless variability. It generates the possibility of problem solving and it also contains the genetic and cognitive memories and mechanisms for anticipating solutions to future problems and for adaptation.**
   - **The two types of memory are likely to involve two different central dogmas. The genetic memory obeys the rules of the first central dogma: DNA $\rightarrow$ RNA $\rightarrow$ Protein(s). A second central dogma, governing the cognitive memory, is likely to exist as well. Its nature is as yet only partially determined, but electrical phenomena are sure to be part if it – and perhaps the actin cytoskeleton as well.**

- **If they are not prematurely eliminated by accidental death, the best problem solvers (= the fittest) are rewarded by a higher level of contentment, and by faster growth and reproductive advantages.**
9. All levels of compartmental organization are subject to evolutionary changes.
10. Life as communication activity, which automatically includes problem-solving activity, is the driving force of its own evolution. The solving of one (set of) problem(s) paves the way for engaging in further problem solving. This may explain why evolution appears to be proceeding in a corkscrew -like mode and at an ever-increasing pace.

CHAPTER 16

PROPERTIES OF 'LIFE'

'INTEGRATIVE' VERSUS 'CLASSICAL' BIOLOGY

COMMUNICO, ERGO VIVO

Keep in mind: There is 'life' after work!

Contents

16.1 Summary of the general properties of 'Life'
16.2 The understanding of 'Life' requires holistic thinking. 'Integrative biology' versus 'classical biology'
16.3 'Good' and 'evil'
16.4 The place of the *Homo sapiens* in the biosphere as a whole
16.5 Biology, ethics and philosophy. *Communico, ergo vivo*
16.6 If Darwin had had a computer...
16.7 **L** = Σ**C**: Towards a biological paradigm?

Essentials

16.1 Summary of the general properties of 'Life'

1. Life is not a machine, it is an ***activity.***
2. This activity involves ***communication*** of which the basic unit is 'the communication act'. ***Communication is transfer of information.*** Information itself is ***immaterial*** but it may require a materialized carrier.
3. A communication system involves hardware and software, each with their own sets of rules for functioning, development and reproduction. This accounts for the '***double continuum** aspect*' of life.
4. At least ***16 levels of compartmental organization*** came into existence in the course of evolution. They are ***hierarchically organized.*** The interest of the higher level outweighs that of all lower levels.
5. The more levels of compartmentalization a communication system has, the more complex the communication activity will be, and the greater the variety of ***problems*** it will be able to ***solve***.
6. ***Life is the total sum of all acts of communication*** performed by a given compartment, from its lowest to its highest level of compartmental organization, at moment t.
7. Because life is a sum of all sorts of communication acts, it has both ***quantitative*** (= a big compartment lives more than a smaller one) ***and qualitative*** (= a baby is as valuable as an adult) ***parameters***.
8. Because life is an activity, it ***requires energy***.
9. Because communication activity uses ***gradients*** (= higher-lower situations), life is far from a thermodynamic equilibrium. To keep on functioning, energy has to be invested all the time. Part of the energy is lost as heat. This accounts for the '***dissipative nature***' of living systems.
10. ***Living systems can solve problems***, non-living cannot. In fact, communication and problem solving activities are almost synonymous. This is not apparent in daily life because we tend to restrict the term 'problem solving' for *conscious* problem solving only. This conscious problem solving represents only a very tiny percentage of the total unconscious problem solving activity going on in any compartment. This property accounts for the ***automaton aspect*** of life.
11. Engaging in problem solving requires ***preexisting signal transduction pathways*** and a ***motivation. Feelings***, whatever their biochemical basis is in the different organisms, can be a tool in problem solving.
12. Life is to a large extent an ***electrical phenomenon***. This is due to the fact that the basic gradient of all cells is an ionic-electrical gradient.
13. Living compartments: the whole is ***more than the sum of the parts***. This is mainly due to the acts of communication that coordinate the successive compartmental levels.
14. Under the proper conditions, communication activity (i.e. life) can ***master any force known in physics***.
15. Because it is an activity, Life cannot be constant. It ***changes all the time***.
16. Life subjects itself to ***self-selection***. The best problem solvers have a better chance for ***growth and reproduction***.
17. ***Evolution alienates***, both genetically and communicationally.

'Life is communication/problem solving activity' acts as an integrative definition that covers all essential properties of 'life' as listed in the multitude of traditional definitions cited in Chapter 5.

16.2 The understanding of 'Life' requires holistic thinking. 'Integrative biology' versus 'classical biology'

At the current time, the teaching of general biology is done with only one central dogma in mind: DNA → RNA → Protein(s). Cells have genes, genes code for proteins, and proteins finally do the work. Furthermore, because genes can change, living systems can evolve. In fact this type of biology deals only with the hardware of living systems. It is single continuum biology. I refer to it as 'classical biology'. The software aspect, which is the most intriguing part, is left to the humanities. My view is that because there are two different types of memory (genetic and cognitive), there should be two central dogmas as well. It should not be so that, because the second dogma that deals with the functioning of the cognitive memory has not yet been adequately formulated, its existence can be flatly disregarded.

The reader will have noticed that this strict separation of disciplines (dualism) is not my cup of tea. I have heard for many years that scientists should endeavor to make their research more interdisciplinary and that teaching should be concentrated more on major principles than on technical methodologies. In practice, these good intentions are seldom translated into coherent novel approaches.

In classical biology, the key questions of biology such as: "What is life? Death? Communication? Information?" are usually not even asked. Hence, unifying concepts are not possible in such an approach.

Communication is a key issue in understanding the very nature of life and death. Therefore, one must face the necessity of introducing into all disciplines of biology the principles of communication to a much larger extent than is the case today, at least where they are relevant. By replacing the 'cell as the basic building block of living matter' by the 'communicating compartment as the basic functional unit', the integration of the different approaches taken by the sciences with those taken by the humanities becomes possible. I propose using the name 'integrative biology' for this approach.

I propose the following two definitions:

> Classical biology is the study of the generation, functioning and perpetuation of organisms and populations. It focuses primarily on the hardware of compartments. It makes use of only one central dogma.

> Integrative biology is the study of the generation, functioning and perpetuation of hierarchically organized carbon-chemistry-based communicating compartments, including both their hardware and their software aspects. It requires two central dogmas.

My proposal broadens the scope of what at present is understood under the term 'integrative biology'. Table 1 lists the major differences between classical and integrative biology.

In the US, some universities (e.g. Berkeley) have laboratories for integrative biology. There is also a Society for Integrative and Comparative Biology.

My preference for integrative biology should not be misinterpreted. I favor holistic teaching, which means emphasizing big concepts and raising questions about the validity of so-called 'established concepts', as I have done in this book. Nonetheless, for experimental work the reductionistic approach is often very rewarding, on condition that it fits well into a given concept. "Think holistically, experiment reductionistically" may be a good guiding principle.

16.3   'Good' and 'evil'

A delicate issue that is not restricted to pure biology is whether it is possible to give a biologically acceptable answer to the questions "What is good?" and "What is evil?"

In my opinion, 'good' is what improves the communication/problem solving activity of the highest level of compartmentalization of a given compartment, and 'evil' is what worsens it. However, what is 'good' for one compartment at a given moment may not be good at other moments, and it can even have deleterious effects on other compartments (e.g. in food chains).

It is not at all easy to come up with a single parameter which allows one to quantify whether the effect of communication is good or not in any circumstance. 'Good' often refers to aspects of expansion due to growth of a compartment in space and time. This expansion can go so far as to include the acquisition of a higher level of compartmentalization and the pinching off of daughter compartments. The family as a whole is doing well when many children and grandchildren are born (expansion). It goes well with a man and a woman (both having compartmentalization level 8: Fig. 4.1) when their mutual communication fits in such a way that they can form a higher order compartment, the heterosexual one (level 10: Fig. 4.1). This level provides to the female the opportunity to achieve compartmentalization level 12 (= baby inside mother, Fig. 4.1) and to generate progeny (expansion). When things go wrong with a couple, they divorce. Divorce means falling back on the scale of compartmentalization (Fig. 4.1), namely from level 10 (heterosexual compartment) to level 8 (individuals with tools).

'Good' and 'evil' are intimately linked to the hierarchy of levels of compartmentalization, just as 'life' and 'death' are. It is the highest level that takes precedence over all subordinate levels. This hierarchy can easily be observed in daily life every time 'higher-order interests' are invoked for one reason or another.

Some examples can illustrate this phenomenon. The highest level of compartmentalization on earth is the planetary. In recent years there has been growing concern that human activities may be reducing biodiversity and may perhaps also be causing climatic changes. Governments are starting to impose restrictions on their citizens to counteract such changes. The citizens are called upon to accept these restrictions in the name of the higher-order interest. In the case of war, what is good for the compartment that is at war (a tribe, a country, etc.) can mean death for individual soldiers. In social insects, what is good for the queen is not necessarily good for the workers. What is good for the family takes precedence over what is good for its individual members. For example, in times when food is scarce, all family members must share. A single individual cannot eat all the available food. The surgical removal of a tumor can be good for a patient but it means the end of the tumor's life.

TABLE I: BASIC PHILOSOPHY OF 'CLASSICAL'- VERSUS 'INTEGRATIVE' BIOLOGY

	CLASSICAL BIOLOGY	INTEGRATIVE BIOLOGY
1. Study object	• living matter	• carbon chemistry-based communication systems (dissipative in nature)
2. Basic unit	• organism	• communicating compartment
3. Smallest unit	• cell (prokaryote or eukaryote?)	• monomembrane communicating compartment
4. Organism	• instrument for metabolism and reproduction • product of its genes and environment • complex because of large number of genes	• instrument for communication and problem solving • product of its genes and communication environment • extremely complex because of large number of genes, protein-protein interactions, and very high number of communication bifurcation points
5. Major functions	• growth, development • reproduction	• communication/problem solving • prevention of transition from order to chaos
6. Number of dogmas and basic rules in biology	• one dogma, namely: DNA $\rightarrow$ RNA $\rightarrow$ Protein(s) for the genetic memory	• two dogmas, one for the genetic- and probably another one for the 'cognitive' memory • the Hodgkin-Goldman-Katz equation (cellular electricity) equally important as first dogma • no life without gradients
7. Basic chemistry	• living matter consists of aggregated fossil stardust • synthesis of organic molecules in 'saline' environment	• the fossil stardust which makes up living matter is creative and self organizing into ionically-electrically compartmentalized units • idem plus bioelectrochemistry
8. Basic physics	• classical linear Newtonian physics • 1st and 2nd law of thermodynamics • time in principle reversible in some processes • mainly deterministic and predictable • 4 dimensions of space-time	• modern Newtonian physics • idem plus non-linear Prigoginean far-from-equilibrium thermodynamics • time in principle always irreversible • less deterministic and predictable • idem plus electrical and immaterial dimensions

9. Information carrier	• nucleic acids • life as a single continuum	• nucleic acids and 'cognitive memory' • life as a double continuum
10. Genetics	• genetics (Mendel, molecular)	• idem plus epigenetics
11. Variability due to	• mutations, meiosis, modifications • changes in *'macromolecular environment'* around the genes	• idem plus bifurcation points • changes in *'macromolecular and ionic environment'* around the genes
12. Evolution	• of life as a single continuum (the *'hardware'*) • neo-Darwinism • changes in genes, evolution of macromolecules, • *'mutated protein force'* • natural selection drives evolution • self-selection hardly mentioned • slow (geological time scale) • 'survival of fittest'	• of life as a double continuum (*'hard- and software'*) • hardware-software theory of evolution • evolution of communication and problem solving strategies • *'gradient/communication/ problem solving force'* • selection is itself a result of something that acted before • self-selection results from life itself and precedes environmental selection • evolution of physical compartments slow but that based on 'software evolution' very fast • best problem solvers grow and reproduce better
13. Definitions of life and death	• Impossible • bodies of organisms are mortal	• possible (this book) • instruments for communication are mortal but communication can have means for survival
14. Scientific approach	• reductionistic experimentation and thinking • sum of parts approximates the whole	• reductionistic experimentation and holistic thinking • the whole is more than the sum of its parts
15. Feedback systems	• for coordinated functioning	• essential for communication
16. Feelings, emotions, freedom in behaviour, consciousness	• largely irrelevant and no conceptual framework to make them fit in	• where present, essential elements in communication
17. Inherent ethical rules	• no ("life is a machine")	• yes, because of interactions among compartments

Complex compartments start to degenerate from the moment that particular lower levels of compartmentalization invoke such high priority for their own good (expansion in space and time) that not enough space, energy or time remains for the normal functioning of all the other constituting subcompartments of the compartment under consideration. The terrorist act of September 11, 2001, which was very much the focus of public and media attention at the time this chapter was being written, is a clear example of this. Terrorism means the same for compartments of level 11 (social compartments) as cancer means for compartments of level 7.

16.4   The place of the *Homo sapiens* in the biosphere as a whole

All living organisms on earth, *Homo* inclusive, are descendants from the very same Progenote. They are all the result of some 3.7 billion years of evolution. Thus, they are all one family. The results of the more than seventy completed genomic projects (early 2002) prove that Darwin's common descent idea is basically correct.

Mammals of the species *Homo sapiens* think that their species is 'superior' over all other species living on earth. Is such a view justified or not? The answer will depend on the criteria that are used to define 'more or less advanced'. This was already briefly discussed in Chapter 2.

In my opinion, the ability to solve problems is a better criterion for complexity, and certainly for fitness, than the number of genes or the size of the genome (for data, see Chapter 2). The studax *Homo sapiens* can be regarded as the most advanced species on earth, because he is able to solve a much greater variety of problems than any other species can do. Apparently, in the course of evolution *Homo* acquired the best combination of genetic and cognitive signal transduction pathways.

In nature it is species for its own. With respect to heterotrophic organisms, the rule is: eat and be eaten. Indeed, such organisms are by themselves not responsible for the fact that some distant ancestor willy-nilly underwent a mutation that resulted in the loss of the capacity to provide its own food, for example by photosynthesis. The only way to survive was to start eating other organisms. There is no rule in nature that says that heterotrophy puts limits on 'the rights' of the prey. Humans should realize that after their death their body will serve as food for bacteria. Bacteria may well look upon *Homo* as one of their farm animals.

The greater problem solving ability of *Homo* does not guarantee that this species will outlive most other species. The rapid expansion of the human population, the exhaustion of natural resources, the man-made pollution, and the aggressive behavior of *Homo* are all serious drawbacks for the success of the future evolution of the *Homo sapiens*.

16.5   Biology, ethics and philosophy. *Communico, ergo vivo*

As a result of the successes of reductionistic experimentation, biology has rapidly evolved into a complex technological discipline. Its vocabulary has become very large and quite often difficult, especially when abbreviations are used which only specialists in the field can easily understand. Such a technological approach to biology alienates people from Nature rather than bringing them to admire it. The sharper focusing of biology on the key issue of communication will lead to a far more harmonious view of the place of man in Nature and in the Cosmos.

Perhaps lawmakers should be more aware of the fact Life carries ethical rules within itself, which I have briefly mentioned in several different chapters, e.g. concerning abortion, euthanasia, self-selection, good and evil, and the higher-order interest principle.

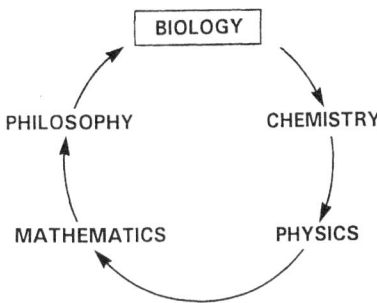

Pope John XXIII (1996) made the statement that Darwin's theory of evolution was right after all when restricted to the material aspect of living matter but that it should not be applied to the mind, because the mind was created by God. Unfortunately, this simply continues the dualistic thinking of Descartes: mind versus body. But the body and the mind are one, not two (cf. "the embodied mind", Varela *et al.*, 1992; and this book). In analogy with Descartes' *"Cogito, ergo sum"* (= I think, therefore I am), we could say: *"Communico, ergo vivo"* (= I communicate, therefore I live).

I always wonder why so many religious people have problems accepting a God who created matter in such a brilliant way that from the moment it came into existence, it was bestowed with the possibility of generating communicating compartments with an inherent immaterial dimension and with the ability to adapt and evolve. I see nothing wrong with this view. To the contrary, it is consistent both with science and religion.

Biologists use chemistry to find explanations for the physiological problems they run up against. Chemists look to theoretical physics for answers to their tough questions. Physicists need mathematics. Mathematicians often touch the boundaries of philosophy. And the philosophers finally end up in biology, and thus in life itself.

16.8 If Darwin had had a computer...

From a scientific point of view, it makes no sense to speculate on the extent to which thinking in contemporary biology would be different if the founding father of the theory of evolution had been familiar with the hardware-software principles of computers. Thinking about evolution in terms of the principles of communication was not even an issue in Darwin's time. Technological innovation and novel insights often go hand in hand. It is up to the present generation of biologists to incorporate the contemporary insights of information science and technology into the theory of evolution. We can be sure that if Darwin had had a computer, he would have hated it and loved it for the very same reasons that most of us hate it and love it.

16.7 L = Σ C : Towards a biological paradigm?

The enormous advances in the biological sciences are causing a true revolution in our perception of the world and the place occupied by man in it. Up to the present time, our worldview has been very much based upon the materialistic concept of living matter that emerged from the enormous successes of physics and chemistry. In other words, the physical-chemical paradigm has prevailed so far. In the near future, the biological paradigm centered on $L = \Sigma\, C$ is likely to grow in importance. In my opinion, communication will be the central issue in the biological paradigm because there can be no doubt that good communication is what really matters in life.

ESSENTIALS

PROPERTIES OF LIFE

INTEGRATIVE BIOLOGY

1. Life in NOT a machine, but rather the ACTIVITY of a special sort of machine.
2. In daily life the terms communication and problem solving are used as if they denote different things. However, if one dissects their underlying mechanisms, it becomes apparent that to a large extent communication is another word for problem solving. Indeed, a sender invariably emits a message in coded form. The first task of a decoder is to solve the problem of how to decode this coded message. Hence, what we call 'communication' in daily life is largely unconscious problem solving. What we call 'problem solving' in daily life is the conscious solving of difficult problems.
3. **Life is the total communication/problem solving activity of hierarchically organized communicating compartments**. *Communico, ergo vivo*: I communicate, therefore I live.
4. The degree of difficulty of the problems that can be solved increases with the number of subcompartmental levels present in a given communicating compartment.
5. Problem solving requires hardware, software, energy and motivation. Hence, all these parameters must be taken into consideration in the theory of evolution. This is done in the 'double continuum theory', as presented in Chapter 15.
6. Communication can master any force of physics. This means that the coming into existence of life some 4 billion years ago marked a dramatic turning point for the universe: it changed from a passive into an active universe, which is able to change itself.
7. Because it is an activity, life cannot possibly be constant. It changes continuously. There are as many different 'lives' as there are different communicating compartments.
8. The interest of the highest level of compartmental organization always prevails over that of the lower levels. Something is 'good' if it improves the communication/problem solving activity at the highest level of compartmental organization. Biology provides the basis for ethical principles.
9. I propose attributing a broader meaning to the term 'Integrative biology' than it presently carries. "Integrative biology is the study of the generation, functioning and perpetuation of hierarchically organized communicating compartments". This approach narrows the gap between the exact sciences and the humanities.
10. Good communication is what really matters.

THE STORY OF CREATION: DYNAMIC VIEW

When space, time and matter did not yet exist, there was perfect symmetry, silence, darkness. And God said, "Let there be a point of asymmetry so a Universe can come into being and life can arise in the fullness of time." At the still point where the symmetry broke there was white heat. In a split second the mass of the total universe exploded into being, expanding at blinding speed and cooling to form time, space and the hydrogen nucleus, primal element of all matter. And God saw that it was good.

Then God said, "Let hydrogen serve as a building block for deuterium and helium and let the expansion of this mass of elementary matter symmetrically proceed for a few hundred thousand years till the time comes that the generation of asymmetry can once more be useful." And it was so. The Universe expanded symmetrically and cooled further. The still naked atomic nuclei captured electrons, which began orbiting around these nuclei, thereby forming atoms. And God saw that it was good.

Then God said, "Let there come points of asymmetry once again in this expanding Universe to trigger condensation of matter, so that the first stars can come into existence." And it was so. At the points of asymmetry, atoms started to cluster, more and more, so that bigger and bigger turbulent clouds of hydrogen and helium appeared. The intense pressure made the inner core of these gas clouds hotter and hotter, to a point where the atoms in the core of the cloud fused with one another, thereby forming elements of higher mass and liberating immense quantities of heat and light. Thus the first stars appeared in the firmament, and God saw that it was good.

Then God said, "Let this process of star formation happen again and again here and there in the Universe so that more and more stars will appear in the firmament of the heavens. In due time, some of them will serve for man as signs of fixed times, days and years. One of these stars, the Sun, will deliver heat and light to sustain life on the planet Earth that will be formed later." And it was so. The fusion reactions yielded elements larger than hydrogen and helium such as beryllium, oxygen and – the crown jewel of all, without which no life is possible - carbon. This process continued steadily until after many millions or sometimes even a few billions of years the stars had burned all their hydrogen. And God saw that it was good.

Then God said, "Let us use the elements which have been formed in the stars as building blocks for other elements with still higher atomic numbers so that finally all the elements that are required for life to come into being sometime and somewhere have been formed." And it was so. In a final burst of activity, the terminal stars exploded as supernovas with such enormous force that a portion of their atoms were knocked into each other and fused, resulting in the formation of heavier elements. The remnants of exploded supernovas were thrown into space and, after a time, some of them came within the gravitational fields of the gas clouds of nascent stars, where they were absorbed. And God saw that it was good.

And God said: "Let these second generation stars go through a similar cycle as those of the first generation so that atoms of still higher mass can be formed through fusion reactions." And it was so. Again, the stars burned their supply of hydrogen and helium, and again they exploded as supernovas, thereby forming heavier elements. The debris was thrown into space, all this occurring in ever continuing cycles. And God saw that it was good. Then God said, "Let nine planets appear at different distances from the star called Sun, and let the moon light the Earth at night." And it was so. Smaller and larger lumps of stardust originating from exploded supernovas started to attract each other at points of asymmetry more than at other places so that aggregating masses of fossil stardust were formed. As these masses grew larger, more pieces were attracted, slamming with increasing force into the growing planetoid, and thereby liberating heat. As more and more pieces impacted, the nascent planet grew larger and larger, and became glowing hot and liquid. And God saw that it was good.

Then God said, "Let conditions be generated so that life can come into being on planet Earth." And it was so. The rain of huge lumps of stardust and ice meteorites decreased, and the surface of the Earth started to cool. After a few hundred more million years, the drop in temperature was great enough that rainwater no longer instantly evaporated upon contact with the Earth's surface, but rather collected in crevices, ponds, lakes, seas and finally in oceans. In this early watery environment, a variety of organic molecules formed, which in due time began forming aggregates. At some moment in this period of prebiotic evolution, certain aggregates got encapsulated by a lipid membrane. And God saw that it was good.

Then God said, "Let the membrane compartments now come to life. Let them use their ability to communicate to solve all sorts of problems so that they can feel well and contented in ever new environments. Let them reproduce and evolve in the course of time so that planet Earth will be teeming with life." And it was so. The membrane compartments started to communicate, their morphology and language became ever more complex. They colonized more and more environments, which thereby became their home. First bacteria were formed. From these emerged cells, which internalized certain smaller membrane compartments. Later, still other types of communicating compartments were formed, ever more complex. The generation of new asymmetries led to male and female, sperm cell and oocyte, sexual reproduction and embryonic development. Some organisms left the water in which they were born and went to live on land, and some even conquered the air. In the course of time, different populations of organisms came to be using such different languages that in the end they no longer understood each other. And God saw that it was all good.

Then God said, "Now, 12 billion years after the moment I created the Universe, time has come on planet Earth, with its millions of living species all speaking different languages, that man should come onto the scene. Let us make man in our image, after our likeness; and let them have dominion over the fish of the sea, and over the birds of the air, and over the cattle, and over all the earth, and over every creeping thing that creeps upon the earth." And it was so. In the line of evolving vertebrates, consciousness increased ever more until a population emerged whose members knew that they knew, and they started thinking about their own origin, their place in the universe and their destiny.

And God said to man, "Behold, I made you in our image, after our likeness. Be worthy of your inheritance. You will have dominion over all organisms that live in the water, on the land and in the air. They may serve you as food, but handle them in a spirit of respect for my creation as a whole. Furthermore, I give you the ability to master all forces operating in non-living matter, but you must use this gift only for the well-being of your fellow man, yourself and my creation." And it was so. And God saw all that he had made, and behold, it was very good.

When God had completed his work, he blessed the seventh day and sanctified it, because in it God rested from all his work which he had done in creation.

- Genesis 1, revised in accordance with the insights of the writers, anno Domini 2002.

Arnold De Loof and Richard Sundahl

REFERENCES

Adams, M. and consortium members (2000) The genome sequence of *Drosophila melanogaster*. Science 287, 2185-2195.

Agard, D.A. and Sedat, J.W. (1983) Three dimensional architecture of a polytene nucleus. Nature (London) 302, 676-681.

Allègre, C. and Schneider, S.H. (1994) The Evolution of the Earth, Scientific American 271 (October, Special volume "Life in the Universe"), 44-51.

Amikura, R., Kobayashi, S., Saito, H. and Okada, M. (1996) Changes in subcellular localization of mtlrRna outside mitochondria in oogenesis and early embryogenesis of *Drosophila melanogaster*. Development, Growth and Differentiation, 38, 489-498.

Anderson, P. and Greenberg, R. (2001) Review: Phylogeny of ion channels: clues to structure and function. Comparative Biochemistry and Physiology B 129, 17-28.

Antoniadis, I., Arkani-Hamed, N., Dimopoulos, S. and Dvali, G. (1998) New dimensions at a millimeter to a fermi and superstrings at a TeV. Physics Letters B 436, 257-263.

Arkani-Hamed, N., Dimopoulos, S. and Dvali G. (1998) The hierarchy problem and new dimensions at a millimeter. Physics Letters B 429, 263-272.

Arthur, W. (2002) The emerging conceptual framework of evolutionary developmental biology. Nature (London) 415, 757-764.

Balinsky, B.I. (1975) An introduction to embryology (4th edition). W.B. Saunders Company, Philadelphia.

Banfield, J.F. and Marshall, C.R. (2000) Genomics and the Geosciences. Science 287, 605-606.

Barritt, G. (1994) Communication within animal cells. Oxford Science Publications, Oxford University Press, Oxford.

Barth, L.J. (1964) Development. Selected topics. Addison-Wesley Publishing Company, Inc., Reading, UK.

Bartusiak, M. (1955) Einstein's unfinished Symphony. Listening to the sound of spacetime. Joseph Henry Press, Washington D.C.

Bate, M. and Martinez Arias, A. (1993) The development of *Drosophila melanogaster* (2 volumes). Cold Spring Harbor Laboratory Press.

Bateson, P. (1983) Genes, environment and the development of behaviour. In "Animal Behaviour". vol. 3.: Genes, Development and Learning. Blackwell Scientific Publications, London, 52-81

Bergstra J.A. and van Vlijmen, S.F.M. (1998) Theoretical Software-Engineering (in Dutch). Zeno, The Leiden-Utrecht Research Institute of Philosophy.

Brenner, C. and Kroemer, G. (2000) Mitochondria – the death signal integrators. Science 289, 1150-1151.

Buxbaum, R.E. (1995) Biological levels. Nature (London) 373, 567-568.

Calvin, W.H. (1994) The emergence of intelligence. Scientific American 271 (October, Special volume "Life in the Universe"), 100-107.

Calvin, W.H. en Ojemann, G.A. (1994) Conversations with Neil's brain: the neural nature of thought and language. Addison-Wesley Publishing Company.

Capy, P. (2000) Is bigger better in cricket? Science 287, 985-986.

Carroll, S.B. (2001) The big picture. Macroevolution. Nature (London) 409, 669.

Cech, T.R. (1986a) RNA as an enzyme. Scientific American, 64-75.

Cech, T.R. (2000) The ribosome is a ribozyme. Science 289, 878-879.

Cech, T.R. and Bass, B.L. (1986b) Biological catalysis by RNA. Annual Review of Biochemistry 55, 599-629.

Chapple, M. (1999) Dictionary of Physics. Fitzroy Dearnborn Publishers, London Chicago.

Clark, W.R. (1996) Sex & the origins of death. Oxford University Press, New York.

Claverie, J.-M. (2001) What if there are only 30,000 human genes? Science 291, 1255-1257.

Cody, G.D., Boctor, N., Filley T., Hazen, R., Scott J., Sharma A., Yoder, H. (2000) Primordial carbonylated iron-sulfur compounds and the synthesis of pyruvate. Science 289, 1337-1340.

Cox, T. (1990) Origin of the chemical elements. New Scientist 29, 1-4.

Cramer, F. (1993) Chaos en Order. The complex structure of living systems. VCH Verlaggesellschaft mbH, Weinheim, New York.

Crick, F. (1994) The astonishing hypothesis: the scientific search for the soul. Simon and Schuster, London.

Crick, F. and Koch, C., (1992) The problem of consciousness. Scientific American 267 (september), 153-159.

Cullen, J.T. (1996) What is Life? A critique of autopoiesis in the light of non-developmental apoptosis. In: Einstein meets Margritte: the Green Book: Man and nature- A world in transition (editor D. Aerts, Kluwer Academic Publishers).

Damasio, A. (2001) Fundamental feelings. Emotion. Nature (London) 413, 781.

Darnell, J.E. and Doolittle, W.F. (1986) Speculations on the early course of evolution. Proceedings of the National Academy of Sciences U.S.A. 83, 1271-1275.

Darwin, C. (1859) The origin of species by means of natural selection. First published by John Murray.

Davenport, R.J. (2001) Getting yeast prions to bridge the species gap. Science 291, 1881.

Davidson, E.H. (1976) Gene activity in early development. Academic Press, New York.

Dawkins, R. (1976) The Selfish gene. Oxford University Press.

De Duve, C. (1991) Blueprint for a cell. The nature and origin of life. Neil Patterson Publishers, Burlington, Carolina, U.S.A.

De Duve, C. (1995) Vital dust. Life as a cosmic imperative. Basic Books, Harper Collins publishers, New York.

De La Cruz, E. and Pollard, D. (2001) Actin' up. Science 293, 616-618.

De Loof, A. (1986) The electrical dimension of cells: the cell as a miniature electrophoresis chamber. International Review of Cytolology 104, 251-352.

De Loof, A. (1992) All animals develop from a blastula: consequences of an undervalued definition for thinking on animal development. BioEssays 14, 373-375.

De Loof, A. (1993a) Schrödinger 50 years ago: "What is Life?" "The ability to communicate", a plausible reply?" International Journal of Biochemistry 25, 1715-1721.

De Loof, A. (1993b) Differentiation: "Keep the genome constant but change over and over again its ionic and/or macromolecular environment"? A conceptual synthesis. Belgian Journal of Zoology 123, 77-91.

De Loof, A (1995) Hormones and the cytoskeleton. Function follows form. Annual Review of Cytology 166, 1-58.

De Loof, A. (1996) :(In Dutch) Wat is Leven? De onstoffelijke dimensie. (First edition). Garant Publishers Leuven (Belgium) and Apeldoorn (The Netherlands).

De Loof, A. and Huybrechts, R. (1998) Review: "Insects do not have sex hormones": a myth? General and Comparative Endocrinology 111, 245-260.

De Loof, A.,Huybrechts, R. and Kotanen, S. (1998) Review: Reproduction and love: Strategies of the organism's cellular defense system. Comparative Biochemistry and Physiology Part C, 120, 167-176.

De Loof, A. and Vanden Broeck, J. (1995) Communication: the key to defining "life", "death" and the force driving evolution. "Organic chemistry-based-" versus "artificial" life. Belgian Journal of Zoology 125, 5-28.

De Loof, A., Callaerts, P and Vanden Broeck, J. (1992) The pivotal role of the plasma membrane-cytoskeletal complex and of epithelium formation in differentiation in animals. Comparative Biochemistry and Physiology 101A, 639-651.

Depew, D.D. and Weber, B.H. (1994) Systems dynamics and the genealogy of natural selection. MIT Press, Boston.

Des Marais, D. (2000) When did photosynthesis emerge on earth? Science 289, 1703-1705.

Dodson, E.O. and Dodson, P. (1976) Evolution: Process and Product. D. Van Nostrand Company, New York.

Dupré, A. (1996) The concept 'time' in physics (in Dutch: Het begrip Tijd in de fysica). In: Lessen voor de XXI eeuw. Davidsfonds Leuven.

Eigen, M. and Schuster P. (1979) The hypercycle. A principle of Natural Self-Organisation. Springer Verlag, Berlin, Heidelberg, New York.

Feynman, Leighton, Sands. Feynman lectures on physics Vol 1. (1977) 7th Edition. Addison Wesley (Reading, Massachusetts)

Finkel, E. (2001) The mitochondrion: is it central to apoptosis? Science 292, 624-626.

Fleishaker, G.R. (1988) Autopoiesis: The status of its system Logic. BioSystems 22, 3749.

Freedman, W.L., Madore, B.F., Mould, J.R., Hill, R., Ferrarese, L., Kennicutt, Jr R.C., Saha, A., Steson, P.B., Graham, J.A., Ford, H., Hoessel, J.G., Huchra, J., Hughes S. M. and Illingworth, G.D. (1994) Distance to the Virgo cluster galaxy M100 from Hubble space telescope observations of Cepheids. Nature (London) 371, 757-761.

Futuyama, D.J. (1998) Evolutionary Biology (3rd edition). Sinauer Associates, Sunderland.

Fry, (2000) The emergence of life on earth. Rutgers University Press, New Brunswick, NJ.

Galas, D.J. (2001) Making sense of the sequence. Science 291, 1257-1260.

Gangui, A. (2001) In support of inflation. Science 291, 837-838.

Geraerts, W.P.M., Smit, A.B. and Li, K.W. (1994) Constraints and innovations in the molecular evolution of neuronal signaling: implications for behavior. In: Flexibility and constraints in Behavioral systems (R.J. Greenspan and C.P. Kyriacou, editors), John Wiley and Sons Ltd, New York: 209-235.

Gestwicki, J.E. and Kiessling, L.L. (2002) Inter-receptor communication through arrays of bacterial chemoreceptors. Nature (London) 415, 81-84.

Gibbons, G. (2000) Brane-words. Science 287, 49-50.

Gilbert, S.F. (1997) Developmental Biology (5th edition). Sinauer Associates, Sunderland, Massachusetts.

Gilbert S.F. and Raunio, A.M. 1997 Embryology. Constructing the whole organism. Sinauer Associates, Inc. Pubishers Sunderland, MA.

Goff, S.A. and consortium members. (2002) A draft sequence of the rice genome (*Oryza sativa* L. ssp. *japonica*). Science 296, 792-103.

Goldman, M.A. (2001) Spandrels of selection (book reviews). Nature (London) 413, 252-253.
Gould, S.J. (1994) The Evolution of Life on Earth. Scientific American, 271 (October, special volume "Life in the Universe"), 63-69.
Gould, S.J. (2002) The structure of Evolutionary Theory. Belknap Press, Harvard.
Graur, D. and Li, W-H. (1999) Fundamentals of Molecular Evolution. Sinauer Associates, Sunderland Massachusetts, 481 pp.
Gray, W. (2000), Mitochondrial genes on the move. Nature (London) 408, 302-304.
Gribbin, J. (1994) In the beginning. The birth of the living universe. Penguin Books Ltd. London.
Gurdon, J.B. and Uehlinger, V. (1966) "Fertile" intestinal nuclei. Nature 210, 1240-1241.
Gurdon, J.B. (1962) The developmental capacity of nuclei taken from intestinal epithelial cells of feeding tadpoles. Journal of Embryology and Experimental Morphology 10, 622-640.
Halder, G., Callaerts, P. and Gehring, W. (1995) Induction of ectopic eyes by targeted expression of the eyeless gene in *Drosophila*. Science 267, 1788-1792.
Harold, F.M. (1986) The Vital Force: A Study of Bioenergetics. W.H. Freeman en Company, New York.
Hawking, S (1996) A Brief History of Time. Bantam books, London.
Hawking S.W. and Penrose R. (1996a) The nature of space and time. Scientific American (July), 44-49.
Hawking, S.W. and Penrose, R. (1996b) The nature of space and time. Princeton University Press.
Hemmerlin, E. (1996) From quantal to material level. In: Einstein meets Margritte: the Blue Book: Metadebates. (Editor D. Aerts, Kluwer Academic Publishers).
Henderson's Dictionary of Biological terms (11th edition, E. Lawrence, Ed.) (1995) Longman Scientific & Technical, Longman Group Ltd, Harlow, England.
Hentze, M. W. (2001) Believe it or not - Translation in the nucleus. Science 293, 1058-1059.
Hopper A.F. and Hart, N.H. (1980) Foundations of animal development. Oxford University Press, New York.
Horava, P. and Witten, E. (1996) Eleven-dimensional super-gravity on a manifold with boundary. Nuclear-Physics B 475, 94-114.
Horgan, J. (1991) In the beginning... Scientific American 264, 103-109.
Huigens, M.E., Luck, R.F., Klaassen, R.H.G., Maas, M.F.P.M., Timmermans, M.J.T.NN and Stouthamer, R. (2000) Infectious parthenogenesis. Nature (London) 405, 178-179.
International Human Genome Sequencing Consortium (2001) Initial sequencing and analysis of the human genome. Nature (London) 409, 860-921.
Jaffe, L.F. and Nuccitelli, R. (1974) An ultrasensitive vibrating probe for measuring steady extracellular electric currents. Journal for Cell Biology 63, 614-628.
Jaffe, L.F. and Nuccitelli, R. (1977) Electrical controls of development. Annual Review of Biophysics and Bioengineering 6, 445-476.
Jaffe, L.F. and Woodruff, R.I. (1979) Large electrical currents traverse developing *Cecropia* follicles. Proceedings of the National Academy of Sciences USA 76, 1328-1331.
James, P., Peebles, E., Schramm, D.N., Turner, E.L. and Kron, R.G. (1994) The Evolution of the Universe. Scientific American (October, special volume "Life in the

Universe."), 29-33.
Jaynes, J. (1982) The origin of Consciousness in the breakdown of the bicameral mind. Penguin Books, Harmondsworth.
Johnston, W.K., Unrau, P.J., Lawrence, M.S., Glasner, M.E. and Bartel, D.P. (2001) RNA-catalyzed RNA polymerization: accurate and general RNA-templated primer extension. Science 292, 1319-1325.
Jones, P.A. and Takai, D. (2001) The role of DNA methylation in mammalian epigenetics. Science 293, 1068-1070.
Kauffman, S.A. (1993) The origins of order. Self-organization and selection in evolution. Oxford University Press, New York, Oxford.
Kendrick, K.M., da Costa, A.P., Leigh, A., Hinton, M.R. and Peirce, J.W. (2001) Sheep don't forget a face. Nature (London) 414, 165-166.
Kestenbaum, D. (1998) Physics: Particle decays reveal arrow of time. Science 282, 602-603.
Kirschner, R.P. (1994) The Earth's Elements. Scientific American (October, special volume "Life in the Universe"), 37-43.
Kobayashi, S., Amikura, R. and Okada, M. (1993) Presence of mitochondrial large ribosomal RNA outside mitochondria in germ plasm of *Drosophila melanogaster*. Science 260, 1521-1524.
Kobayashi, S. and Okada, M. (1989) Restoration of pole-cell forming ability to UV-irradiated *Drosophila* embryos by injection of lrRNA. Development 107, 733-742.
Koch, C. and Crick, F. (2001) The zombie within. Unconsciousness. Nature (London) 411, 893.
Langman, J. (1976) Medical embryology. The Williams and Wilkins Company, Baltimore, (First edition in 1966, many later editions).
Leaky R.E. and Lewin, R. (1977) Origins. E.P. Dutton, New York.
Lee, D.H., Granja, J.R., Martinez, J.A., Severin, K. and Ghadiri, M.R. (1996) A self-replicating peptide. Nature (London) 382, 525-527.
Lin, E.C. and Cantiello, H.F. (1993) A novel method to study the electrodynamic behavior of actin filaments. Evidence for cable-like properties of actin. Biophysical Journal 65, 1371-1378.
Lovelock, J. (1995) Gaia. A new look at Life on Earth. Oxford University Press.
Løvtrup, S. (1974) Epigenetics. J. Wiley and Sons, London, New York.
Luna, E.J. and Hitt, A.L. (1992) Cytoskeleton-plasma membrane interactions. Science 258, 955-964.
Lyko, F., Ramsahoye, F. and Jaenisch, R. (2000) DNA methylation in *Drosophila melanogaster*. Nature (London) 408, 538-540.
Margulis, L. (1981) Symbiosis in cell evolution. W.H. Freeman and Company, San Francisco.
Margulis, L. (1998) The symbiontic planet. A new look at evolution. Weidenfeld & Nicolson, London.
Margulis, L. and Sagan, D. (1995) What is Life? Simon and Schuster, New York.
Mathog, D., Hochstrasser, M., Grünbaum, Y., Saumweber, H. and Sedat, J. (1984) Characteristic folding pattern of polytene chromosomes in *Drosophila* salivary gland nuclei. Nature (London) 308, 414-421.
Maturana, H.R. and Varela F.J. (1980) Autopoiesis and Cognition: The realization of the Living. D. Reidel, Publishing, Dordrecht, Holland.

Maynard Smith, J., and Szathmary, E. (1995) The evolution of complexity. W.H. Freeman and Company, New York.
Melino, G. (2001) The Sirens' song. Apoptosis. Nature (London) 412, 23.
Meulemans, W. and De Loof, A. (1992) Changes in the cytoskeletal actin patterns in the Malpighian tubules of the fleshfly, *Sarcophaga bullata* (Parker) (Diptera: Calliphoridae), during metamorphosis. International Journal for Insect Morphology and Embryology. 21, 1-16.
Miller, S.L. (1987) Which organic compounds could have occurred on the prebiotic earth? Cold Spring Harbor Symposia on Quantitative Biology, 52, 17-27.
Miller, S.L. and Orgel, L.E. (1973) The Origins of Life. Prentice Hall, Englewood Cliffs, N.J.
Mitchell, P. (1979) Compartmentalization and communication in living systems. Ligand conduction: a general catalytic principle in chemical, osmotic and chemiosmotic reaction systems. European Journal of Biochemistry 95, 1-20.
Müller, W.E. (2001) Review: How was the metazooan threshold crossed? The hypothetical Urmetazoa. Comp. Biochem. Physiol. A 129, 433-460.
Murphy, M.P. and O'Neill, A.J. (1995) What is Life? The next fifty years. Speculations on the future of biology. Cambridge University Press, Cambridge, New York.
Nakaseko Y. and Yanagida M. (2001) Cytoskeleton in the cell cycle. Nature (London) 412, 291-292.
Nelissen, M. (2000) (In Dutch) De bril van Darwin (= Darwin's glasses). In search for the roots of our behavior. Lannoo, Tielt, Belgium.
Newman, A. (2001) RNA enzymes for RNA splicing. Nature (London) 413, 695.
Nuccitelli, R. (1986) Ionic currents in development. Progress in Clinical and Biological Research, Volume 20, Alan R. Liss Inc. New York.
Nüsslein-Vollhard, C. (1994) Of flies and fishes. Science 266, 572-574.
Orgel, L.E. (1994) The Origin of Life on Earth. Scientific American (October, special volume "Life in the Universe"), 53-61.
Papaj, D.R. (1993) Automatic behavior and the evolution of instinct: lessons from learning in parasitoids. In: Papaj, D.R. and Lewis, A.C. (eds.) Insect learning. Chapman and Hall, New York: 243-272.
Peach, K. (1998) Time's broken arrow. Nature (London) 396, 407 408.
Pease, R. (2001) Brane new world. Nature (London) 411, 986-988.
Pedersen, S.F., Hoffmann, E.K. and Mills, J.W. (2001) Review: The cytoskeleton and cell volume regulation. Comparative Biochemistry and Physiology Part A, 130, 385-399.
Pennisi, E. (2000) Hardy microbe thrives at pH 0. Science 287, 1731-1732.
Penrose, R. (1994) Shadows of the mind: on consciousness, computation, and the new physics of the mind. Oxford University Press, Oxford, New York.
Pestic-Dragovich, L., Stojiljkovic, L., Philimonenko, A., Nowak, G., Ke, Y., Settlage, R.E., Shabanowitz, J., Hunt, D.F., Hozak, P. and de Lanerolle, P. (2000) A myosin I isoform in the nucleus. Science 290, 337-341.
Pierce, M.J., Welch, D.L., McCure, R.D., Van den Bergh, S., Racine, R. and Stetson, P.B. (1994) Hubble constant and Virgo cluster distance from observations of Cepheid variables. Nature (London) 371, 385-389.
Porath, D., Bezryadin, A., de Vries S., Dekker, C. (2000) Direct measurement of electrical transport through DNA molecules. Nature (London), 403, 635-638.
Postlethwait, J.H. and Hopson, J.L. (1991) The nature of life. McGraw-Hill Publishing Company, U.S.A.

Press, F. and Siever, R. (2001) Understanding Earth (3rd Edition). W.H. Freeman and Company, New York.
Prigogine, I. (1980) From Being to Becoming. Time and Complexity in the Physical Sciences. W.H. Freeman and Company, New York.
Prigogine, I. and Stengers, I. (1985) Order out of Chaos. Man's new dialogue with nature. Flamingo (Harper Collins Publishers).
Prigogine, I. and Nicolis, G. (1971) Biological order, structure, and instabilities. Review of Biophysics 4, 107-148.
Raven, P.H. and Johnson, G.B. (1996) Biology (4th edition). Wm. C. Brown Publishers, Dubuque.
Riordan, M. and Schramm, D.N. (1993) The Shadows of Creation. Dark Matter and the Structure of the Universe. Oxford University Press, Oxford, New York.
Rizotti, M. (Ed.) (1996) Defining life, the central problem in theoretical biology. University of Padova, Italy (Seminario di Scienze Biologiche, via Venezia 5, 35131 Padova).
Robert, F. (2001) The origin of water on earth. Science 293, 1056-1058.
Ronshaugen, M., McGinnis, N. and McGinnis, W. (2002) Hox protein mutation and macroevolution of the insect body plan. Nature (London) 415, 914-917.
Rosen, R. (1991) Life itself. A comprehensive inquiry into the nature, origin, and fabrication of Life. Columbia University Press, New York.
Sauman, I. and Berry, S.J. (1994) An actin infrastructure is associated with eucaryotic chromosomes: structural and functional significance. European Journal for Cell Biology 64, 348-356.
Schejter, A. and Agassi, J. (1994) On the definition of life. Journal for General Philosophy of Science. 25, 97-106.
Schrödinger, E. (1946) What is Life? Cambridge University Press, Cambridge.
Seife, C. (2001) Big Bang's new rival debuts with a splash. Science 292, 189-190.
Sklar, L. (1974) Space, Time and Space-time. University of California Press, Berkeley.
Smart, J.J.C. (1964) Problems of space and time. Problems of Philosophy series. Paul Edwards, General editor. The MacMillan Company, New York.
Smith, C.A., McClive, P.J., Western, P.S., Reed, K.J. and Sinclair, A.H. (1999) Conservation of a sex-determining gene. Nature (London) 402, 601-602
Sneden, C. (2001) The age of the Universe. Nature (London) 409, 673-674.
Solvay, E. (1894) Le rôle de l' électricité dans les phénomènes de la vie animale. Hayez, Brussels.
St Johnston, D. and Nüsslein-Volhard, C. (1992) The origin of pattern and polarity in the *Drosophila* embryo. Cell 68, 201-219.
Stiller, S. (2001) Sexual selection and the maintenance of sex. Nature (London) 411, 689-692.
Stringer, C. and Davies W. (2001) Those elusive Neanderthals. Nature (London) 413, 791-792.
Strobel, S.A. (2001) Repopulating the RNA world. Nature (London) 411, 1003-1006
Stynen, D., and De Loof, A. (1982) Sugar binding properties of the vitellogenic proteins of the Colorado beetle, demonstrated by hemagglutination and precipitation experiments. Wilhelm Roux's Archives of Developmental Biology 191, 159-162.
Sutovsky, P., Moreno, R. D., Ramalho-Santos, J., Dominko, T., Simerly, C., Schatten, G. (1999) Ubiquitin tag for sperm mitochondria. Nature (London) 402, 371-372.
Szathmáry, E., Jordán, F. and Pál C. (2001) Can genes explain biological diversity? Science 292, 1315-1316.

Templeton, A.R. (2002) Out of Africa again and again. Nature (London) 416, 45-51.
True, H.L. and Lindquist, S. (2000) A yeast prion provides a mechanism for genetic variation and phenotype diversity. Nature (London) 407, 478-483.
Turner, P.C., McLennan, A.G., Bates, A.D. and White, M.R.H. (2000) Instant Notes Molecular Biology. BIOS Scientific Publishers Ltd., Liverpool.
Valadkhan, S. and Manley J.L. (2001) Splicing-related catalysis by protein-free snRNAs. Nature (London) 413, 701-707.
van den Ent, Amos, L.A. and Löwe, J. (2001) Prokaryotic origin of the actin cytoskeleton. Nature (London) 413, 39-44.
Van Dooren, P. (1993) Life. Origin and future of a unique phenomenon (in Dutch: Leven. Ontstaan en toekomst van een uniek fenomeen). Pelckmans Publishers, Belgium.
Van Poecke, L. (1996) Non verbal communication (in Dutch: Nonverbale communicatie). Garant Publishers, Leuven (Belgium), Apeldoorn (The Netherlands).
Vanden Broeck, J., De Loof, A. and Callaerts, P. (1992) Electrical-ionic control of gene expression. International Journal for Biochemistry 24, 1907-1916.
Varela, F. Thompson, E., and Rosch E. (1992) The embodied mind: cognitive science and human experience. MIT press, Cambridge, Massachusetts.
Venter, J.C. and consortium members (2001) The sequence of the human genome. Science 291, 1304-1351.
Vlaar, M. (1997) Creation (in Dutch: De Schepping). De Bezige Bij, Amsterdam.
Wächtershauser, G. (1988a) Before enzymes and templates: theory of surface metabolism. Microbiology Reviews 52, 452-484.
Wächtershauser, G. (1988b) Pyrite formation, the first energy source for life: a hypothesis. Systematical and Applied Microbiology. 10, 207-210.
Wächtershauser, G. (2000) Life as we don't know it. Science 289, 1307-1308.
Wahl, G.M., de Saint Vincent, B.R. and De Rose, M.L. (1984) Effect of chromosomal position on amplification of transfected genes in animal cells. Nature (London) 307, 516-520.
Watson, J. D. and Crick, F.H.C. (1953) Molecular structure of nucleic acids. A structure of deoxyribose nucleic acid. Nature (London) 171, 737-738.
Weinberg S. (1994) Life in the Universe. Scientific American 271 (October, special volume "Life in the Universe"), 22-27.
Whitrow, G.J. (1959) The structure and evolution of the universe. Harper Torch Books Science library, Harper and Brothers, Publishers, New York.
Whittaker, R.H. and Margulis, L. (1978) Protist classification and the kingdoms of organisms. BioSystems 10, 3-18.
Wolpert, L. (1998) Principles of Development. Oxford University Press, Oxford.
Wood, W. and consortium members. (2002) The genome sequence of *Schizosaccharomyces pombe*. Nature (London) 415, 871-880.
Woodruff, R.I. en Telfer, W.H. (1980) Electrophoresis of proteins in intercellular bridges. Nature (London) 286, 84-86.
Yao, S., Ghosh, I., Zutshi, R. and Chmielewski J. (1998) Selective amplification by auto- and cross-catalysis in a replicating peptide system. Nature (London) 396, 447-450.
Yu, J. and consortium members (2002) A draft sequence of the rice genome (*Oryza sativa* L. ssp. *indica*). Science 296, 79-92.
Zeleny, M. (ed.) (1981) Autopoiesis: A Theory of Living Organization. Elsevier North Holland, Amsterdam, New York.

INDEX

abortion 289
accretion 26
actin 75
actin, bacterial 236
actin, electricity conduction 103
active transport 185
adaptation 313
agglutinin, -ation 262
aggregate 100
Aids virus 201
alienation, evolution 310
alive, being 96
allele 296
amino acid, structure 47
amino acids, coding 49
analogy 39
animal, definition 103, 278
apoptosis 123, 133
Archaea 40
asymmetry 24
asymmetry, double, principle 288
automation 81
autopoiesis 123
bifurcation point 162, 171
big bang 20, 233
big hello 231
biocoenosis 109
biodiversity 57, 92
biogenetic law 292
biology, classical 324
biology, integrative 324
blastula 278
Branchiostoma 276
brane theory 30
BSE, prion disease 201
budding, external 103
budding, internal 107
C. elegans 259
Caenorhabditis 259
cancer, benign 260
chaos 195
choice, making 162
cholesterol, precursor 265
chromosomal skeleton 75

classification system, 16 levels 95
classification systems, classical 39
cleavage 277
cloning 282, 290
cogne 305
colony formation 105
coma 138
communicating compartment 87, 94, 114
communication act, definition 88, 143
communication and time 82
communication system, 71, 72
communication vs interaction 82
communication, definitions 70
communication, purpose 80
communication, self- 210
compartment 134
compartment, communication 87
compartmental organization 95, 312
complexity, defining 59
computer, life 206
concentration chamber 186
consciousness 301, 303
contentment 85, 86, 309
continent, formation 27
continuum, life, double 83
Creation, story, classical 32
Creation, story, dynamic version 331
Creationism 315
cultural evolution 161, 298, 301
Cyanobacteria 27
cytoplasmic bridge 102
cytoskeleton 81, 97
cytoskeleton, origin 235, 238
Darwinism, overview 296
death 133
death, definition 137, 140
death, duality 135
dementia 138
depolarization 188
deuterium 22
development, embryonic 287
developmental biology 275
Dictyostelium 105

diffusion 242
dissipative system 122, 321
DNA methylation 267
DNA world 233
DNA, electricity 193
DNA, properties 74
DNA, structure 46
dogma, central, first 48, 82
dogma, central, second 84
Dolly 283
drive, of life 173
Drosophila 101
Drosophila, development 280-281
Drosophila, gene number 59
duration, time 221
electric field, extracellular 191
electricity 181
electricity, biological 182
electrophoresis, miniature chamber 192
electrospherization 108
embryonic development 273
emotions 85
epigenetics 161
epithelium formation 278-79
epithelium, compartment 103
equilibrium, punctuated 299, 318
ethical considerations 138, 289, 326
eukaryotic cell 99
evil, when 323
evolution, cultural 300, 304
evolution, driving force 299, 304
evolution, general 295
exon 53
extremophiles 40, 236, 241
feedback, communication 71
feedback, spiral, evolution 313
feelings 85
female, the choosing- 269
fertilization, *in vitro* 289
fitness 312
force, driving evolution 313
fossils, oldest 28
free will, evolution 315
freedom, in acting 175
Gaia, compartment 109
gamete formation 255

gap junction 102
gastrulation 278
gender 263
gender, genes involved 264
gene, expession, electric 194
gene, expression 285
gene, expression, ions 286
gene, maternal 282
gene, numbers 59
Genesis, dynamic version 331
Genesis, Opening account 19, 32
genome, human 59
genome, size 60
germ cell line 101, 254
germ cells, primordial 256
good, when 323
GP-Triple S principle 240, 313
gradient formation 72
gradient, ionic 236
gradients 311
gradients, in evolution 304, 309
gradualism, phyletic 299, 318
Haeckel 292
hardware, definition 70
hardware, evolution 299
headquarters, installation 112
helium 14, 22
heterosexual compartment 106
heterosexuality 93
heterosexuality, origin 112
hierarchy 91, 115
holistic thinking 322
homology 39
hyperpolarization 188
immune system 261
imprinting 267
impulse conduction 194
indeterminism 195
infection, bacterial, hypothesis 257
information, definitions 79-80
instinct 174
interaction, vs. communication 82
intron 53
ion channel 164, 185, 186
ion flux 190
ion pump 164, 183, 184, 189, 190

ionic environment, differential 286
ions, inorganic 182
iron 24
Kingdoms, five 40
kinship, degree 61
life, a machine 125
life, artificial 202
life, classic definitions 118, 127
life, definition, requirements 143
life, double continuum 83, 297
life, electrical phenomenon 196
life, electronic 211
life, new definitions 155-156
life, origin 233
life, properties 321
life, symbolic notation 151
life's drive, impulse 173
light, measuring time 225
light, speed 225, 226
lipids 45
macro-evolution 57
males, superfluous 267
man, descent 62
Margulis, theory 44, 58, 93, 100
mega-evolution 57, 94
mega-evolution, mechanisms 110
meiosis 114, 255-258
membrane potential 183, 236
membrane, excitable 194
membrane, permeability 188
membrane, physiologist 204
meme 305
memory, cognitive 74, 163
memory, prerequisites 74
Mendeleyev, table 24
micro-evolution 56, 296
mitosis 251
modification 55
Monera 40
monoepithelium 103
monoorganismal compartments 97
morula 101
motion, time 224
motivation 88
multicellularity 100
muscle contraction 145

music, analogy 165
mutation, types 55
Na+K-ATPase 145
natural selection 298, 303,304
Neanderthal man 62
necrosis 133
neo-Darwinism 296-298
neurotransmitter 144
nucleic acids 46
nurture 297, 299, 311
organisms, models 275
orgasm 269
osmosis 242
oviposition 263
oxygen, formation 24
oxygen, in atmosphere 26
paradigm, biological 327
parthenogenesis 265
passive transport 186
PCR 29, 124
philosophy, biology 326
phyletic gradualism 297, 316
planet, formation 26
plate tectonics 27
pole cells 256
polyepithelium 103
polyorganismal compartments 105
population, compartmentalization 107
potential gradient 186, 187
primordial soup, 235
prion 202
problem solving 81
problem solving, evolution 304
Progenote 40, 238, 249
Progeny, goal vs.gift 268
Prokaryote, level 1 97
prokaryotes 40
protein synthesis 48
proteins, structure 47
Protista, Protoctista 40
punctuated equilibrium 297, 316
receiver-decoder 149, 150
recorder, example 165
regeneration 253
remembering 73
reproduction, asexual 252

reproduction, sexual 253, 255
reverse transcriptase 54
revolution, definition 95
RNA structure 46, 52
RNA world 54, 233
segmentation 104
selection 315
self selection 240, 242, 288
self-communication 210
self-electrophoresis 192
sender-encoder 147
sex-hormones 265
sex-reversal 265
sex-steroids 260, 266
social compartmentalization 106
software, characteristics 76-77
software, definition 70
software, evolution 76, 299, 302
solar system 25
somatic cell line 101
somatoplasm 254, 260,
soul 63
soup, primordial 235,

space-time 218
speciation 107
splicing 54
stardust, fossil, 19
struggle for life, 304, 310
supernova 24
symbiosis 93
symbiotic theory 44
synapse 144
syncytium 101
thermodynamics, laws 121
time, arrow 227
time, classical definitions 218
time, definition 223
tip cell 259
toolization 104, 105, 210, 301
transmission channels, types 146, 148,
twins, identical 161, 284, 299
Urkaryote 251
virus 201
water, origin 26
whistle, example 163
yolk, formation 261

www.ingramcontent.com/pod-product-compliance
Lightning Source LLC
Chambersburg PA
CBHW071229230426
43668CB00011B/1357